THE BUSINESS OF
Genocide

THE BUSINESS OF
Genocide

THE SS, SLAVE LABOR, AND

THE CONCENTRATION CAMPS

MICHAEL THAD ALLEN

The University of North Carolina Press Chapel Hill & London

Set in Minion and Meta types by Keystone Typesetting, Inc.

Manufactured in the United States of America

Publication of this work was aided by a generous grant
from the Lucius N. Littauer Foundation.

The paper in this book meets the guidelines for permanence
and durability of the Committee on Production Guidelines for
Book Longevity of the Council on Library Resources.

Library of Congress Cataloging-in-Publication Data
Allen, Michael Thad.
The business of genocide : the SS, slave labor, and the
concentration camps / by Michael Thad Allen.
   p. cm.
Includes bibliographical references and index.
ISBN 0-8078-2677-4 (cloth: alk. paper)
1. Nationalsozialistische Deutsche Arbeiter-Partei. SS-Wirtschafts-
Verwaltungshauptamt—History. 2. World War, 1939–1945—
Conscript labor—Germany. 3. World War, 1939–1945—
Concentration camps—Germany. 4. Genocide—Germany—
History—20th century. 5. Holocaust, Jewish (1939–1945).
6. Forced labor—Germany—History—20th century. 7. Holocaust,
Jewish (1939–1945)—Germany. I. Title.
DD253.6 .A65 2002

940.53′18—dc21   2001041474

06 05 04 03 02   5 4 3 2 1

CONTENTS

ILLUSTRATIONS

ACKNOWLEDGMENTS

The research and writing of this book have been supported by grants from the National Science Foundation, the Fulbright Kommission, and the Social Science Research Council as well as travel grants from the Deutscher Akademischer Austauschdienst and the Deutsches Museum.

Countless people have aided this work along the way. When I had the great pleasure to enjoy the hospitality of the United States Holocaust Memorial Museum, where I held a research fellowship in the fall of 1998, I was greatly aided by the advice of Keith Allen, Peter Black, Betsy Anthony, Martin Dean, and Severin Hochberg. I finished the final editing of this manuscript while in residence at the Zentralinstitut für Wissenschafts- und Technikgeschichte, Munich, as an Alexander von Humboldt Fellow. During my stay in Munich, my host, Ulrich Wengenroth, provided critical comment, as did Helmuth Trischler, Stephan Lindner, Jonathan Petropoulos, Matthias Heyman, Jörg Hermann, Martina Blum, Luitgard Marschall, Thomas Wieland, and Falk Selinger.

The bulk of this writing was done in Berlin. Without the help of the late Arno Mietschke and his wife Franciska Mietschke on my arrival in Berlin I would scarcely have found my way. Jonathan Wiesen, Mats Fridlund, Gabriel Finkelstein, and Julia Sneeringer all read drafts and donated their time to discussions. My special thanks also go out to Jefferson Chase, a constant source of what Paul Fussel (as opposed to R. P. Blackmur) has called "fresh idiom," and to Christian Hufen, whose quiet but uncompromising self-reflection makes East Berlin a much better place. Omer Bartov gave insightful comments at the 1996 German Studies Association meeting, where I presented some of the work here in a panel organized by Paul Jaskot. And no conversations were more profitable than those I shared with Paul; rarely have I ever disagreed more with *and* learned more from any individual. My colleagues at the Georgia Institute of Technology have continually provided the kind of support that assistant professors need most. In addition, Seth Fein, Therkel Straede, and Matthew Payne always had time when pressed to read this or that last or next-to-last alteration. All of these people care about good prose and good ideas, and without them this book would be much, much less than it is.

Several people I met in Berlin were kind enough to give up scarce time and offer their help as longtime professionals. I especially thank Peter Hayes, Raymond Stokes, Heinrich Volkmann, Michal McMahon, and Herbert Mehrtens. Karin Hausen also gave useful comments during a presentation at the Social Science Research Council's seminar series in the Program for Advanced German and European Studies. Michael Neufeld and Jens-Christian Wagner aided me with crucial facts of the V-2 rocket narrative. The wizened Miroslav Karny, survivor of the Nazi ghettos, the concentration camp Theresienstadt, and communist jails after the Prague Spring, also receives my sincere gratitude for helping me locate archival materials in eastern Europe. And on research trips in Poland I benefited from the untiring energy and meticulous knowledge of Tomasz Kranz at the State Museum of Majdanek as well as helpful tips of Wojciech Płosa at the State Museum of Auschwitz.

The Arbeitskreis around Professor Wolfgang Scheffler at the Institut für Antisemitismusforschung at the Technische Universität–Berlin also provided a clearinghouse for good ideas and research leads. My special thanks go out to Andrej Angrick, Martina Voigt, Peter Klein, Birget Jerke, Alexandra Wenck, Christian Gerlach, and Marcus Gryglewski; likewise to Karin Orth and Christoph Dieckmann of the Institut für die Forschung des Nationalsozialismus in Hamburg. It has also been my good fortune to work with editors who have seen above the pettiness of much Holocaust historiography—which is rocked about every five years by one academic scandal after another, imagined or real. Chris Wickam, Ronald Smelser, Miriam Levin, John Staudenmaier, Kenneth Barkin, and Chuck Grench all pushed me to make this project better than I could have made it alone. Smelser also organized a panel at the 1998 Lessons and Legacies of the Holocaust conference in Boca Raton, Florida, at which George Browder helped refine this work with his comments, as did Rebecca Wittman and Fran Sterling.

There are also those without whose support this book would have suffocated in its crib: Wolfgang Scheffler and my dissertation advisers Thomas Parke Hughes, Edward Constant, and Frank Trommler. When I first proposed this project almost a decade ago, many told me that it was "impossible." Many asked why anyone would be interested in such a topic in the first place. When I took the project to these four, they said simply, "You should go ahead and do that." They gave me all possible help and encouragement even when it was inconvenient to do so, as when Frank Trommler wrote a steady stream of stipend recommendations during his sabbatical. Here they receive my deepest gratitude now that the work is done.

Last, my wife, Helen Rozwadowski, was always there when I needed her most.

ABBREVIATIONS

The following abbreviations are used in the text. For additional abbreviations used in the notes, see pages 287–88.

A-4	Aggregate-4 (V-2 rocket)
AG	Aktien Gesellschaft (joint-stock company)
BI	Bauinspektion (Building Inspection)
DAF	Deutsche Arbeitsfront (German Workers Front)
DAW	Deutsche Ausrüstungswerke GmbH (German Equipment Works)
DESt	Deutsche Erde- und Steinwerke GmbH (German Earth and Stone Works)
DWB	Deutsche Wirtschaftsbetriebe GmbH (German Commercial Operations)
GmbH	Gesellschaft mit beschränkter Haftung (corporation with limited liabilities)
HAHB	Hauptamt Haushalt und Bauten (Main Office for Budgets and Building)
HSSPF	Höhere SS- und Polizei Führer (Higher SS and Police Führer)
IG	Interessengemeinschaft
IKL	Inspektion (Inspekteur) der Konzentrationslager (Inspectorate [Inspector] of Concentration Camps)
NSDAP	Nationalsozialistische Deutsche Arbeiterpartei (National Socialist German Workers Party)
ODBS	Ost-Deutsche Baustoffwerke GmbH (East German Building Supply Works)
OSTI	Ost-Industrie GmbH (East Industries)
RFSS	Reichsführer SS (Heinrich Himmler)
RKF	Reichskommissar für die Festigung deutschen Volkstums (Reichskommissar for the Reinforcement of Germandom)
RSHA	Reichssicherheitshauptamt (Reich Security Main Office)

SA   Sturmabteilung
SS   Schutzstaffel
SSPF   SS- und Polizei Führer (SS and Police Führer)
TexLed   Textil- und Lederverwertung GmbH
            (Textile and Leather Utilization Ltd.)
V-2   Vergeltungswaffe-2 (rocket)
VuWHA   Verwaltung- und Wirtschaftshauptamt
            (Administration and Business Main Office)
VW   Volkswagenwerke
WVHA   Wirtschaftsverwaltungshauptamt
            (Business Administration Main Office)
ZBL   Zentralbauleitung (Central Construction Directorate)

THE BUSINESS OF
Genocide

INTRODUCTION

Before January 1944 less than a fifth of all Allied bombs dropped throughout the entire course of the war had fallen on Axis targets, but in just the next six months, between January and July, the total tonnage increased by almost half again as much. The pace and ferocity of bombing only increased from that point onward, leaving the famous "rubble mountains" in every major German city. The previous summer the Red Army had lured the Germans into a trap at Kursk, after which Hitler's armies never again mounted any major offensive against the Red Army. And yet in the spring of 1944, to the Allies' great consternation, German war production continued to rise. Moreover, the Allies were yet to land at Normandy; the Soviets had yet to launch the major offensives that would lead them on to Berlin; Wehrmacht officers had yet to stage their abortive assassination of Adolf Hitler; and Hitler and his leading paladins were increasingly enthusiastic about "wonder weapons" like the V-2 rockets and the V-1 cruise missile. These proved vain hopes, but especially for those who wished to remain blind, obvious signs of utter collapse were still several months away.

German engineers and midlevel managers were chief among those who refused to give up. Not the least of their contributions was the oversight of millions of forced laborers who had come to make up one-fourth of Germany's total work force. To German management fell the daily task of reconfiguring modern production around these laborers in a last-ditch effort to match the Allies tank for tank and plane for plane.[1] Foreign civilians made up the majority of this compulsory labor force. Limited recruitment campaigns for foreign workers had started as early as 1940, but after March 1942 a special "General Plenipotentiary for the Labor Action" began large roundups of "Eastern Workers" to ship west to German factories. Over 700,000 concentration camp prisoners labored under the most brutal conditions, and even if they formed only a small part of the overall German war economy, by 1944 hardly a single locale with any factory of note lacked a contingent of prisoners. Every morning columns of somber workers, starving and bruised, could be seen marching from fenced enclosures down the streets of ordinary German towns. By 1944 Heinrich Himmler's SS (Schutzstaffel) was parceling out these inmates by the thousands for everything from aircraft factories to chain-gang-style construction.

This book is about the managers of that process. They worked in a special division of the SS called the SS Business Administration Main Office (Wirtschaftsverwaltungshauptamt, or WVHA). This office spread a network of slavery across German-occupied Europe. From its pool of prisoners came the bulk of the work force for the V-2 rockets as well as other "wonder weapons." Most concentration camp prisoners, however, worked under the WVHA's elite corps of civil engineers, which specialized in breakneck construction projects, among them the conversion of underground tunnels into factories such as the eerie caverns where V-2 rocket assembly took place.

By the spring of 1944, these efforts were reaching a climax. At the time, a relatively obscure midlevel manager, Kurt Wisselinck, like so many other officers of the WVHA, was working longer and harder hours trying to squeeze production out of desperate and expiring prisoners in this system of slavery and murder. Introducing Wisselinck is perhaps a good way to introduce the WVHA as a whole, for he was a compulsive doodler and left a clear image of how the WVHA viewed itself and its mission. Amid his work Wisselinck took the time to sketch a handsome, square-jawed man on office stationery. The man gazes sidelong down a string of telephone poles with focused intensity. Rings around his eyes betray fatigue, but his determination is undimmed. He holds a telephone to his ear, and it is impossible to say whether he is giving or receiving orders, but the pose—ready for action—portrays virtues that the SS's industrial managers wished to see in themselves. The man is dynamic, the master of modern technology, and, with his high forehead and perfectly straight nose, he is a model of Teutonic racial fortitude.[2] At the margin of this sketch Wisselinck also scribbled an almost unreadable note about some kind of reimbursement for petty cash. Taken as a whole, this curious artifact bears witness to both a heroic ambience in managerial tasks imagined by SS officers like Wisselinck as well as the trivial paper pushing that filled their days, even as they presided over the life and death of human beings. The latter, the inane details of administration that made the Nazi genocide possible, has preoccupied historians since Hannah Arendt first published *Eichmann in Jerusalem: A Report on the Banality of Evil* in 1963; yet Wisselinck's steady-eyed, vigorous Nordic hero hardly squares with the image of Adolf Eichmann, who had become a vacuous, middle-aged man with thick spectacles when he was put on trial in Israel.

When Arendt wrote her biography of Eichmann, she created much more than a portrait of one desperate ex-Nazi indicted for his crimes. She fixed, for the next forty years and likely more, popular conceptions of the Nazi bu-

Sketch by Kurt Wisselinck, drawn sometime in 1944.
U.S. National Archives Microfilm Collection T-976, Roll 18.

reaucrat. Here was a failed vacuum oil salesman who had become one of civilization's all-time greatest killers. Indeed, it was almost as if the foolish Willy Loman of Arthur Miller's *Death of a Salesman*, by some horrible accident, had escaped fiction and become the engineer of the Holocaust. Although Arendt by no means trivialized Eichmann or his crimes, she brought out his pathetic bathos in the same way that Miller, the playwright, had made the empty life of the Western "organization man" the subject of his drama. The major difference, of course, is that Eichmann killed other people while Loman killed himself.

Arendt's portrayal of Eichmann tapped into a widespread tendency to view midlevel managers in modern society as the twentieth century's numb and inane one-dimensional men. Eichmann was, in other words, the classic, atomized "organization man" or what Lewis Mumford called the "penny-in-the-slot automaton, this creature of bare rationalism."[3] Arendt's famous book asks us to see an utter emptiness in Eichmann's conscience and, worse, a complaisance in that emptiness. That is, she did not condemn Eichmann for having stupid, inconsistent, or condemnable ideals but for being "thoughtless," for having no ideals whatsoever. In consequence, SS men like Eichmann are not, as Arendt falsely promises, condemned for enacting evil but for being amoral; for being "perfectly incapable of telling right from wrong"; for being afflicted by an "inability to think."[4] Thus the human engagement of Nazis in bureaucratic function—the kind that emanates from the shrewd eyes of Wisselinck's sketch—has receded from view.

A paradox has always rested at the center of Arendt's judgment, a contrast between the Kafkaesque torpor of bureaucracy and frenetic genius. First there is the miserably stupid Eichmann, afflicted by an "utter ignorance of everything that was not directly, technically and bureaucratically, connected with his job."[5] Yet simultaneously Eichmann has been accorded a perverse intelligence as vast as his worldly conscience was small. In Vienna in 1939, as he confronted the monumental task of cataloging all the Jews of Austria for deportation, he proved so innovative that contemporaries and historians alike have marveled. "This is like an automatic factory. . . . At one end you put in a Jew . . . and he goes through the building from counter to counter, from office to office, and comes out at the other end without any money, without any rights."[6] The paradox is resolved by attributing managerial creativity to the very source of its banality: "This use of human beings," as James Beniger notes about modern organization, "not for their strength or agility, nor for their knowledge or intelligence, but for the more objective capacity of their brains to store and process information."[7] Max Weber is perhaps most renowned for casting this enduring image of administrators in the famous metaphor of the iron cage. Bureaucracy supposedly constrains because it imposes cultural meaninglessness and renders the individual impotent to resist its imperatives. Thus rationality generates power precisely by driving humane sensitivity to the margins. In Nazi Germany this meant the failure to oppose the genocide.[8]

Weber introduced his metaphor of the iron cage by comparing the disenchanted but efficient bureaucrat to the universal humanity embodied in Goethe's Faust, and the comparison is instructive. At the end of Goethe's play, Faust takes part in a kind of Holocaust. We find him embarked upon the construction of a perfectly ordered society, and he directs Mephisto to remove an elderly husband and wife who have settled in the path of one of his massive social engineering projects. Unbeknownst to Faust, Mephisto murders the couple and provokes the protagonist's final grief. By contrast, the genocide was hardly such an absentminded distraction and could not have issued from any isolated individual decision. At every stage institutions, especially the SS, mediated the horror. Furthermore, the SS did not need Mephisto's supernatural smoke and mirrors. It could rely upon midlevel managers, and these, unlike Faust, rarely repented their deeds.

Historians have documented the willing and energetic identification of individuals with the new organizational milieus of the late nineteenth and early twentieth centuries, and Wisselinck's sketch confirms that he did not feel imprisoned in an iron cage. He felt empowered. Therefore, as this book examines the management of slave labor and murder, it questions not how modern structures divested Nazi "technocrats" of moral agency but rather how perpetrators endowed their institutions with personal significance.[9] Much scholarship that seeks to understand the barbarity of the SS begins by asking the question, Why were those involved not repulsed by their actions? or, to quote Hans Mommsen, "Why did so many who participated in the series of events that led directly and indirectly to the extermination of the Jews fail to withdraw their contribution either through passive resistance or any form of resistance whatsoever?"[10] Entire books are dedicated to explaining how Nazi perpetrators were able to overcome repugnance for their deeds, which presupposes that they indeed found them repugnant.[11] Historians essentially ask why SS men did not have the good sense to act as we hope we would have acted in their position, that is, as moral, upstanding citizens who would have saved fellow human beings. But the SS confronts us with a world of murderers, not good citizens; more precisely, SS men were the model citizens of a murderous regime. Instead of asking why SS men did not feel what they did not feel or why they failed to act as they might, should, or could have done, this book poses the question, Why did they believe it was the right thing to do?

We may use Arendt's biography of Eichmann as a point of departure. First, historians have long overturned her portrait of the miserably blinkered Eichmann. Hans Safrian, with much broader access to evidence than that available to Arendt, has documented the conscious moral dedication—anything but a banal "inability to think"—of Eichmann and the officers gathered around him.[12] The slave-labor moguls of the WVHA were dedicated in equal measure. Second, Arendt's picture of the perversely brilliant Eichmann, the manager of industrial genocide, invites further inquiry on one smaller point: namely, Eichmann never managed a factory in his life but made his career as a police administrator (in the SS Reich Security Main Office, or Reichssicherheitshauptamt). Unlike Eichmann, our sketch artist Wisselinck worked in real rather than metaphorical "factories of extermination." He and his co-workers shifted prisoners to labor sites across the breadth of Europe and collected their broken bodies for liquidation when this "human material" (as WVHA correspondence put it) had been used up.[13] Looking back upon the twentieth century, in which genocide now seems more likely to recur than it did to Arendt in the 1960s, some have claimed that the bureaucratic and technological nature of the Nazi genocide is the sole feature that distinguishes it from Bosnia, from Stalin's collectivization campaigns, from Rwanda, or from Pol Pot.[14] WVHA engineers arranged for the "stationary crematoria, incineration stations, and execution installations of various kinds" built in the camps after

1942.[15] Yet while Eichmann and Reinhard Heydrich have become common names of infamy, who recognizes the leaders of the WVHA: Oswald Pohl, Wilhelm Burböck, Gerhard Maurer, or Hans Kammler, let alone the obscure Kurt Wisselinck?

Why did the SS set out to broker hundreds of thousands of prisoners to Hitler's war industries? Many speculate that the SS wanted to gain "control over the economy." To me this answer is unsatisfactory, for it discounts any real motivation. The image of banal careerists immersed in the office work of murder too often dovetails with such an image of institutions in which a purely pragmatic "will to power" supposedly eclipsed decisions about moral right and wrong. Although the Third Reich, like any complex state, played host to numerous conflicts, we should not be too hasty to label it, as did Franz Neumann, as a Behemoth, eaten up from within by a war of all against all in a raw bid for power. Who would ever deny that the Third Reich was exceptionally fragmented? Neumann's great service was to point this out. Two executive organs existed for agricultural policy; there were two justice systems (SS and civilian), two armies (Waffen SS and Wehrmacht), two chancelleries (party and state). Sometimes three or four institutions overlapped, and they fought each other incessantly. But histories of the Third Reich have dwelt too much on struggles for power; likewise, they have too readily attributed inefficiency and conflict to what is commonly known as "polycracy," defined as the "rule of many" and first established by the German historian Peter Hüttenberger. The historian Peter Hayes once remarked that, on one hand, we are led to believe that Hitler's Germany was polycratic and thus incapable of concerted organizational effort because everyone struggled against his fellows; on the other hand, this small country in central Europe kept the entire world at bay well through 1942, even into 1943, while losing about the same number of soldiers in combat (3–4 million) over the course of the whole war as the number of Red Army prisoners the Wehrmacht captured in the first six months of the Soviet invasion.[16]

Beyond the Nazis' startling efficiency at many different tasks, it is in the very nature of multiple, overlapping institutions that they created as many venues for cooperation as for infighting. I would argue that "polycracy" relied on cooperation, and that this followed ideological consensus precisely because—with so many agencies—the historical actors had to constantly exercise their initiative and conscious choice. Motivation mattered more, not less, due to the higgledy-piggledy nature of National Socialist organizations. In fact, the progressive rationalization of the camps could not have proceeded without the help and encouragement of Reich ministries, private industrialists, and civilian managers.[17]

WVHA officers also made their careers in the midst of a curious generational break. They mostly came from a relatively new class, the white-collar workers whose numbers began to swell at the end of the nineteenth century and were beginning to dominate the twentieth. Often they had grown up in old-middle-class families; their fathers had been farmers, shopkeepers, or countless other petty tradesmen. The white-collar workers had deserted these backgrounds to enter the novel work-world of the factories and large urban firms with their branching managerial systems. This new class departed from Weber's (or Arendt's) image of modern managers as much as Wisselinck's determined Aryan at the field telephone departed from Kafka's pusillanimous bureaucrat. For that matter, Kafka's *Castle* describes a world of traditional administration from which modern management differed just as Ford's factories differed markedly from craft or traditional batch production. For example, as richly described by Reinhart Koselleck, the small cadre of Prussian civil servants and bureaucrats at the beginning of the nineteenth century relied on prose reports; in fact, many disdained statistical shorthands for their duties. Accordingly they sank beneath mountains of paper that recall Kafka's Sordini, whose "every wall is covered with pillars of documents tied together" and whose workroom reverberated with the thunder of falling tomes.[18] By contrast, in 1944 Wisselinck called for the WVHA to rationalize the management of Gross-Rosen and, needless to say, did not call for pillars of ledgers. He used the language of charts and graphs and imposed the terse statistical surveillance of input and output. Indeed, it is little known that scrupulous tables to tot up the "fit," "unfit," and the dead—statistics almost synonymous with the coldly efficient Nazi temperament—appeared in the concentration camps only after 1942, when WVHA officers began to take charge in an effort to serve modern industry.[19]

Kurt Wisselinck again serves as a brief introductory example of how this modern management operated within the SS. First, there is no denying that power struggles marked Wisselinck's career with the WVHA, as Neumann and Hüttenberger would quickly point out. In 1944 Wisselinck was Chief Factory Representative (*Hauptbetriebsobmann*) within the SS to a rival institution, the German Workers Front (Deutsche Arbeitsfront, or DAF). The DAF had crushed the German trade unions in the first year of Hitler's seizure of power, but it also demanded social welfare programs and appointed representatives like Wisselinck to enforce its decrees. Private industry disliked many of these policies. Management often viewed DAF representatives as usurpers who sought to trample its prerogatives. The DAF had also founded industries of its own, which posed unwanted competition to private corporations. (The most famous was Volkswagen, organized to manufacture the Beetle, the "people's car.") Beyond claiming the right to place representatives in SS companies, the DAF had set up its Volkswagen plant as a model of technological efficiency and National Socialist principles.[20] The WVHA—which managed state rather than private corporations—had cause to fear the DAF's encroachment, for the WVHA's own corporations also posed as beacons of Nazi ideals. Robert Ley, head of the DAF and by all accounts an ambitious and fanatic Nazi, might well have perceived the SS as a threat to his own industrial empire and vice versa. And if it came to a pitched bureaucratic struggle, Wisselinck, as SS officer and DAF representative simultaneously, held key leverage. He might undermine the SS from within; on the other hand, he might act as an agent of the SS and undermine DAF intervention. Yet when conflict erupted, as it did in February 1944, the issue did not turn on the extension of bureaucratic influence but on ideological principle, and the outcome differed from that which orthodox interpretations of "polycracy" might lead us to expect.

Wisselinck had heard of misconduct at the SS Granite Works of Gross-Rosen. The Granite Works had started as one of the SS's first large-scale industrial projects in the concentration camps, founded to provide stone for the Nuremberg Party Rally Grounds. After the advent of total war, the SS also tried to convert its facilities to take in armaments production. Wisselinck routinely visited such SS factories on rotation, but this time his trip was different. He had put through a special request to the WVHA to inspect Gross-Rosen because he sensed a severe transgression of DAF policies toward civilian employees. Since its inception, the quarries had worked prisoners to death, but this was not what Wisselinck considered unjust. Rather, his interest had been piqued by rumors that Gross-Rosen was not exploiting the prisoners enough. Thus, when he heard that no one had distributed clothing confiscated from Jews to SS manager-trainees as specified, he acted quickly to make the factory conform to Nazi policy. What he learned upon arrival further appalled him. Trainees complained that their instructor was a drunk and was sleeping with his secretary. Wisselinck also suspected embezzlement. In addition, he alerted the headquarters of the WVHA that one cook seemed to favor the prisoners: "The apprentices complain that they are being withheld additional portions of potatoes with the justification that there are no more left, while it can be observed that the prisoners receive the food as additional rations."[21]

Anyone who has seen documentary footage of the camps knows what the prisoners began to look like in 1944, emaciated skins stretched over skeletal bodies. Mortality statistics, which fluctuated wildly, were running at about 10 percent a year at Gross-Rosen. The WVHA tracked them carefully.[22] Something as simple as an occasional potato could have made the difference between surviving the last year of war or perishing of starvation and disease within a few weeks. Furthermore, a steady diet—even of leftovers—might have even benefited production. Gross-Rosen's managers in fact mentioned this in their own defense. They complained that Wisselinck's presence endangered efficiency. To no avail. The head of the entire WVHA, Oswald Pohl, personally communicated his "sharpest disapproval" to the Granite Works.[23] The message was clear: nothing should be given to starving prisoners. So important were the issues involved that the chief of the WVHA backed Wisselinck, a DAF representative, against others in the SS's own management.[24]

If a war of all-against-all defined the Nazi period, does this explain Pohl's and Wisselinck's deeds? After all, polycratic interpretations of National Socialism can account for cooperation. Institutions often worked together to eliminate mutual rivals and thus increase their own influence. Alliances of convenience were indeed common. But Wisselinck's trip to Gross-Rosen did nothing to enhance the DAF's or the WVHA's authority, nor did the pursuit of power seem to define his purpose. Wisselinck neither extended the DAF's reach nor advanced the WVHA's factory operations. At stake was not the expansion of bureaucratic authority; rather these events proceeded along ideological lines in which the DAF and WVHA shared common commitments. Within the WVHA—and other Reich institutions as well—most believed that the bereavement of concentration camp prisoners was just. Although one cook at Gross-Rosen saw things somewhat differently, Wisselinck mobilized the entire apparatus of SS bureaucracy against her: he wrote reports, compiled statistics, called in his superior. In the end he actually demanded the "rationalization" of the camp's kitchen. More strict bookkeeping would ensure that such "embezzlement" could not happen again.[25] Moreover, if Wisselinck had "just followed orders" as a man constrained to an "iron cage," he might have overlooked the camp entirely. His visit was a matter of personal initiative, and he comported himself not like Arendt's banal Eichmann, but as an interventionist manager who thought it necessary to act on principle. He went out of his way to ensure that Gross-Rosen's management *did not* help prisoners to survive, to insure that it *did* give confiscated Jewish belongings to SS recruits; and he carried through with his inspection in spite of complaints that he was actually endangering efficiency.

Skeptical readers will doubt that Wisselinck was really ideologically engaged in the whole matter. Might he not simply have been striving for his superiors' attention? Yet at about this time, at no one's bidding, he wrote a lengthy memorandum with no apparent reader other than himself: "The business undertakings of the Schutzstaffel are the best means to breath new life into National Socialist ideals, to let them become reality, to blaze new trails in the area of applied socialism. We must live socialism as the deed! Our example must spur other corporations forward to emulate us in order to see the growth of a healthy, satisfied, and happy *Volk*." Wisselinck operated neither as an agent of the DAF nor as an agent of the SS but as both, for he went on to express spontaneous enthusiasm for a plexus of ideologies that formed the raison d'être of the SS's business enterprises *and* those of the DAF as well. Every SS company should offer its German employees generous social benefits (programs championed by the DAF). In turn, he connected these to Nazi racial imperialism. Affordable SS housing should encourage Aryan families "rich in children" and tie them to their "Motherland." "Blood and Soil" should unite the Nazi homestead and further garner loyalty to the factory community, a microcosm of the larger national community of Nazism. Before Hitler's rise to power, Wisselinck claimed, "primitive housing" had proved a "breeding ground of immorality" and a "feeding trough [*Nährboden*] of Marxism."[26] He also blamed banks, thus condemning communism and capitalism in the same breath. Ideals like these could make the distribution of leftover potatoes seem like an issue of national security and cultural renaissance.

As extraordinary as Wisselinck's manifesto may sound, it was by no means unusual within the WVHA, and we will have ample opportunity to encounter other midlevel managers as ideologues. Of special interest here, Wisselinck was fixated not by one monomaniacal drive but by many, mingling them in his manifesto to the point of incoherence. Much has been written of single ideas that caused the worst crimes of National Socialism. If Arendt or "polycratic" interpreters of Hitler's Germany have erred by underplaying agency and motivation, others err by attributing the Holocaust to one "crisis of German ideology" and one alone, whatever it may be. Most prominent among them are "anti" ideologies: anti-Bolshevism, anticapitalism, anti-Semitism, and antimodernism.[27] Similarly, some authors attribute the violence of Nazi hatred to a pathological "fear" of the Third Reich's victims.

While no one should discount the Nazis' rabid suppression of communism, their hatred of Jews, their fantasies of a romantic German past, and their intervention in the national economy, Wisselinck did not apply himself so energetically merely because he feared this or that. It is well to remember the words of Richard Evans regarding right-wing violence: "The murderers' actions, and the brutal language accompanying their deed, suggest that it was not fear, but loathing and contempt, which motivated them."[28] Wisselinck's proactive assertion of identity filled his prose and his actions. This was Nazi activism, not reaction: "The SS . . . must be an example and ever again an example in social policy."[29] Even in 1944, when the Reich was already beyond saving, he still saw himself in the vanguard of social change.

I have used the admittedly awkward term "plexus of ideologies" because I believe the image of a complex network, one with branching, even partial systems of ideals, provides a better understanding of how organization men like Wisselinck worked. They operated within a broad current whose tenets sometimes ran together, sometimes followed parallel courses, and sometimes collided. Wisselinck did not become an accomplice to murder by following any single tributary but by working within the whole. He felt competent to switch and modify his course continually and was encouraged by the National Socialist emphasis upon passion and activism over logic and consistency, upon syncretism over synthesis. On the other hand, this did not mean that the WVHA, or any other National Socialist institution for that matter, acted arbitrarily. The organization as a whole tended toward efficient action when multiple ideals, individuals, and institutions reinforced each other. In the case of Gross-Rosen, Wisselinck was able to mobilize his superior officers through the WVHA's bureaucratic edifice in favor of DAF policies. The outcome was no accident. Shared ideals reached into constituencies outside the WVHA and lent coherency to this collective action. A useful metaphor is perhaps a river delta in which currents may eddy or alter direction but nevertheless eventually and inevitably issue into the ocean. Precisely because of the importance of consensus, Nazi ideology issued, finally, in one massive sea of blood. Understanding the multiple valence of ideological tributaries, their conjunctions and contradictions, best explains why SS men like Wisselinck chose to do what they did.

Questioning why they did what they did brings us directly to the junction of modern organization and ideological motives. This book argues that ideology is embedded in the quotidian tasks of bureaucratic operations because it lies at the root of collective identity and consensus. The function of consensus is best understood by considering the nature of modern management, whose techniques transform local, particular experiences and artifacts into fungible information amenable to collation, interchangeability, and abstract transfer. Above all modern organizations do so through statistics. At issue here is the role of consensus in evoking individuals' identification with impersonal institutions and abstract information. The most banal statistics have always depended upon the input, trust, and collective work among white-collar workers.[30] At the juncture of personal as well as collective trust among managers, ideological consensus has always played an indispensable role by helping render the information they worked with fungible. Information could be more readily transferred when SS officers trusted each other than when they had cause to doubt each others' motives. Ideals also served a function by animating large bureaucratic hierarchies—which are otherwise impersonal and even alienating—with a sense of individual purpose, a sense of personal mission. SS men worked harder and information within the WVHA flowed better when they believed in what they did. Again, I do not argue that one motive inspired all SS men; likewise no one individual needed to identify with each and every principle of the SS. Above all, ideology cannot be reduced to a single-minded goal, which organizations then set about to instrumentalize. But the WVHA functioned best when it succeeded in evoking the active identification of its officers—for whatever reason—with elements of its social cause. Once officers identified with the institution and their fellows, if only with fragments of larger, grander visions, their specialized skills could be mobilized in unison for the whole.[31]

Managing concentration camp industry involved three separate professional communities, each with its own distinct style and career patterns. Their interconnections, conflicts, and consensus all shaped the brutality of the concentration camps and slave labor. Two developed internally to the SS. First were the managers of the WVHA—businessmen, accountants, and engineers, among them Kurt Wisselinck. Second were the Kommandanten, the leadership core of the Inspectorate of Concentration Camps. The third community was external to the SS and entered only when the German economy plunged into total war late in the winter of 1941–42, namely, the state planners and industrialists within the Reich Ministry for Armaments and Munitions. Each community had its own vision of how to foster managerial teamwork and marshal "organization men." In general, these overseers of forced labor considered themselves idealists and wished to convert their visions into reality. Their ideals were manifold, as Wisselinck's manifesto has already hinted.

Some currents developed uniquely within the SS. First, the SS consciously set out to remake Europe in its own image. Police surveillance of the private and public lives of citizens in the name of "German values" was only one aspect of this drive. The SS also wished to build what came to be known as the "New Order," a program both to extirpate "unworthy" races from eastern Europe and to place model Nazi communities in their stead. Wisselinck's manifesto referred to this program when he wrote of "settlement houses . . . to maintain a perpetual stream of fresh [Aryan] blood."[32]

Second, a strong commitment to the Führer principle—a doctrine of Nazi leadership and national unity—drove decision-making and organizational structure. The Führer principle was Janus-faced, as much a communitarian ideology as a spur to "internecine strife," for it stressed unity and individual initiative at one and the same time. It did so by emphasizing that individual leadership grew out of collective identity. Every "Führer" conceived himself as the manifestation of the "will" of his subordinates; likewise, he conceived himself as a man in confluence with the will of his own "leaders." Adolf Hitler sat at the top, nothing less than the supposed embodiment of the historical mission of the German will. The Führer principle prompted individuals both to act spontaneously *and* to close ranks obediently, to act out *but also* to act in communion with other like-minded men. This dovetailed with the very structures of modern, centralized bureaucracy, which depend on the creative initiative of organization men but which focus that initiative upon collective endeavor in order to accomplish what no single individual can do alone. As Ronald Smelser has elegantly put it, "One could hope for success not as an isolated atom in a highly individualistic society, where failure or bad luck could bring precipitate social destruction, but rather as an integral part of a dynamic organization reaching out in an almost chiliastic fashion for total transformation of the world."[33] Oswald Pohl, the chief of the WVHA, made the very structure of his institutions and corporations reflect this goal.

Third, as we have seen, the SS emphasized the socialism in National Socialism. WVHA managers wished their businesses to serve goals of Nazi community without regard for pecuniary gain. They resented the threat that international markets posed to homogeneous "German" culture. Whether profitable or not, the SS wanted to manufacture a National Socialist renaissance, and they suspected businessmen of being loyal only to their purse strings. If we were to describe this as an anti-ideology, cultural anticapitalism is perhaps least awkward. It differed from anticapitalism of other stripes. The SS did not oppose monopolies or joint-stock companies, as did many liberal critics, because of the threat they posed to individualism; nor did the SS wish to redistribute the means of production to the working class, as did many socialists and communists; rather the WVHA opposed capitalism because of the threat that it posed to a homogeneous German culture.

For this doctrine, "productivism" serves as a better label. Productivist ideology meant that companies should not so much do business and make products as make Germans and Germanness. It promised to make the factory floor into a system with which to stamp managers and workers alike with an indelible national harmony. In industrial terms, this meant an elevation of factory organization and technology as a supreme concern over consumption, marketing, or distribution, which both Richard Overy and Mary Nolan have noted from quite different methodological approaches.[34] As Detlev Peukert pointed out, "The consumer-goods market promotes the individualism and freedom of movement that the political system [attempted] to obliterate."[35] This was another reason why liberal capitalism disgusted SS men like Wisselinck: it had spawned the "vulgar" street life of the Weimar Republic, materialistic pursuits, and a corresponding proliferation of tastes. When Nazis imagined revolutionizing consumption, rigidly standardized products like the VW Beetle or the Volks Radio were the result. These foresaw little room for consumer choice. The national organization in charge of distribution in Nazi Germany actually advertised its services as delivering the "IMPULSES OF THE ECONOMY to the daily life of the people" (emphasis in original).[36]

By contrast to the varieties of consumer impulse and expression, a well-run factory displayed unified organization and bent the material world to a collective human will. Specialized machine tools, standardization, and assembly lines had captured the fantasy of Hitler in the 1920s. The SS followed this lead. In 1924, just a month before delivering his first political speech, Heinrich Himmler had written to a close friend, "So you're reading Henry Ford . . . one of the most worthwhile, weighty, and most spirited predecessors in our fight."[37] It is significant that Himmler praised Ford for his "spirit" and not his wealth. To many ardent National Socialists, Ford's River Rouge was not so much a business as a manifestation of supreme will and the harbinger of a new world. To the WVHA, production was the forge of national identity, not first and foremost an act of economic output. (The WVHA actually pooh-poohed the dictates of economic rationality.) Wisselinck was tapping into this productivism when he proselytized for the modern factory as the locus of ideal German community.

By praising Ford, who had popularized these techniques, Himmler was merely echoing the widespread enthusiasm for the visionary potential of modern production. Nazi productivism reinforced a strong current of modernization, a fourth mission within the multivalent ideology of the WVHA. Even in seemingly old-fashioned industries like stone quarrying, the SS tried to introduce modern machines, despite their unsuitability to the conditions of forced labor. Modern factories may be defined by their operations, which took in raw materials and yielded finished product in a continuous stream, displacing traditional, small-scale batch or craft production. New technologies had made this possible by substituting machines for the work of human hands as well as for the human regulation of labor. As a vanguard of National Socialism, the SS wished to claim such futurism for itself. Jeffrey Herf has coined the phrase "reactionary modernism" to describe this impulse. In his view, the Nazis sought to reject Enlightenment doctrines of reason and individualism while using technological rationality in order to pursue their preposterous, irrational dreams. Who could deny this was true? But Herf also proposes that this represented a burdensome ideological contradiction. How can one reject the Enlightenment, equated more or less straightforwardly with modernity, but then champion technological prowess? This is a dilemma, however, only if we mistake the Enlightenment for modernity in general and mistake technological rationality as the pinnacle of all human reason derived from the Enlightenment. By and large, National Socialists were not among those who indulged in these assumptions. Moreover, the supposed contradiction between technical rationality and romanticism never bothered industrialists in Germany or anywhere else. As such it was never unique to the German engineering profession or National Socialism and seems to have manifested itself already during the French Revolution as well.[38] At least in the way that SS men spoke and acted, modernity had less to do with eighteenth-century political philosophy and more to do with a claim to futurism staked in terms of their mastery of the machines and modes of organization new to the twentieth century.

Ernst Jünger, for example, captured this fascination more than fifty years ago in terms that professional historians and sociologists would summarize again in the "modernization" debates of the past three decades. With near exaltation, he wrote: "Here the following must be named: the technological engagement of industry, economy, agriculture, traffic, administration, science, public opinion—in short, each special substance of modern life in a self-enclosed and elastic space, inside of which a common character of power manifests itself."[39] Jünger was a novelist and a man of letters, but the phenomenon he celebrated filled the imaginations of quite ordinary German managers and engineers. Witness Wisselinck's sketch. Their organizations and the novel technology that they commanded, not to mention the new social group of white-collar workers to which they belonged, had changed the visage of what all recognized as the "modern age." The SS officers at the focus of this book were no different. They consciously sought to articulate and construct a Nazi modernity and heralded their institutions and technological systems with no less enthusiasm than Jünger, even if they did so in much worse prose.

One last ideal appears in this study only in context with the other four, not because it is of lesser significance but because it pervaded all other ideological currents. Its separate treatment could not do justice to its influence. Namely, those who led SS technological enterprise shared a deep-seated belief in their own racial supremacy. They therefore believed in the legitimacy of murder and the forced labor of Jews and whoever did not count as "Aryan." Describing their sentiments as racial supremacy in no way downplays their anti-Semitism. Rather Nazi racial supremacy was much more prodigious in generating contempt for human life than an anti-ideology alone can account for. If we hold, as does Elie Wiesel, that "those who speak about the 11,000,000 [total victims of Nazi extermination] do not know what they are talking about . . . it is 6,000,000 Jews," we can never explain why the WVHA worked myriad concentration camp inmates to death and not only Jews.[40] There were differences of degree and number, to be sure, but slave labor was not a discriminating business in the fate that awaited the SS's victims.

Neither racial supremacy nor an unswerving belief in modernization was in any way peculiar to the SS. These two ideals were discussed in normal managerial communities throughout Germany and the West even before the Nazi era.[41] The SS rarely proved creative except in radicalizing general sentiments, and here again ideological consensus comes to center stage. Only rarely did outsiders, even powerful ones, oppose SS aims. As the movement to write the "history of everyday life" (*Alltagsgeschichte*) in the 1980s has shown, the Nazis proved successful in prosecuting only those policies that encountered no inveterate resistance in the population at large. Citizens on the sidelines seldom rose to impassioned activism in the Nazi cause; for this, only a few institutions like the SS were necessary. But the Nazis' most fanatic policies proved most successful when citizens had nothing against them or passively acquiesced.[42] What was true of Nazi society at large was no different in the microcosm of concentration camp industry. On one hand, the power of convictions could stir some individuals to resistance, but this happened rarely. On the other hand, ideological consensus moved others to passive toleration and cooperation. Some individuals opposed the SS on certain issues while complying on others. Put simply, at every stage in the WVHA's history, whether it lost or gained influence, ideology and its multivalent content mattered.

Within the WVHA a distinct community of officers inhabited each departmental division. Each depended on its members to solve problems with expert knowledge and to forge a working consensus—that is, on the ability to act on ideals as well as mere issues of problem solution. Failure could result from a deficit of either sound business skills or consensus, while neither alone sufficed in and of itself to ensure success. The WVHA's construction corps presents an example of managerial success in which both consensus and skill coincided. Here Oswald Pohl recruited a tightly knit cadre of civil engineers and architects who had already worked together in a parallel branch of the state, the German Air Force. They represented the highest concentration of technically trained officers within the WVHA and, in all likelihood, within the SS as a whole. Not insignificantly, they came from an engineering tradition with the longest-standing connection to state service and the military. Further, a significant proportion had overt commitments to the Nazi cause. Many were activists. Their chief, Hans Kammler, was able to inspire their cooperation, and their success during the war was horrific for its brutal efficiency. Here "extermination through work" became a reality as civil engineers managed productive labor and genocide on the same projects.

If the WVHA's corps of engineers fostered concerted action through shared goals, other managerial echelons proved dysfunctional due to ideological strife. For example, during the general shortage of all white-collar personnel in Germany after 1936, Oswald Pohl had to look for competent factory managers outside his close circle of SS ideologues but failed to elicit their dedication to the plexus of ideological goals embedded in the SS's prison factories. As a result, dedicated managers soon complained that they could not work with the newcomers; meanwhile, the newcomers complained about the hostility of the old guard. No one worked together, and their enterprises fell apart accordingly.

It was part of the absurdity of Nazi Germany that prisoners paid most dearly for such mismanagement. Among Kommandanten, the commanders of the concentration camps, and their staffs a core had formed who shared a homogeneous sense of purpose, albeit one that demanded the brutalization of prisoners. On the other hand, the Kommandanten had few managerial skills, administrative or technical. They excelled only at terror and wreaked havoc on industrial production. In fact, industries proved successful in utilizing the SS's slaves only when they removed concentration camp guards from the direct technical supervision of production (dealt with in chapters 6 and 7). Most SS factories crumbled along fault lines of managerial inconsistency and conflicting commitments. They failed, however, not because rational managers cannot operate under fanatic ideological influences, as is commonly held. Rather, SS industry broke down because SS managers came to loggerheads over antinomic issues, issues in which they believed.

This book is organized narratively and begins with the origins of the WVHA in the SS administration of the early thirties. It ends with the utter collapse of the Third Reich, which brought the SS's empire down with it. Throughout, however, the emphasis falls on what it meant to be an SS "organization man" and how SS managers strove to establish their own identity as members of a modern German "race" by dehumanizing the Third Reich's outcasts. This is not to "adopt" the viewpoint of the criminal but to lay bare the capacity of otherwise normal, modern organizations for barbarity. The SS Business Administration Main Office is of twofold importance. First, it reveals the historically unique use of modern means in slave labor and genocide. Second, this institution operated with the same basic structure as any other modern organization in the West. In this sense its managers were "ordinary men." Because the WVHA worked in such familiar ways, here the Nazi catastrophe cannot be conceived as an unfathomable exception in Western history.[43] "I suppose you would feel better if I told you all those who implemented the holocaust were demented," Raul Hilberg once remarked.[44] Had they been, the task of maintaining a just and equitable society would certainly

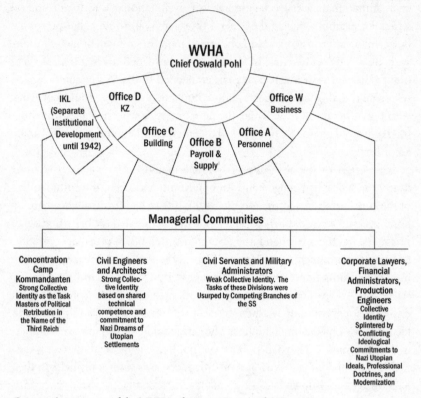

WVHA
Chief Oswald Pohl

IKL
(Separate Institutional Development until 1942)

Office D
KZ

Office C
Building

Office B
Payroll & Supply

Office A
Personnel

Office W
Business

Managerial Communities

Concentration Camp Kommandanten Strong Collective Identity as the Task Masters of Political Retribution in the Name of the Third Reich	**Civil Engineers and Architects** Strong Collective Identity based on shared technical competence and commitment to Nazi Dreams of Utopian Settlements	**Civil Servants and Military Administrators** Weak Collective Identity. The Tasks of these Divisions were Usurped by Competing Branches of the SS	**Corporate Lawyers, Financial Administrators, Production Engineers** Collective Identity Splintered by Conflicting Ideological Commitments to Nazi Utopian Ideals, Professional Doctrines, and Modernization

Community structure of the WVHA after 1942. Drawing by Steve Hsu.

be much easier. We would only have to round up the cretinous madmen who conform to the caricatures of evil presented as the bad guys in American Saturday morning cartoons. Sadly, however, the task is much more difficult and requires understanding how men and institutions that differed little from those found in any other modern industrialized society became the eager tools of genocide.

CHAPTER 1
ORIGINS OF THE SS
THE IDEOLOGY IS THE
MODERN ORGANIZATION

The SS began as a small clique of Nazi street fighters in the SA (Sturmabteilung, or Storm Division) and swiftly became the dominant organ of executive power in the Third Reich. It ballooned after 1933 into a nationwide organization almost overnight. Many historians speak of a "pragmatic" and a "fanatic" side of National Socialism, of coldly rational, officious functionaries and a lunatic fringe of propagandists. But the dividing line is hard to draw between such pragmatic and fanatic "sides" in the SS. Its founders sought to make the very structure of their organizations embody National Socialism. At first it may seem contradictory to note that Nazis set out to institutionalize their self-proclaimed dynamism in hierarchical institutions. What other name for such institutions is there than bureaucracy, which the Nazis so loudly derided? But the SS identified its own hierarchies with the regeneration of "German" values. "Vision" was supposed to set them apart, and this was no empty rhetoric. The SS's conception of an ideal society shaped the personnel that Heinrich Himmler recruited no less than the offices they inhabited. Conversely, incoming individuals with a strong sense of purpose could shape the SS. The relation was mutual, especially in the flux of the early, formative years when the SS achieved rapid successes. Once the SS became more than a street fighters' club, the coordination of even the most rudimentary information necessarily demanded modern bureaucracy. Here too ideology and organization ran together because the will to "be modern" itself motivated SS men (though it was hardly unique to them).

This chapter focuses on three aspects of the SS's early push: its financial administration under the leadership of Oswald Pohl, Himmler's first industrial ventures, and the foundation of the concentration camps by Theodor Eicke. Eventually, the exploitation of slave labor after 1942 brought them all together, but in the mid-1930s they remained distinct. By 1936 Pohl promptly built a modern administrative corps after Himmler recruited him to oversee the SS's expanding budgets. At the same time Himmler founded the fledgling SS companies within his own Personal Staff that were the precursors to the SS's industrial empire. Initially they were run in such a dilettante manner that, by 1938–39, Pohl had to step in and take them over completely. From 1933 to 1936 the concentration camps grew exponentially as Theodor Eicke established their legitimacy as new prisons, secured funds for their expansion, and recruited officers with a self-proclaimed mission to protect the Nazi body politic; his guards, specially named the Death's Head Units, established a firm identity as the Third Reich's appointed punishers.

From the very beginning the particular ethos of how to be Himmler's "organization man" within such diverse branches of the SS was often contradictory and even led to genuine internecine strife. Pohl's men prided themselves as modern administrators. When it came time to manage prison industries, they clashed with camp guards who, true to their self-conception as punishers, often beat and killed prisoners even when this undermined productivity within the SS's own corporations. In all cases, however, these were not empty disputes about authority or mere power. Conflict erupted over the SS's vision and mission in Nazi Germany and often gained in bitterness when all involved agreed about lofty goals of National Socialist fundamentalism but could not agree about implementation. Whenever mismanagement, "polycratic" strife, or incompetence arose, however, prisoners ultimately bore the brunt as its victims. Sadly, one point of solid consensus among all SS men involved was that prisoners should and must suffer and that SS bureaucracy should serve this end.

Because production became the focus of later disputes over contradictory SS goals, persons recruited within the camps as technological managers warrant special attention, for they alone possessed the training to organize the complex conjunction of workers, raw materials, and machinery in factory systems. We would therefore expect to find them in Himmler's SS companies, but that is where competent technological management was almost wholly lacking, a major cause of the bungling that would prompt Pohl's entrance into these enterprises in the first place. Initially, the SS recruited only one significant group of engineers, namely construction personnel, primarily in the concentration camps. Factory managers were notably absent. Nevertheless, the SS civil engineers began to form a corps of technically competent and ideologically dedicated personnel who later propelled Pohl's organization to the center of the Nazi war economy by efficiently working prisoners to death.

Modern Men: The New Administrative Officers of the SS

Six months before the first Nazi attempt to seize power in the Beer Hall Putsch of 1923, Adolf Hitler asked two of his closest henchmen to organize a small guard. With typical narcissism, he dubbed these men the Shock Troop Adolf Hitler. After the abortive putsch, the Nazi Party was banned, and the members of this shock troop either went underground or fell out of Nazi circles until the party reemerged in 1925. Nevertheless, despite the obvious discontinuity, the SS claimed the Shock Troop Adolf Hitler as its point of origin. The mature SS came to include a massive, professional secret police force, the nation's civil police precincts (the *Ordnungspolizei*), elite military battalions, medical institutes, schools, government ministries, and an industrial empire. The later hydra-headed organization owed little except its myth of foundation to the period before 1925. Most important SS leaders, including the Reichsführer SS Heinrich Himmler, had never belonged to the Shock Troop Adolf Hitler, yet the myth was important, for Hitler himself attributed the origins of the SS to the early days in Munich when he had needed "a small guard, whose men are sworn to do exactly what they are ordered without resistance, even if it means standing against their own brothers."[1] With the exception of Himmler's belated and ridiculous attempt to forge a separate peace with the Allies in mid-April of 1945, the SS never deserted Hitler's mandate to protect, with unquestioning loyalty, the person of the Führer. Further it assumed the duty to "protect" all Germandom, that is, to police the German body politic in Hitler's name.

The SS first began to take real shape in 1927 when Heinrich Himmler came from the Nazi hustings of Lower Bavaria. There he had worked in the propaganda office of Gregor Straßer, one of the most charismatic Nazi pundits who campaigned on a populist brand of anticapitalism: Jews and big business were squeezing the virtuous "little man" out of German society; spiritless urban living had removed German citizens from healthy, community life; and, worse still, the manipulative markets of mass consumer society were rotting German "culture." Himmler had also been in the Artamanen Bund, a band of young visionaries who dreamed of leading German expansion eastward as colonial "warrior-farmers." They drew their inspiration from Richard Walther Darré, like Straßer, a charismatic populist who preached that the nation could preserve its racial mettle only by bringing German citizens into communion with the German soil and, further, that the nation itself needed new living space (*Lebensraum*). The young Himmler imagined himself as an Aryan crusader on his way to reclaim eastern Europe from uncultured Slavic "races." Although his policemen later purged the Nazi celebrities around Straßer in 1934 and deserted Darré in 1938, there is no reason to doubt Himmler's personal identification with the ideals these men had espoused. Rather Himmler believed that the mature SS had to achieve a truer, purer expression of its earlier visions.[2]

Aside from formative ideals, the most important thing that Himmler brought with him was his capacity for organizing. After January 1929, when he took charge as Reichsführer SS, his hand is unmistakable. The SS had had around 1,000 members; on the eve of the Nazi seizure of power it had 50,000. The SS quickly tripled, then quadrupled, and then burst all bounds as membership in all Nazi organizations began to grow exponentially. More important than the statistics was the kind of man the SS called to its rolls. In the 1920s the SS had scarcely differed from the SA, a group of thugs who liked to brawl and did the party's street fighting. The ascendance of the SS began precisely because Himmler was able to create a supple, responsive organization in contrast to the Nazis' mercurial roughnecks. SS men wore black instead of the "Brown Shirts" in the SA. Of course, especially early on, the Black Corps had their share of street bullies too, but by the 1930s recruits began to conform to middle-class values that Himmler himself reflected in his punctiliousness, his prudishness, and his abstinence from alcohol.[3]

A new character resulted from effective recruitment among middle-class professions and white-collar men, which intimately linked modern organization and ideology. As white-collar workers started joining the bread lines in the heart of the world depression, Germany was full of men looking for alternatives, especially for organizations peddling a sense of purpose and a vision of the future. The SS told them that Germany's economic chaos confirmed what the NSDAP had foretold all along. The SS offered the aura of political activism, national pride, and the chance to build new institutions. The SS succeeded in establishing itself as a "new aristocracy" in German society; it drew in those of high stature and simultaneously conferred status to members seeking social advance. Because "modernity" is generally equated with dynamic social mobility, the SS also seemed "modern" in this regard. The SS sold itself as an organized elite open to the capable and the energetic, a home to men who wished to build something new. Here the SS's white-collar recruitment proved one of its most valuable assets. Its expansion was driven by an influx of precisely those people who had the knowledge and experience to control modern administrative structures: clerks, bookkeepers, midlevel managers, and lawyers.[4]

One of the practical problems faced by the SS was its burgeoning budget needs. Before the 1930s budgeting had involved little beyond postal fees for pamphlet distribution. SS men had also paid dues of fifty Pfennigs into self-help funds. After 1930, as membership climbed into the tens of thousands, the SS needed a way to coordinate its budget. In 1931 Himmler appointed a treasurer (*Geldverwalter*) to his Personal Staff, the head of a new Department IV. At the same time, every regional SS Section (Abschnitt) had to erect a parallel Department IV for the management of dues and fees. The SS treasurer was quickly overwhelmed by the increasing scope of his tasks, for the SS once more doubled its membership in the five months between January 1933 and Hitler's birthday on 20 April. After this, the SS doubled yet again, achieving 200,000 members by 1934. Only the German air force would hold a greater appeal to young German men seeking elite formations. At this point, Himmler closed membership temporarily. Nominally he did this to protect the mystique which he had carefully built up around his Black Shirts. "Newbies" (*Neulings*) were seen as opportunists, and they posed a danger to the ideals of loyalty upon which Himmler had built the SS's legitimacy. Beyond these concerns, however, the burgeoning SS had outstripped his staff's ability to control its growing complexity.[5] The SS had to maintain the cherished activist image of its early years while erecting an impersonal, hierarchical organization to contain and then channel it.

During this period, Himmler consciously began to recruit top administrative talent, and he looked first for military officers. Since assuming the title Reichsführer SS he had envisioned a paramilitary organization, a fantasy evident in the SS's myth of origin as an elite "shock troop." Glorification of the military had much to do with a general adulation of soldiers in Germany, especially the veterans of the First World War. Himmler had been too young to serve but shared with many of his generation a romantic nostalgia for the trenches, which represented a perished time and place of common resolve, glory, and discipline—where men had united to protect kith and kin in unambiguous devotion to the Fatherland. Himmler grafted these imagined ideals onto his organization. His officer corps was to be a military community in which all subordinated their personalities to the interest of national renewal and a higher "good." It typifies Himmler's elite pretensions that he aspired to even more than the competence and discipline of the traditional German army; he wished his men to display ideological fervor above and beyond that of the Wehrmacht. He believed that political conviction would prepare the SS to sacrifice more, to fight more fiercely, to work harder in the name of Nazi Germany than regular troops. Therefore he actively sought out military administrators who believed in the Nazi movement in order to transfer their expertise, élan, and organizational practice to the SS.[6]

Thus it was not the first time or the last that the SS tapped the talent of established career officers when, in the spring of 1933, Himmler asked a navy paymaster, Lieutenant Captain Oswald Pohl, to help him build a new administration from the ground up. In fact, high-ranking military men sympathetic to the SS aided Himmler. This was, in fact, the other side of the "polycratic" coin minted by Hitler's regime. National Socialism spawned a wild growth of overlapping institutions, but, as often as not, the result was eager cooperation rather than knee-jerk rivalry. The armed services were still chafing under the restrictions of Versailles, and often welcomed new paramilitary, nationalist organizations. Cooperation with Himmler in this case was wholly voluntary, and it took place in the *gemütlich* (charming) atmosphere of beer gardens. Himmler came across Pohl after Admiral Wilhelm Canaris mentioned him as an especially energetic officer known to be a dedicated Nazi. A letter to Himmler from Pohl's immediate superior also gave a glowing recommendation, "I especially wish to call your attention to navy Lieutenant Captain Pohl. He was already active in the [National Socialist] movement in 1925, and he is a first-rate man in every way."[7] Perhaps Pohl elicited this letter, for it reached Himmler on the same day that both men met outside a casino in Kiel. Himmler had little trouble convincing Pohl to dedicate his career to the SS. The chummy tone of Pohl's follow-up letter shows that the navy officer could hardly wait:

> You know quite well that I have an absolutely secure life in the navy, a career that is quite envious in its outward splendor, its benefits, and this or that advantage. I find, however, that my professional activities do not offer me sufficient spiritual satisfaction. They do not offer me a vent for my lust for accomplishment and for my fury to work. I can and will work until I collapse. If you think you could use such a fellow, then I will accept in spite of the insecurity this would imply for my future career.[8]

Pohl was unusually old for the SS; he would even acquire the status of a "gray eminence" in wartime. But he was old only by comparison to the overwhelming youth of the SS officer corps. When he and Himmler sat down together in that Kiel beer garden, Pohl's forty-first birthday was little more than a month away. Nevertheless, he was looking across the table at a man ten years his junior. Pohl in his turn would seek out young, energetic men. In 1942 his leading officers would be, on the average, just thirty years old.

Pohl's father had been a foreman in the August Thyssen Steel Mills. His childhood was financially secure, if not opulent, and he had attended the *Realgymnasium* (a modern secondary school, as opposed to the *Gymnasium*, whose curriculum was based on classic texts, Latin, and Greek). After the war, Pohl claimed that he had wanted to study science but that his father had lacked the means to send him to a university. This is probably true, but Pohl's social standing was sufficient for him to forge ahead into an elite navy career, a branch of the German armed forces known for its relative openness to social advance as well as technical and organizational modernization. He joined in 1912, and when he could afford to pay the tuition of thirty Reichsmarks a month, he entered the navy's administrative course of studies. He served out the First World War as a paymaster and began legal studies at the University of Kiel after demobilization. He quickly abandoned these when he was again offered a navy commission. This made him one of the few soldiers to weather the drastic cutbacks demanded at Versailles. By 1934 he presided over a staff of 500 men and associated with a group of young officers who promoted reform. When Himmler offered him a commission in the SS, the chance to create an entirely new organization undoubtedly held an irresistible attraction. Pohl admitted as much after the war, even when he had begun to lie about his past to save himself from the gallows: "I saw a wonderful opportunity to carry out the ideas expressed in [my] reformist tendencies. . . . They were my own ideas. I accepted the offer [from Himmler]."[9] This shows a level of ambition beyond mere concern for power or career promotion. Pohl yearned for the challenge of new tasks and the satisfaction of creating and mastering the complexity of modern institutions.

Most SS administrative officers would share Pohl's restless spirit of administrative reform, and, moreover, most had also received systematic training in some aspect of modern administration. In fact, almost eight out of ten received training above the secondary level and over 40 percent of these held higher academic degrees, most typically in economics, business law, or—in the future construction corps organized by Pohl—civil engineering. This was over twice the proportion one could expect in the SS in general, and the SS officer corps was, as a rule, more highly educated than the German population at large. Typically Pohl's men had sought pathways to mobility through the new administrative and managerial occupations of the twentieth century. Out of 126 officers who left accurate information, 27 percent had come out of old middle-class families to work at office jobs. In other words, they were the sons of farmers or small-time proprietors who entered the burgeoning world of white-collar labor, the world Siegfried Kracauer called "the model of the future."[10] In general, very few SS officers moved from working-class backgrounds into the officer corps; nevertheless, at least seven officers under Pohl made this jump. The largest single group of men in SS administration had entered from backgrounds already recognized as elite—having fathers, for example, who were doctors, lawyers, entrepreneurs, or higher civil servants. On the other hand, the absence of people threatened by declining social status is notable. Given the most skeptical interpretation, roughly a quarter (23 percent) might be counted at risk of falling in social status before joining the SS, but almost all of these were young men who had merely "fallen" from the elite stature of their fathers into lower white-collar positions early in their careers. Those on the rise outnumbered them. Most in this corps of officers had already made career choices that manifested a "modernizing" drive, and few seem to have joined the SS out of financial desperation.

Generally, they also held deep ideological convictions. Here too Pohl is typical. He had long belonged to various German ultranationalist organizations as well as the Free Corps Löwenfeld. (The Free Corps were right-wing, ultranationalist organizations of demobilized troops that met, drilled, and clandestinely sought to preserve Germany's military might after the crippling cutbacks demanded at Versailles.) He joined the NSDAP in 1922 and remained active in a cover organization after the party was banned (the Volksbund Uwe Jens Lornsen). By 1926 he ran as a Nazi candidate in local elections. In short, he was a political activist. "I was a National Socialist before National Socialism came into being," he declared in 1932.[11] He wanted to be in the vanguard and sacrificed no small portion of his free time to this end.

Private life plays only a small role in the history told here, but it is nevertheless interesting to note that, for the most part, SS administrators married active women. Here again they showed themselves to be "modern" men. Usually they had met their wives in the new work world of the twentieth century and its connate white-collar culture. Pohl himself divorced his first wife to marry Elenore von Brüning, a graphic designer. Of his top officers who offered information in their personnel files, fewer than a third identified their wives under the traditional category "housewife," the rest claiming some kind of profession or trade. Pohl officially enforced gender stereotypes in his offices, forbidding, for example, secretaries to wear pants or makeup and ridiculing those who did as "little men-women" (*Mannweibchen*), "cannibals," and "Papuas."[12] But were not these wives the active, new women of the 1920s officially despised by SS rhetoric? Were not these the "flappers," the "feminists" with bobbed hair as shown in some SS marriage pictures? This brief contrast points up another character trait that appears again and again among SS administrators, businessmen, and engineers. If they felt they had a deep need to institutionalize their ideals, like so many fundamentalists of any stripe, they also possessed a nearly limitless capacity for hypocrisy.

Pohl's pathway into Himmler's SS was marked indelibly by circumstances peculiar to the man, yet he embodied traits of life-style, ideological vision, and professional training in common with those he sought to recruit. What kind of institutions did such men build? When Pohl officially joined Himmler's staff in February 1934, he found almost unlimited tasks at hand. SS headquarters still lay in Munich but would transfer to Berlin in the new year. Expansion quickly threatened to burgeon out of Himmler's control. The Reichsführer subsequently split SS functions among three new "Main Offices": the Race and Settlement Main Office, which planned SS cultural and racial policy; the Security Main Office, which controlled the SS political police; and, last, a coordinating bureau simply named the SS Main Office. Parallel to these, Himmler continued to expand his own Personal Staff.

Before funds had come mostly from membership dues, but now the Reich government and party treasury began covering ever larger portions of the SS's budgets. To get state money, Himmler knew that he needed a top organization man who could lend an aura of fiscal fidelity, and Pohl was to make good on his promised "fury for work" in this capacity. In January 1935 he took over the Office IV–Administration, soon to move to Berlin with the rest of Himmler's SS offices. Pohl's tasks were twofold. On one hand, he had to justify expenditures in the eyes of state and party bureaucracies. On the other hand, Pohl knew that he had to professionalize the SS's bureaucratic structure to accomplish this. Himmler created a new position throughout the regional SS Sections, the Administrative Führer, to coordinate accounts and bookkeeping. The SS had long attracted men with bookkeeping skills to these sections. Now members who had handled such work in their free time received promotions and full-time jobs. Some of Pohl's most loyal officers would rise from this low level. At their head, Pohl was plying the skills he had learned as a civil servant over the past twenty years in the navy. He succeeded in molding informal and amateurish accounting practices into a standardized, homogeneous system so that the SS could stand up to public audit. National agencies began to respect the SS accordingly.[13]

So pleased was Himmler that he promoted Pohl, then an SS colonel, to brigadier general, a forward leap of two full ranks. Simultaneously, Himmler made him the official liaison to the Nazi Party treasurer and bestowed a new title, SS Chief of Administrative. Pohl was elated and initially found the news too good to be true. On learning of his new titles, he first wrote to his personal friend, Himmler's adjutant, Karl Wolff, and asked if there might be some mistake. Wolff assured him that it was true, and that the Reichsführer SS expected his administrative chief to master the increasing complexity of his tasks.[14] Pohl eagerly greeted the challenge by forming a "V" office within the Personal Staff ("V" stood for *Verwaltung*, administration). From now on, all party funds flowed through his hands and were disbursed through the chan-

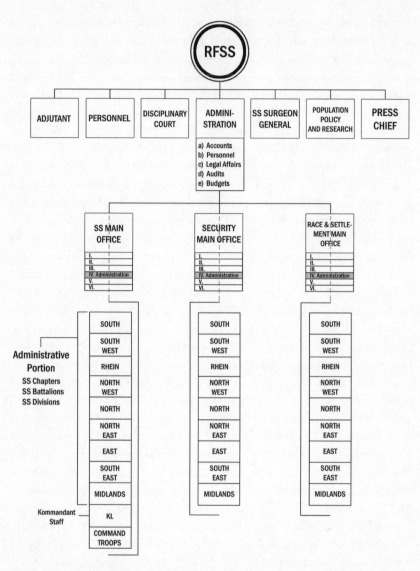

SS administration at the time of Pohl's first administrative reforms, based on reproduced chart from BAK NS3/555 (also reproduced in Tuchel, Konzentrationslager, *259). Drawing by Steve Hsu.*

nels he had defined. His office coordinated parallel administrative bureaus in all SS main offices.[15] In short, Pohl had quickly created a nationwide, impersonal modern administration.

In June 1936 Hitler had just made Himmler the Chief of German Police, granting the Reichsführer SS overarching authority for all state security forces.

This marked a consolidation of SS power. It also gave Himmler justification to demand Reich funds for his police formations, funds Pohl managed. Pohl also helped to tap the coffers of Reich ministries for Himmler's fledgling military battalions (the future Waffen SS) and, in 1936, for the concentration camps.[16] He seems to have helped quietly, working up budget proposals that others then submitted to the Reich Interior Ministry through the SS Security Main Office. He also helped recruit capable officers to the concentration camps' administrative divisions.[17]

By late 1938 Pohl's internal system of centralized control over nested administrative offices was beginning to run smoothly. Increasingly he deputed financial administration to one of his most trusted subordinates, August Frank, who, like Pohl, was energetic and ambitious. He had come to the SS from the Bavarian civil service, the state police administration (*Landpolizei-Verwaltung*), and had done the bulk of Pohl's tedious work, mostly projecting and compiling budgets. Yet, typical of the spirit Pohl had sought to foster, Frank did not find this work banal; rather he felt himself to be the kind of self-made man whom the SS had given a chance to rise. "From recruit up to general every administrative Führer can make it under his own power. In 1933 I was myself just a simple SS Man," Frank boasted when he became the second administrative officer to make general's rank (brigadier general) in 1940.[18]

The Führer Principle

Meanwhile, as Pohl left daily routine in the hands of such capable subordinates, he began to seek new vents for his "fury to work." A restless quest for tasks characterized Pohl's leadership as he consciously followed the Führer principle, at once a Nazi doctrine of personal leadership and national community. The Nazi Party had achieved its electoral gains by condemning the factionalism and bickering of Germany's parliamentary democracy, which seemed to go hand in hand with hyperinflation and world depression. It was easy to believe that both democracy and liberal free trade had delivered Germany to the predation of international interests while the nation itself had been left adrift and emasculate. In a startling feat of irrationality, Jews were considered to be behind all evils. In all cases the Nazis promised national unity as an antidote, to be crystallized by a Führer who harmonized the will of all Germans. Conflicts of labor and capital, of state government and opposition, of local and national identity, all would be swept aside, obviated by the manifest power of the Führer to embody the will of all. Debate would yield to resolve; the economy would spring to the firm hand of command. Such dreams of order had wide appeal across all segments of German society, with perhaps an unusual success among middle-class, white-collar workers.[19] The SS benefited from this popularity. Many of its most creative and energetic officers joined after the Nazis' first big electoral gains of the early 1930s precisely because they believed that a national community (*Volksgemeinschaft*) united by Hitler could reverse Germany's demoralization by fiat.[20]

No one believed such rhetoric more deeply than Oswald Pohl, whose managerial philosophy reflected his tutelage in SS ideology. He tended to mistake action for innovation, effort for results, and congenial agreement for consensus on substantive issues. This was unsurprising in an organization in which "restless realization" (*restlose Durchführung*) was a watchword and "ruthless" (*rücksichtslos*) larded praise of work well done. Pohl's ideals led him and his subordinates to diagnose the symptoms of managerial inefficiency in paralysis, disharmony, and lack of commitment. He and his officers prescribed a cure: action, united resolve, and political dedication. In the midst of reforms in 1939, one memo appealed directly to this elite sense of community:

> The industrial tasks of the SS should and must be carried out by the SS man, for he brings more discipline and the dedication which is an absolute requisite for this work. More can be demanded of him than from a civilian; he is more worthy than those . . . who simply wish to put on the uniform of the Black Corps and strive after an influential job in the main offices of the Reichsführer SS—whenever possible immediately—out of material interests.[21]

Instead of materialistic pursuits, Pohl expected submission to a "higher cause." As hard as it is to believe from the vantage of the present, the SS supplied a sense of mission in abundance through its hypernationalism and racial supremacy. Its task was nothing less than the reinvention of all Germandom. "[The SS] is above all else an institution of the NSDAP [National Socialist German Labor Party] as the dynamic element in the state," one officer declared.[22]

The Führer principle told its believers to rush headlong in quest of new problems to solve, new Gordian knots to cut, and it tended to absolve its followers from any demand for consistency. Believers in the Führer principle tended to trust their instincts. Men who aspired to be Führer administrators believed that their leadership abilities emanated naturally from a personal embodiment of their subordinates' will, likewise of the German will writ large. Thus once Pohl assigned officers to new tasks, he expected to see them constantly generating initiatives as he himself did. Pohl often showed a talent for locating skilled personnel to fill subordinate positions, like Frank; but he would also fail, as when he rushed into the management of prison labor after 1939. In 1938, however, it was not yet clear in what new direction Pohl would storm off to next; it was only clear that he felt himself to be a revolutionary administrator, spoiling for action.

Needless to say, this general style did not encourage a rigid "iron cage." It bred hierarchy, to be sure. If the Führer principle encouraged managers to act decisively, this doctrine never meant to unleash unbridled individualism, for which Hitler, Himmler, and the SS had nothing but scorn. They associated individualism, in the traditional sense, with the old-fashioned ideas of Anglo-American liberalism. Although the Führer principle encouraged individuals to take initiative, it also led them to interpret their acts in the framework of collective mobilization. In short, it fit, however perversely, the essential tension between impersonal hierarchy and individual command and control at the heart of dynamic modern organizations.

There was one crowning irony in Pohl's "restless" initiative, however. Since 1934 he had never really undertaken anything radically new. True enough, he had been engaged in constant reorganization, and the scale of his activities had increased steadily; nevertheless, he had done nothing other than methodically and competently apply the professional skills taught to him by the navy: modern financial administration based on statistical, standardized accounting. Shortly, Pohl would divert his energies to industrial production. This called for technical management skills foreign to the navy paymaster, namely the statistical surveillance not only of accounts but of machines, labor processes, and the material world of the factory. Before examining this in detail, however, it is necessary to introduce the original SS companies as well as the first managers of prison labor in the concentration camps. Both sprang into existence independently of Pohl.

Heinrich Himmler's Favored Industrial Projects

Himmler began dabbling in corporate ventures almost immediately after the Nazi seizure of power, first founding a publishing house in the last weeks of 1934. The Nordland Verlag brought out around 200 publications over the next ten years. These ran the gamut from training manuals, political tracts, collected speeches, pseudoscientific propaganda, and diatribes against Jews, Freemasons, and organized religion. The press also brought out novels in which heroes persevered by virtue of superior German culture and the Nordic race. The SS could have easily contracted with existing publishers, but Himmler conceived of the SS as a rescue organization for Germany's cultural soul.

This mission lay at the root of all SS institutions, including its budding industrial empire. Nordland promised to "bring the SS world view [*Weltanschauung*] to the SS membership and to the people," a purpose too lofty, too important to trust to private industry.[23]

Permutations of racism appear in the title of a Nordland anti-Soviet brochure, "The Subhuman" (*Untermensch*).[24] But racial supremacy was only one ideological current. One of the most pronounced values in SS entrepreneurship was its productivism, defined here as the belief that industrial and economic activity should be bent to the service of national identity rather than sordid profit gains. To the SS, the efflorescence of German culture, the forging of spirit—not the prosperity of citizens or economic utility—constituted the highest purpose of the factory. Therefore a strong, centralized state should seek to control industry in such a way as to preserve the sanctity of Germandom, a cause in which mere cost could not be allowed to set the ultimate bottom line. SS business managers would cling to this ideal even as the Third Reich began to crumble. An SS lawyer who later directed the acquisition of SS industries in occupied Poland even wrote a manifesto on this theme:

> Why does the SS pursue business? This question is thrown at us especially by those who think in purely capitalist terms and look unfavorably on public enterprise or at least on enterprises that have a public character. The time of liberal economics promoted the primacy of business. That is, first comes the economy and then the state. In contrast, National Socialism stands by the point, the state commands the economy; the state is not there for the economy, but the economy is there for the state.[25]

The lawyer, Leo Volk, prepared this tract to indoctrinate a new generation of SS businessmen in an officer school. By serving the state, SS companies were to manufacture, literally, German unity and Nazi values.

SS men were wont to arrogate to themselves the mantle of pure culture, and they believed that a German renaissance required a new industrial order. This feeling often superseded any conception of what form, in the end, German culture should actually take, but it was no less passionate. A low-level memorandum by an SS security officer in Munich captures well the SS's mixture of pretension and resentment:

> The top executives think only of their balance sheets. . . . The new [National Socialist] development of economics carries with it a spiritual precondition: the willingness of top executives to dedicate their creativity to the nation. . . . They do not understand it. We are therefore called upon as an organ of the state to dethrone the monopoly power of business [*konzernge-*

bundene Eigenunternehmung] and to break its hold on private initiative. We must win over the soul of the top executive so that he can act according to our point of view.[26]

This SS officer did not so much want to topple monopoly power, secure the rights of the working class, or equalize the distribution of goods and services throughout German society as other critics of capitalism often do. Rather, he was upset about the corporate soul; he was worried that businessmen just did not exhibit the correct "feelings," the right enthusiasm for national identity and the SS "point of view." Himmler wanted SS companies to provide the necessary lever for cultural policies that could not be trusted to the mere businessmen who simply "did not understand it." SS industries were to serve as beacons of soulful uplift.

On the other hand, productivism strongly discouraged systematic formulation of financial goals. Talk of money was beneath these men's cultural dignity, and this sentiment goes a long way toward explaining the chaotic nature of the first SS companies. Himmler rushed into a mishmash of different industries. Rhetoric of unified will notwithstanding, a concomitant mishmash of legal and financial organization was the result. By the summer of 1937 the SS had acquired a 55 percent interest in a photographic studio, FF Bauer GmbH, based on the personal friendship between Himmler and two brothers who had started the firm, Friedrich and Franz Bauer. Like Nordland, the studio was supposed to dedicate itself to "cultural and ideological [*weltanschauliche*] tasks." It was not required or even expected to run in the black.[27] The SS had also invested in a building cooperative, the Cooperative Dwellings and Homesteads GmbH (Gemeinnützige Wohnungs- und Heimstätten GmbH) to build idyllic SS communities.[28] "These things interest us," Himmler stated of a similar undertaking to preserve German monuments "because they are of great importance in ideological and political struggle. It is my goal that every SS regional division [*Standarte*] shall have one cultural center in which German greatness and the German past can be displayed . . . in such a way that is worthy of a cultural people."[29]

At the beginning of January 1936 Himmler entered manufacturing for the first time with the founding of the Allach Porcelain Manufacture GmbH, which remained forever after his industrial darling. Like the Bauer brothers, its founders had befriended the Reichsführer SS and secured a starting capital of 45,000 Reichsmarks. This firm too displayed the SS's special obsession with culture. Himmler wanted "art in every German home, but first of all, in the houses of my SS men" and intended Allach to provide it.[30] The founders were all artists: the sculptors Theodor Kärner and Bruno Galke, and the painters

Franz Nagy and Karl Diebitsch. Here again profit was not the main goal, and the firm booked a loss equal to almost 90 percent of its starting capital in the first year, partly because Himmler insisted on giving away its wares to SS dignitaries.[31]

Allach Porcelain Manufacture also indulged pretensions to technical wizardry typical of the SS. "It was the will of the Reichsführer SS to distinguish the selected elite of the SS on a cultural level," remarked Karl Wolff, who headed Himmler's Personal Staff, a point that he tediously belabored to his captors after the war:

> [Allach was] a kind of billboard for the SS's cultural representation. Only highly valued, artistic porcelain was produced, so distinguished that it overcame the greatest technological difficulties. These consisted in producing a figurine horse with one rider supported only by the two thin legs of the horse without the usual support of the heavy body of the horse under the belly with an allegorical tree trunk, branch, or flower. Even the other famous German manufacturers like Meissen, Nymphenburg, and so forth could not manage this technical achievement. It was the will of the Reichsführer SS.[32]

Despite high pretensions, Allach produced kitschy statuettes not unlike those peddled in second-rate gift shops in any tourist town, yet breast-beating in the name of culture and technological progress were no less real for all this hypocrisy. Productivism—a quest for cultural purity in the act of making things— gave free rein to a fascination for high technology *and* a faith that this expressed culture, especially that which seemed to embody the future or elite status. At the very least, it is hard to construe the SS's ethos in such matters as "antimodern" or "anti-intellectual." That Allach consulted with former Bauhaus students such as Wilhelm Wagenfeld and the director of the State Porcelain Gallery of Dresden is just further evidence of this.

Visions of fabulous inventions fueled Himmler's imagination, although he himself was no engineer. Beyond producing statuettes, in these early years the drive to incorporate modern technology, innovation, and rational design (distinct from the actual practice of mastering such technics) issued in the shabby enterprise of the Anton Loibl GmbH in September 1936. Anton Loibl had been one of Hitler's chauffeurs and styled himself as an inventor. As had previous SS entrepreneurs, he approached Himmler as a personal friend. In the mid-1930s he claimed to have developed and patented, of all things, an advanced bicycle pedal light. Himmler allowed Loibl an absurd entitlement to 50 percent of profits in a new corporation, although the "inventor" carried none of the risks (normally inventors receive no more than 5 percent). Despite the mundane nature of this technology, we should not forget that such gadgets were relatively new in the 1920s and 1930s. The founder of Panasonic, in fact, had got his start by selling bicycle lights in Japan little more than a decade earlier, and this was a time when electricity commanded the same fascination that computers do now. By 1944 Himmler's gullible mesmerization by "gee-whiz" technology would drive SS interest in the so-called miracle weapons, but in the salad days of the mid-1930s it led to a genuine expectation that the Loibl GmbH, with its pedal-light patent, would blossom as a research and development house for "technical articles of all kinds."[33] Loibl ran the company in a dilettante manner, and in 1939 Pohl's executives exposed his patents as indefensible and removed the chauffeur altogether. Similar to Allach, which aspired to cultural and technical artistry but produced kitsch, the Loibl GmbH aspired to high-tech research and development but produced bicycle lights.

These early endeavors remained small, if not utterly ridiculous. The largest investment before 1938 was tied up in a nonprofit foundation, the Society for the Promotion and Care of German Cultural Monuments, and amounted to a loan of 13 million Reichsmarks from the Dresdner Bank. Though supposedly not-for-profit, the society proceeded to funnel much of this capital into shady SS business adventures.[34] Bruno Galke, one of the cofounders of Allach Porcelain Manufacture, did gain appointment to Himmler's Personal Staff as the head of a special department for "business aid" (*wirtschaftliche Hilfe*) but exercised little control over these disparate enterprises. They were managed so amateurishly that one of Pohl's administrative officers later referred to them as "worse than wretched."[35] In addition, the SS was also flirting with graft. Himmler authored a law requiring all German police bicycles to purchase Loibl's pedal lights, and Allach, for instance, sold its wares to members of Himmler's personal staff at 40 percent of cost.

There is little evidence of Oswald Pohl's direct involvement in the SS companies before mid-1938. Pohl undoubtedly knew of Himmler's dabbling as a CEO of business enterprises and served as an adviser.[36] Possibly such projects had already piqued Pohl's special interest in the early days, but more likely the massive reorganization of administration during the SS's most formative years demanded almost all of his time. From 1935 to 1938 almost 500 million Reichsmarks fueled the SS's paramilitary buildup, and the "V" Office within the Personal Staff doubled its size to cope with this expansion.[37] The handful of SS companies were paltry by comparison. Pohl's enormous appetite for large-scale, complex tasks could have never been sated by a few charitable foundations, a porcelain manufacturer, a building cooperative, a photo studio, and a book press; nor, one suspects, could his drive for modernization be satisfied by ceramic figurines or bicycle lights. In the future Pohl placed SS industry at the center of his activities, and to do so he even neglected the traditional military administration that originally drew him into the SS.[38] But only the SS's move to install large-scale modern industry in the concentration camps after 1937–38 first created this change of heart in the SS top administrator. Before that the SS companies remained a sideline, cultivated by Himmler almost at random.

"We Are No Pencil Pushers!": Theodor Eicke's Total Institution and the Primacy of Policing

While Pohl was erecting an administrative system for Himmler's SS, slave labor in the concentration camps began in a completely different institutional context.[39] Long before Pohl helped to set up large-scale prison industries in the late 1930s, the camps' commanding officers, the SS Kommandanten, had already become the SS's first slave drivers, not the young officers of Pohl's modern administration. The camps have gained a historical mystique of demonic efficiency, but, even more so than the initial SS companies, camp management was chaotic from the beginning. Its dissolution reflected the professional training and ideals of the Kommandanten no less than the order of Pohl's organization reflected his own modernizing spirit. If SS management could be reduced to mere formal rules as an "iron cage," the interplay of "material self-interest" alone, or the rational development of "absolute power," phenomena such as heritage, personality, and esprit de corps would not matter, but they did. When Pohl's officers eventually began to take over prison industry, dissonance started to rend their managerial work precisely because they had to confront an entrenched conception of what it was to be an "organization man" among the Kommandanten. These confrontations were not made any less bitter by the fact that each group believed it was speaking on behalf of the SS in general. The history of the camps and their leadership therefore warrants some close attention.

The formation of professional concentration camp officers begins with Theodor Eicke, whom Himmler had appointed Kommandant of Dachau in June 1933 in the midst of a scandal. Like many police detention centers, Dachau had been founded shortly after the Nazi seizure of power, in this case in an old powder factory near Munich. Its initial organization by Nazi rowdies was ad hoc, amateurish, and lacked clear administrative hierarchy. The SS had taken over on 11 April, but this had changed little. SS guards celebrated by getting drunk, tormenting prisoners, and selecting Jews for special torture. By the next night they had already killed their first inmate. The torture of Jewish prisoners and more or less random violence continued with renewed zeal until the Bavarian police arrested the first Kommandant, Hilmar Wäckerle, for the blatant murder of political prisoners. To Himmler, Eicke seemed a competent strong man who could impose discipline where Wäckerle had not, and Eicke did not let his chief down. He immediately developed his "Rules of Discipline and Punishment for the Prison Camp" and "Service Regulations for Watchmen" to govern the conduct of guards and the camp's administrative structure. Camp personnel still meted out brutality and murder, but under Eicke they began to conform to predictable methods.

Rudolf Höss, future Kommandant of Auschwitz, recalled the changed tenor of work created by Eicke:

> I remember exactly the first punishment by beating that I saw—according to Eicke's order at least one company of the [SS] troops had to be present. Two prisoners, who had stolen cigarettes in the canteen, were condemned to received 25 lashes. The troop formed in an open square under arms. In the middle stood the whipping blocks. Both prisoners were led forward by the Block Führer [SS noncommissioned officer].[40]

This was ritual violence. It enforced discipline among the guards as much as it punished their victims. Nevertheless, Eicke's regulations quickly mollified outside authorities, who began to accept the camp as a legitimate penal institution. They had objected less to the brutality of Dachau than to the capriciousness of its misrule.[41]

Himmler was pleased with Eicke, in whom he saw a man whose vision and expansive aspirations corresponded to his own. Over the course of May and June 1934 Himmler made him the Inspector of Concentration Camps (IKL, or Inspekteur der Konzentrationslager) and expanded Eicke's authority outward from Dachau. Eicke took charge of all long-term political detention centers in Germany and quickly consolidated them into five main camps: Esterwegen, Lichtenburg, Moringen, and Sachsenburg along with his original Dachau. It was time to take up the broom and the hammer at once, to sweep out the old institutions of the Weimar Republic and erect the foundations of a new era in one decisive act. The camps of the mid-1930s, which at the end of 1936 held little more than 4,500 inmates, remained tiny in comparison to the network to come. Auschwitz alone would hold more by an order of magnitude. Eicke's achievement lay not in the scale of his system at this time but in the precedent he established. The SS set out to create new rules of incarceration and new standards of punishment, but perhaps most important, at least to Eicke, the

SS had to create new men to direct the system. "We are not mere prison watchmen," Eicke said of his guards; "we are political soldiers, and as such we are body guards of our Führer."[42]

The Shock Troop Adolf Hitler had now passed into Nazi legend, and the SS claimed to be its heir; Eicke was merely projecting the SS mandate as body-guard of the Führer onto the entire body politic that Hitler supposedly personified. Only the best National Socialists could assume the task of protecting the Reich, he argued, and therefore he had to be in the business of recruiting perfect Nazi citizens. No less than among Pohl's men, a spirit of modernity, of being in at the foundation of new and dynamic institutions radiated through Eicke's organization. "In just 11 months," he crowed in the summer of 1936, "I have reorganized 5 concentration camps. . . . I have built them up and clarified relations between them." Then he added, "A new, great, and modern concentration camp is under construction at Sachsenhausen."[43] And Eicke had organized Sachsenhausen on the principles of a panopticon. As Himmler observed, the SS would build "a completely new, modern concentration camp for a new era," and many outside the SS shared this impression as well.[44]

Conflict with Pohl arose out of what exactly "modernity" meant in different branches of the SS. According to Eicke's self-conception, modern society meant regulation by principles of efficient, military discipline and a state prepared to carry out its political dictates through swift, executive force. His specially named Death's Head Units stressed the prestige of the uniform, the power of their weapons, spit and polish, and a "snap-to" readiness for action. Eicke's vision also contained a deep irony: he aspired to be modern and yet harbored a latent hostility to modern bureaucratic administration. He was a paragon of that archetype that once prompted Christopher Isherwood to characterize Germany as a nation of noncommissioned officers. Eicke himself tended to ignore office work in favor of discipline and drill. In 1939 he would leave the camps to follow his true passion and lead a Death's Head troop in combat, after which he never again concerned himself with prison administration. He was killed in action in 1943. His greatest organizational innovation was an image rather than unified administration, an ability to demand commitment and receive it and to serve as a model of inspiration.

Eicke began by creating a tightly knit group. One of his first reforms was to force all Dachau guards into the SS, and he imparted to them a sense of belonging and mission. His men called him "Papa Eicke," and there is every reason to believe that they personally identified with his cause, especially in the early 1930s, when many of the later Kommandanten and top officials got their start (e.g., Rudolf Höss, Arthur Liebehenschel, Richard Glücks). Höss later characterized this as the "Dachau School" where officers imbibed Eicke's vision. Eicke was a giant, barrel-chested man, personal and jovial with even his lowest-ranking soldiers. He invited recruits into an inner ring of comrades, to whom he offered the confidence of a wizened patriarch. On the other hand, he cursed his enemies outside with a brutal vulgarity that shocked even Himmler at times. To devotees this only reinforced the image of a man who would stop at nothing to fight for what he believed and who refused to toady before traditional bureaucrats. He also brought his reputation as a tough, embattled Nazi who had taken part in the early years of struggle. He had joined the movement in 1927 and owed his professional career to the SS after political agitation had cost him a secure job as a watchman during the world depression. Before Himmler had called upon him to reform Dachau, Eicke had led a wayward, broken life, including arrests and a brief tenure in a mental institution. With the appointment to Dachau, he received the chance to shape not only his own future but also the future of National Socialism; to him, the two were inextricable. From his subordinates he expected the same: he called upon them to quench their personal identity in the mission of the Death's Head Units.[45]

As much as the Death's Head Units rested on Eicke's charisma, they were also built up into a very homogeneous group by conscious policy. With a mixture of bombast, bravado, and deadly earnestness, the IKL encouraged their sense of privilege and a corresponding sense of betrayal, should anyone leave the fold. The original Death's Head guards were mostly young men, usually cut off from other sources of firm identity. Karin Orth refers to them as the "camp SS" to distinguish them from the combat formations of the Death's Head Units later in the war (who rarely crossed over to camp duty). Over 90 percent were unmarried compared to 57 percent in the SS in general. More often than not, they had frustrated or stagnant careers behind them, like Eicke himself, which they blamed on the liberal democracy of Weimar, international business conspiracies, the communists, or the Jews. They generally possessed mediocre education, and at the height of their civilian careers they might have been skilled laborers, farm workers, or petty salesmen, if not simply unemployed. Few had received the formal training requisite for new white-collar jobs and had often found their way to social mobility blocked. Such common background meant that new recruits saw many faces in the IKL that reminded them of their own life, men whom they could understand. Indoctrination and training reinforced similarities of background. The concentration camps provided for all their needs. Units slept and ate with each other in common barracks, camp canteens provided for their leisure hours, the sick bay attended to their health, and, at the same time, their mutual experiences and extraordinary work enforced an attendant alienation from the outside world. Eicke personally designed their uniforms to set recruits apart from all other SS units so that even the smallest insignia announced their distinction from other Nazi formations. In addition, secrecy bounded their world, for although Eicke encouraged his men to discuss matters candidly, they had to keep IKL affairs among themselves.[46]

Eicke took great care to ensure that his soldiers found camp life to be an experience of honor and empowerment. He told his men that careful genetic selection and rigorous training had marked them as a natural aristocracy and also appealed instinctively to their low-brow resentment of upper-class privilege to awaken pride and dedication. He excoriated digressions from duty as "bourgeois weakness."[47] "To political agitators and subversive intellectuals let it be known: you must hide yourselves. If we find you we will pick you up by the scruff of your necks and bring you to silence after our own recipe!"[48] In keeping with the resentment of traditional elites fostered by Eicke, within the Death's Head Units boundaries between higher and lower social class were more permeable than in the state bureaucracy or military. Unlike Wehrmacht officers, there was no prerequisite for education, and enlisted men of no standing could rise to officer rank. For many young men, this was their first independence from home and a chance to "make something of themselves." Officers under Eicke also had to take their mess with enlisted men and often used the informal *du* instead of formal *Sie* to address subordinates.

Death's Head identity was also politicized, and ideological dedication was in turn held up as a badge of status. For example, Eicke forbid the Death's Head Units to wear civilian clothes, even on vacation. They were required to greet everyone, even girlfriends or family, with a straight-armed Hitler salute, regardless of where they met. Inside the camp fence, compassion counted as emasculate, and "hardness" was, in turn, endowed with political meaning as national service: "Tolerance is weakness. Once you realize this, you will be ready, without giving it a second thought, to attack when and where the interest of our Fatherland demands it."[49] Eicke told his men to consider prisoners as the enemies of the Nazi movement and, as such, to make them personal enemies. It would be exaggerated to say that the entire camp SS believed this as passionately as Eicke; nevertheless, zealous brutality was easy and free of risk; it was one of the most public ways to show identification with Eicke's cause. As one of the most thorough biographers of concentration camp personnel has observed, Eicke "wanted his men to know why they were serving in his camps, and he often overlooked willful mistreatment of the prisoners as long as he was convinced that the misconduct was the result of an inner identification with the tasks of the concentration camps."[50]

This systematic molding of collective identity should not be confused with sound organization. The same appeals that allowed Eicke to kindle his men's devotion also fomented contempt for competent management. "Bureaucrats become comfortable, fat, and old," he boasted. "As fighting men, we remain healthy and vital. As bureaucrats we would be bound to the dead letter of the law; but as political soldiers we deal in the founding law of revolutions."[51] Thus Eicke institutionalized an esprit de corps that associated scrupulous record keeping with pusillanimous civil servants. In practice, Richard Glücks, Eicke's adjutant, slowly usurped his superior's duties as Inspector of Concentration Camps. Far from a play for power, Glücks simply took over routine paperwork that Eicke scorned. Glücks eventually became Eicke's successor, yet his rise shows Eicke's ineffectual management more than Glücks's own energy and competence (he was quite unlike August Frank under Pohl). A pattern of succession in lazy administration soon repeated itself within the IKL. Glücks's own deputy, Arthur Liebehenschel, dutifully took over administrative details that his chief preferred to ignore. In all cases, officers in the IKL central office remained in Berlin and seldom intervened directly at individual camps. On occasion Glücks or Liebehenschel did issue sweeping agendas, but these rarely contained details for implementation or the quintessentially modern techniques of statistical surveillance of camp affairs.

The IKL's lax administration actually encouraged the ablation of its own functions. Other SS organizations absorbed camp duties by creating a superior administrative apparatus, which then had no problem displacing the IKL. In June 1936 the official medical core of the IKL was erected under the SS Sanitation Office, an outgrowth of Himmler's Personal Staff.[52] The offices of Oswald Pohl began to intervene in camp construction as early as 1935, and, by mid-1937, Pohl took charge of the IKL's budgets.[53] By wartime, the IKL had lost authority over hiring its own guards, so that, even before war began, the IKL received only second-rate men for camp details.[54] Generally there is little evidence that the IKL resented this slow erosion of its administrative control. In 1938 Eicke eagerly welcomed and even praised the officer whom Pohl chose to direct construction in the camps.[55] Eicke was somewhat mercurial, but after 1939 Glücks allowed the piecemeal usurpation of his duties with passive disinterest.[56] The Death's Head Units idealized military bearing, discipline, and the decisive act; as a corollary they disdained desk work as the pursuit of idlers and do-nothings.

Because Pohl eventually subsumed the IKL, his relationship to Eicke deserves some scrutiny. After the war, Himmler's adjutant testified to their vexed and stormy encounters:

Pohl wanted administrative order and wanted to be able to extend a reasonable amount of influence. However, Eicke refused to comply with this. The Reichsführer ordered both men to come to him. . . . He ordered both men to discuss the matter in my anteroom until they had reached an agreement. Only then was I permitted to take them to Himmler. Since both of them were very strong-headed, the discussion lasted for three hours. I was not allowed to give them any water or refreshments. After three hours they were allowed to see the Reichsführer.[57]

Unfortunately we do not know exactly when this conversation took place or the issues at stake. It appears a classic case of a bureaucratic power struggle, and has been interpreted as such. To the contrary, however, Eicke needed Pohl, and their cooperation deserves as much attention as their "polycratic" conflict. Note that Himmler's adjutant stressed the exceptional nature of this encounter *and* the fact that both men eventually agreed to work together.

Most confrontations were often of little substance and focused on mere venality. Eicke typically resisted the SS administrative chief only when Pohl threatened the personal perquisites built into the camp system. For instance, when Pohl sought to audit the entertainment funds disbursed at Dachau, Eicke exploded. Pohl's auditor, Karl Möckel, as Eicke complained, "is incompetent . . . [and] has probably already suffered a nervous breakdown."[58] This was the kind of choleric outburst that made Eicke famous among his admirers and notorious among detractors. Yet however indignant Eicke's insults, he actually had little knack for daily fiscal organization. He contentedly allowed Pohl's administrators a free hand in other matters of common interest, and the IKL was often glad when Pohl assigned quality officers. Needless to say, the very officer Eicke described as incompetent later returned without further incident—only one indication that petty clashes over authority alone rarely produced lasting effects. Enduring conflicts would arise only from different values and ideals of the SS's proper mission, in particular over slave-labor management. The mere fact of contested power is less significant than how deep, institutionalized differences emerged within the plexus of SS intentions.

Slave labor became a sticking point because it served a vital role in constituting the Death's Head Units and their prisons. Before Eicke came to Dachau the Gestapo had already proposed to use inmates to drain the moors of Emsland, but these initiatives came to nothing. While Germany still suffered from the widespread unemployment of the world depression, however, officials both within and outside the Nazi Party opposed prison labor. To them it seemed a crime in itself to give work to Germany's political "opponents" instead of reserving all possible jobs for civilians. Yet the professional identity of the Death's Head Units fostered another, equally important reason why plans like draining the Emsland moors were so easily pushed aside. Namely, the initial "pragmatic" role of work in camp security had little to do with meaningful, economic labor. Inside the camps, work conformed to the primacy of policing—defined as the enforcement of discipline—not to the primacy of production. Work details kept prisoners constantly working and continually exhausted. Menial tasks like digging ditches made them amenable to control, less likely to mount resistance or plot escape. One official of the Reich Ministry of Justice formulated this functional role quite bluntly: "One of the most valuable tools for securing the safe incarceration of the criminal is [to make him] work—all day long, from morning to night, every week, month, and year of his imprisonment. This leaves him no time for stupid thinking [*dumme Gedanken*] and, as an added bonus, helps to raise discipline within the institution."[59]

Beyond the efficacy of control, concentration camp labor also served as a means of inscribing political identity, on both prisoners and guards. Very early on, the SS designed work details chiefly to punish political enemies. The first Kommandanten assembled labor brigades according to prisoners' political transgressions, not their skills or abilities. The camp SS reserved the hardest labor for the most unforgivable political or racial enemies (communists and Jews, for instance). In the original Dachau before Eicke, such prisoners were to receive "a hard bunk and, as provisions, a warm meal consisting of one-quarter the normal rations."[60] Eicke's service regulations were less draconian but sprang from the same punitive spirit: "Prisoners, without exception, are obligated to carry out physical labor. Status, profession, and background will not be taken into account."[61] Like the original concentration camp guidelines, Eicke showed little interest in the exploitation of prisoners' skills; in fact, he told Kommandanten to pay no heed to professional criteria. He conceived of labor as a tool with which to impose the might of National Socialism upon prisoners. Death's Head watchmen kept their eye less on productive organization than security and punishment and typically left the details of work to "Kapos," specially chosen prisoner-supervisors who usually came from the genuine criminal elements of the camps' population. Some Kapos earned notoriety for acts of violence even more vicious than their keepers'. A drawing by the survivor Maurice de la Pintière captures their menacing presence.[62] The camp's overseers chose them precisely because of their brutality, not because they were skilled foremen.

All aspects of the primacy of policing in forced labor—both practical *and* ideological—are evident in the testimony of Oskar Mühlner, a survivor of a

Sketch by the survivor Maurice de la Pintière: "Écrasés sous la botte de la brute sadique."
KZ-Gedenkstätte Mittelbau-Dora, Nordhausen.

labor camp associated with Buchenwald. In a postwar interview he told of his
restlessness: "After weeks of inactivity . . . I was genuinely happy on the first
day of work." In practical terms, this meant that during his "weeks of inactivity" guards would have found Mühlner uncontrolled. Menial labor offered a ready solution to this problem. Upon his first day of work, Mühlner

actually recalled, "With alacrity [I] sunk my shovel and pick deep in the earth. I filled one cart after another. Without pause, but at a more moderate pace, I worked for a full eight hours without apathy."[63] Labor imposed a routine, one with which the prisoners were often willing to comply. Yet the IKL wished to extract more from Mühlner. Even though he returned at the end of the day with blisters on his hands, he narrowly escaped a beating for being "lazy." The concentration camp did not demand the quality or quantity of his labor, or his mere passivity; work was a means of demoralization.

Individual guard details proved ingenious at dehumanizing prisoners. Consider this example of dredging near the concentration camp Neuengamme:

> The prisoners stood in a row at a distance of about one meter between them. . . . Other prisoners in a similar compact formation pushed carts one after the other past this row of prisoners. . . . The prisoners with the carts were not allowed to stop at any time. They were ordered to receive one shovel-full of dirt from each prisoner in the row, from the first to the last. It did not matter if the cart was full or not. When the carts were overfilled, the excess dirt simply fell back to the ground.[64]

Another survivor's testimony, that of Fritz Schmidt, presents an evil parody of Frederick Taylor's classic tale of "Schmidt," selected according to principles of scientific management to carry pig iron in time-motion studies:

> Long, heavy iron beams . . . were not carried, say, on wagons or carts but rather on the shoulders of the prisoners. The supervisors and Kapos had experimented to establish exactly how many men could carry one beam. Not one additional man was assigned to ease the burden for support or assistance, although plenty of men were available. Whoever fell or collapsed received a beating with the nightstick, rubber clubs, or from the fists of the Kapos until he got back up and moved on or remained lying unconscious.[65]

As Daniel Goldhagen has pointed out, such cruelty required a large degree of initiative and creativity from low-level guards and functionaries. In this he is correct, for, as morally disturbing as it may be, the two selections here show a "rational" supervision of labor designed to maximize brutality. But Goldhagen also claims that capricious violence focused almost exclusively on Jews and was motivated by German anti-Semitism alone.[66] In fact, Schmidt's work detail was of mixed prisoners. Fellow prisoners, the Kapos, also took initiative in cruelly innovating new punishments along with the camp SS. Organized brutality stemmed more from the Death's Head Units' affirmation of their identity as Hitler's punishers (an identity in which racial supremacy was certainly a large part) than it did from any single anti-ideology. In both examples, the camp SS maximized the demeaning nature of labor, not only by beating prisoners but by making a mockery of their toil, an insult to their integrity. Ultimately the camp SS broke them bodily as well.[67]

Lax management only reinforced such wanton violence in Eicke's system. Long before war imposed real shortages of food and provisions, supplies to camp inmates were abominable both in quality and quantity. Chronic overcrowding, a related problem, plagued Dachau from its conception and continued to plague the entire camp system throughout its history. Conditions made it bodily impossible to sustain long periods of heavy manual labor, and as a result mortality due to sickness alone was atrocious, achieving rates between 8 and 11 percent per month by 1942 and even higher in the winter. Because production yield was not a central goal, such phenomena carried on without qualitative or quantitative controls. Production lay under jurisdiction of the highest officer, the Kommandant, who remained unmonitored from Berlin. Few had any formal knowledge of the techniques necessary for modern production: the statistical surveillance of input, output, and throughput. They were familiar with neither the organization of labor nor the efficient operation of machines necessary to control steady-flow manufacture; and, in short, modern management was generally beyond them. In daily operation, this meant that they had the power to intervene in production without being required to understand or promote it. What productive labor there was was usually dedicated to the personal enrichment of higher camp officials. Skilled woodworkers, tailors, or jewelers made personal effects for them and their wives; prisoners tended gardens at their houses. Indeed, prisoners coveted such positions, because personal service made their lives more valuable and thus increased their chances of survival.[68]

When the IKL eventually merged with large-scale SS industry, Pohl's administrators confronted an array of concentration camp practices that the Kommandanten had actively perpetuated through the training and indoctrination of new officers and guards. The result would be bitter conflict whose ultimate outcome increased the murderous suffering of prisoners. Before talking about that clash in mid-1939, however, we should first introduce the differences between the IKL and the SS Administrative Office precisely because they were subtle and often came clothed in the same vocabulary of Nazi boosterism. Like differing shades of the same color, it is necessary to view the two in the right light from the right angle to see where they meet and then diverge.

Perhaps the similarities and differences are best described by the contrast between Oswald Pohl and Theodor Eicke, as both fairly accurately represent the officers who served them. On the surface, they had much in common. Both were older men—even gray eminences of the SS—but if their age set them off from their younger subordinates, it had not dulled their energy. Both were aggressive organization builders who enjoyed the direct ear of Himmler. Nevertheless, after the First World War Pohl had overseen a large staff in the navy. Eicke had been fired from job after job. Similarly, the successful careers of men in Pohl's service generally distinguished them from concentration camp personnel. By contrast to the truncated ambitions and false starts that many concentration camp Kommandanten had suffered as civilians, Pohl's top-ranking officers often came from flourishing careers as attorneys, businessmen, or civil servants. Instead of the low-brow origins of the Death's Head Units, many had come from comfortable middle-class or even upper-class homes, and their fathers had often sent them to Germany's best universities. Either through formal training or work experience, many more of them had learned the modern tools of statistical bookkeeping, personnel administration, engineering, or law.

Pohl strove to impart a sense of elite comradeship among his administrative managers no less than Eicke had done among his "political soldiers." Many SS administrative officers believed that they, as SS men, were more energetic, more forceful, and tougher than those who sat in Reich ministries or the boardrooms of private industry. They believed in their ability to rush in and cut through Gordian knots where others would simply dally, complain, or make excuses. They would, in the words of the SS's official historian, "provide the genuine military tradition, the bearing and breeding of the German nobility, and the creative efficiency of the industrialist."[69] "For this is my goal," Pohl himself wrote to a trusted friend, "—to replace the bureaucrat once and for all with an administrator of soldierly bearing."[70] They would not shrink from desk work. By contrast, Eicke shouted, "We are no pencil pushers [*Bonzen*]!"[71]

Here again the differences emerge only against a field of basic similarities. Like the Death's Head Units, for instance, the future SS corporate managers believed that prisoners had to be driven to work. Perhaps because of Pohl's inexperience in production, he tended to conceive of labor as a more or less homogeneous entity, subject to intensification only by lengthening the amount of time spent working. He believed that force and discipline, not coordination, were the keys to efficiency and did not differentiate rational processes of labor exploitation.[72] Yet, when the time came to found large-scale concentration camp industries, officers under Pohl nevertheless conceived of slave labor as a resource in modern production, even when they mismanaged it; on the other hand, to Eicke's militarized Death's Head Units labor mattered only as an occasion for brutality or as a tool of camp discipline.

One last characteristic, shared by both the IKL and Pohl's managers, deserves mention. Pohl's administrators often bungled production because technically trained industrial managers were conspicuously absent among them (Pohl's engineers were mostly civil engineers). Only committed production engineers could have rendered the complexities of manufacturing into abstract statistics so that administrators higher up Pohl's hierarchy might accurately evaluate the material world of production from afar. Factory operations demanded a fusion of information management (skills which Pohl's officers had) with the management of labor, machines, and material (skills which they lacked). As much as they idealized modern management, in practice SS business executives ignored the nuances of the factory floor, including the human component of labor. As a direct result, they continued to rely upon Eicke's "political soldiers," setting a course for conflict that had murderous results for the prisoners. Concentration camp personnel brutalized working inmates even to the detriment of industrial operations.[73]

Origins of the SS Construction Corps

Civil engineering represented one exception to the general dearth of technical expertise. Institutional niches within Eicke's system began to collect construction engineers and architects in the early 1930s, long before Pohl took charge of concentration camp affairs, and these niches notably contained officers who brought both modern managerial skills *and* enthusiastic dedication to the SS cause. In the 1930s, as now, significant differences existed between the career paths of engineers and architects, but historically in Germany they have nevertheless shared the same aspirations and often the same jobs. This book treats the two professional categories in common. Especially on big, complex projects like those of the SS, the architects acted less as aesthetes than as managers amid the stream of constant building; likewise SS civil engineers took over design functions such as those required by steel-reinforced concrete. Such new, modern materials required more science-based training than older wood or masonry construction, and here again architects and civil engineers held overlapping positions when it came to modern building sites.

In short, both became modern technological managers. They confronted tasks not unlike those of a symphony conductor: they had to call a scheme or system of production into existence for the moment, elicit the diverse talents of many for one performance, and then pack it all up when the job was done. The tasks of SS engineers resembled those of the conductor rather than the design engineer or artist-architect. Some may no doubt object that such a metaphor aestheticizes the work of these SS engineers, for they were essentially murderers. But the metaphor is instructive. SS productivism led them to see technological labor as a cultural act, one that helped unify and glorify the nation. They were constructing more than barracks; they were constructing Nazi identity, and many conceived themselves as creators, again, not unlike the conductor. To recognize this aesthetic moment in their work is not to embrace it, and, as perverse as it may seem, they could believe in it even as they worked human beings to death. In any event, it is far less misleading than presenting them as apolitical "technocrats," for here again ideology and organization became one and the same thing.[74]

Ideology and "rational" management did not contradict but rather reinforced each other. Every concentration camp possessed a Construction Directorate (Bauleitung) to take care of small-scale projects and maintain the camp's physical plants. The directorates used prisoners in building commandos that were grueling, exhausting, and demeaning. This work never spared prisoners the bludgeon of guards or Kapos, but here engineers did manage the labor site directly to assure that the proper work got done at the proper time. Under their direction, work could be at once dehumanizing *and* productive. They used their engineering knowledge to find innovative solutions to the supposed contradiction between slave labor and sound, modern technological systems. To point this out is not to celebrate their "successes" but to dispense with the common conception that something latent in modern institutions makes them incompatible with ideological lunacy like that of the SS and its slave regime.

Of course, the capabilities of the early Construction Directorates should not be exaggerated. They remained decentralized and effectively uncoordinated up to mid-1941, and a personnel list from 1940 shows that not all camps were able to fill their civil engineering slots.[75] Even if a camp did have its one trained engineer, not all were competent. At some camps, they hearkened to the Eicke style and ran affairs without systematic oversight. Nevertheless some did begin to keep accounts of building materials, labor timetables, and contracts, and they constituted a kind of functional elite: by controlling specialized knowledge, they could command a degree of independence. Regardless of rank, the Kommandanten had to rely upon them to get practical tasks done. The nature of construction also lent the directorates additional latitude. Building necessarily took place outdoors and on site, so that daily work required engineers to supervise disparate locations, away from the Kommandanten who usually lacked both the initiative and the knowledge to monitor even simple projects.[76] Construction Directorates also began to look for masons, carpenters, electricians, roofers, and other skilled workmen among the camps' population to serve as Kapos instead of merely selecting criminal thugs, or they commonly sought out criminal thugs who possessed technical skills. Although unskilled laborers derived little benefit from construction projects, SS engineers usually shielded their *skilled* prisoners from the worst of the camps' brutality. Construction officers therefore began to form an important center of gravity within the SS. Those who were capable began to draw competent people into their ambit, and they began to draw ever more important projects into their hands.

Gerhard Weigel is a typical SS engineer of this early period, a man whose small beginnings hardly hint at his future sinister duties as a top manager of slave labor. In 1941 and 1942 he, more than any other officer, helped organize the SS Building Brigades, what were essentially mobile concentration camps. They rushed prison labor crews to urgent building sites and commanded specially equipped railcars for this purpose. Weigel personally led a brigade of over 350 guards and 2,500 prisoners in the construction of fortifications on the Channel Islands. He routinely sent transports of his sick, exhausted, and dying prisoners, up to 200 at a time, to be liquidated.[77] Yet before 1942, his career was inauspicious. His father was a petty official who inspected taxis, and perhaps because of his family's limited means and stature, Weigel's education had been erratic. After grade school he took apprenticeship as a metalworker and installed heating and ventilation systems. Probably through night classes, he attempted to better himself by attending an advanced vocational school (Höhere Fachschule Chemnitz). It was the kind of school that trained solid engineers to fill a gap between technical foremen and the elite polytechnic graduates. Such vocational-school graduates were becoming the mainstay of technological industries: the white-collar workers at routine problem solving, factory oversight, or drafting.[78]

Nevertheless Weigel could claim the title of heating systems engineer when he finished his studies in 1930, but his timing could not have been worse. Long-term investment in construction prospers only in stable, growing economies, and the 1920s and early 1930s offered no such thing. Like many Death's Head officers, his curriculum vitae lists no steady employment from 1930 to 1933 except as an assistant at a sport camp.[79] He was not the only one to fall on hard times. These early officers had suffered a higher incidence of unemployment than the later SS corps of engineers (21 percent as opposed to only 10 percent of those who entered SS service after 1939). But one did not need to be unemployed to feel threatened. Despite a reputation as an "apolitical" profession, the Weimar Republic fostered resentment among many German engineers that quickly took on an ideological hue, especially in the building trades. Democracy seemed incapable of keeping construction up and running, even though Germany suffered from a housing shortage that was plain for all to see. Some estimates claim that up to 60 percent of technical-university graduates went without finding work in their profession in 1933. This had the greatest effect on architects and construction engineers, who together represented the largest portion of students in those years. As Gerd Hortleder has shown, engineers in general began to doubt the future; those in the prostrate building sector must have doubted more than most. Whereas big corporations could often guard their managerial talent when short-term earnings did not justify it, no such protection existed in the fragmented building sector. It did not take many engineers long to come to the conclusion that a society that had failed to harness their technical skills was incapable of progress.[80]

In such an inhospitable climate, one might easily conclude that frustrated engineers like Weigel sought out National Socialism just to get a job, but Weigel's motives were anything but opportunistic. His activism suggests that National Socialism fulfilled his need for a vision of a better future as well as steady work. He had joined the SS long before his career prospects faded and proudly declared his leading role as a volunteer administrator in the first SS Section of Chemnitz. He was also willing to back up his belief with his fists. "I have been injured several times in street and salon battles [i.e., with political opponents]," his curriculum vitae claims, "twice critically."[81] Shortly after Hitler's seizure of power, Weigel volunteered for SS "Special Commandos" that acted as auxiliary police forces; sometimes officially deputized, sometimes not, they led the sweeping arrests of the Nazi's political enemies early in 1933. Through his own commando Weigel first came into contact with the nascent concentration camp system, and by 1935 he was working full-time for the Construction Directorate at Sachsenhausen. Like Weigel, many of the first technical officers of the Death's Head Units were early activists. We cannot know what they said to each other in local bars or casual conversations, where political judgment or racial slurs were most likely to be bandied about, but even in the detritus of their institutional records twelve of thirty-three early SS engineering officers left some evidence of their commitment, and, of those, nine were activists like Weigel, willing to go far out of their way to promote the Nazi cause. Among them broken careers were not uncommon, but knowledge and training often distinguished them from other drifters. Thus compared with most low-brow Death's Head recruits, they represented a small managerial elite.

Men like Weigel could manage large building projects and manage them well, but few had experience in national civil-engineering hierarchies, the kind it would take to oversee modern organization. As a rule the early IKL engineers were not organization builders and could hardly make up for

Eicke's lack of unified national administration. Furthermore, not all directorates were headed by men like Weigel; many bumbled projects and misallocated funds. It was Pohl, not the likes of Weigel, who negotiated with Hitler's building czar, Fritz Todt, or Hermann Göring's Office of the Four Year Plan for cement, lumber, and steel. Pohl also controlled the acquisition of property, which was essential for the foundation of new camps. Pohl's influence was strongest in exactly such matters of administrative protocol. He also set about creating a core of technical competence in an SS construction bureau independent of the camps, and in the IKL his touch could be seen in the progressive assignment of key officers who slowly strove to put Eicke's house in order. Eicke, for his part, was glad to get the capable men.[82]

Perhaps some SS engineers considered their job to be like any other, but the long association of most early construction officers with the Nazi cause suggests otherwise. On the average, they had joined both the SS and the Nazi Party before the decisive seizure of power in 1933. Furthermore, the SS was proud of its buildings, sought to gain publicity and attention through them, and often built on property confiscated from Jews or from the Nazis' political enemies. For example, Himmler's Race and Settlement Main Office showcased a settlement in the Krumme Lanke suburb of Berlin. The SS patterned the housing development on the garden city movement popular among Western architects and urban planners before and after the war. It expounded a new kind of urban world in which street bustle and industrial sites could be separated from residential districts located close to nature. This movement was not against the modern city but sought to reconfigure it, and the SS also sought to vest its settlements with its own ideals of social order.

An article in a glossy architectural journal described exactly how the SS expected its members to live and how homes and community buildings were supposed to shape their identity as Germans. It planned to settle residents with room for a garden in each plot. Family cottages would constitute each SS man's piece of the Third Reich, a constant reminder of what National Socialism provided and why its "mother soil" was worth defending. Through such value systems, the settlement sought to replace private life with a structured harmony as the central experience of ideal citizens: "The tenants' souls will be filled with true comradeship. Divisions among them by fences are superfluous. They will take up residence in a beautiful spirit of comradeship under the tree tops of the local pines."[83] In terms of urban planning, this meant that each family's roadway wended toward public buildings—where political meetings were held, for instance. These were the "central point of the settlement, the unifying idea that imprints the new garden city with character."[84] Meanwhile, architects proposed standardization down to minute details of kitchen layout. Typical of the SS's productivism, design choices were put forward not as a means to cost savings but in order to create unity, just one indication that even in building the SS expressed its drive to construct community.

Productivism further informed a smug critique aimed at the former development firm from which the SS had taken over the Krumme Lanke site: "The integrity of artistic idea in the blueprints [of the former firm] is completely absent. Here one tried only to build with all possible parsimony. . . . When one looks at their street layout, one must think of an iron grill upon which a piece of living Nature is to be roasted, cooking down each parcel of land."[85] The SS assured its readers that National Socialism would build for political community and cultural ideals instead of such capitalist greed, and the blueprints called for ostentatious use of cut stone, because Himmler believed that "natural" materials of the German soil better portrayed the German spirit than new materials like concrete. Stone also cost more and demonstrated the SS's insistence on putting "cultural" considerations first. The act of construction, in this case, was to produce identity, "artistic idea," and eschewed business sense.

Perhaps it is conceivable that bureaucratic routine made men like Weigel and other SS engineers blind to such ideals in the very structure of the buildings they worked on, but it seems unlikely. Although IKL engineers did not develop the Berlin, Krumme Lanke settlement, every concentration camp had similar officers' houses. The very details of their architecture were supposed to demonstrate the kind of community the SS strove to build. Both Paul Jaskot and Robert Jan van Pelt have shown that concentration camp architecture utilized the same ostentatious materials. Nowhere is the contrast starker than between the serried rows of standardized cottages for SS men outside the gates of Sachsenhausen, for example, and the equally standardized cramped barracks for concentration camp inmates, which, van Pelt notes, engineers carefully projected to cost no more than seventeen Marks per inmate and which provided, on the average, no more space than "the surface dimensions of a large coffin or the volume of a shallow grave."[86]

A further hint that Nazi architectural vision was not incidental but central to SS engineers can be found in a volume coauthored in 1934 by the future chief of the SS construction corps, Dr. Engineer Hans Kammler. Kammler was then working for the city of Berlin and would not join Pohl's organization until 1941, but he was already consulting with the SS Race and Settlement Main Office, likely on the SS project in Krumme Lanke, where he was assigned as a city official. The title of his book (written with a docent at Berlin's Technical University) suggests little more than a presentation of managerial techniques in dry, technical prose: *Foundations of Price Calculation and Orga-*

Kameradschaftssiedlung der ℋ in Berlin-Zehlendorf

Kameradschaftssiedlung der ℋ in Berlin-Zehlendorf

Exhibit of the SS "Comrades House" in Berlin, Krumme Lanke, a model of the entire settlement, and a site plan drawing of the community building. Hans Gerlach, "Die Kameradschaftssiedlung der SS am Vierling in Berlin-Zehlendorf," Siedlung und Wirtschaft 19 (1937): 701, 703.

nization of Construction Firms for Dwelling and Settlement Construction in City and Country. Indeed, the book delivered exactly what it promised, offering reviews of recent technological innovations and recommendations for how to run modern managerial hierarchy. Yet the Nazis' penchant for productivism meant that ideology and organization could be one and the same, and the authors went on to link this to Nazi settlement ideals:

> [Under liberal capitalism] houses and settlements were seen as a commodity. . . . The production and maintenance of this commodity became a good business. . . . Man and the soil did not stand at the center of these measures but rather materialism, bureaucratic technicalities, legalities, and the salesman's point of view.
>
> The policies of National Socialism are now dedicated to the firm connection of the man to the soil through hearth and home as the basic

foundation of the people [*Volk*] and the state. The German man's hereditary health and the hereditary health given by the German soil therefore stand at the focal point of the German Reich's program of renewal.[87]

Kammler wished to elevate settlement development above cost accounting, which put price tags on German homesteads. He and his coauthor thus held up garden city development as an imperative of national identity, akin to an absolute prerogative, and proclaimed the romantic notion that contact with native soil might necessarily elicit the dedication of each citizen to a strong, unified Germany. The affinity between a modern, rationally trained, technical professional and the romantic, irrational doctrines of National Socialism may seem contradictory, for what could be more alien to someone who had to read blueprints and calculate static equations than Nazi fundamentalism? And yet the productivism on which it was based made perfect sense to activists: it made their work an act of cultural devotion and simultaneously promised to release them from preoccupation with cost, supply, and demand.

Kammler, unlike Gerhard Weigel, had studied at Germany's elite polytechnic universities, where German technical education had always nurtured such a quest for a "calling" along with engineering competence. Werner Durth, for example, has shown that German architects were taught to believe in the importance of their contribution to culture; likewise, they were taught to seek out Germany's social problems and apply their skills to solve them. As masters of *technological* progress, engineers had long learned to associate their professional tasks automatically with *social* progress and the betterment of humankind. The urge to serve a better society was not, in itself, ignoble; condemnable were the ideals of a "better society" that engineers sought to turn into social reality. The student bodies of the technical universities were overwhelmingly National Socialist.[88] Engineers like Kammler saw no contradiction between notions of blood and soil and the methods of modern organization and technology. He wished to place the best means of modern organization at the disposal of National Socialism; and among National Socialism's ideals was the glorification of modern means in the name of productivism.

Civil engineers were perhaps more attracted than others to a strong state that promised to let loose the full force of modern organization, for such ambitions had been continually stymied in the building sector. Unlike many employers of electrical, chemical, or mechanical engineers, construction firms remained small; the centralized organization of mass production seldom benefited their need to adapt to individual customers' changing demands at specific locations. In fact, the building trades have almost always benefited most from other technological trajectories, such as a flexible production, in which adaptable machines mobilize a wide range of materials depending on the needs of the moment. Standardized mass production associated with modern technology tended to remain in the realm of materials supply—like bricks and cement—which could then be adapted at individual sites. The new mechanized factories of the 1920s and the large, hierarchical organizations new to the twentieth century, so successful in the automobile industry, nevertheless exerted an almost irresistible draw. This was true even when mass production was unsuitable, and the SS was not alone in this fascination. One of Albert Speer's leading architects dreamt of "a kind of house-building machine through which a five-story building can be produced as if by a stamping press."[89] One future SS construction officer had also experimented with mass-produced housing.[90] These visions appealed to the SS's productivism precisely because SS ideals glorified standardization as cultural unity and hierarchical organization as a sign of racial potency. The SS carried its modernizing ambitions even into areas where other forms of organization and technology would have been more "pragmatic." This became all the more clear after 1938–39 when the SS's industry, its administrators, and its concentration camps merged under Oswald Pohl.

CHAPTER 2

A POLITICAL ECONOMY
OF MISERY
THE SS "FÜHRER" CORPORATION

From 1936 to 1938 two simultaneous developments brought Oswald Pohl and slave-labor industry together. First, as early as April 1936 the Reich Ministry of the Interior agreed to provide annual funds for concentration camp guards, and then, by 1938, it extended this support to the entire camp system. This huge gain for Himmler's growing police apparatus demanded a more modern administration to handle the new scale of operations. Concentration camp accounting had formerly been dissolute, if not appalling, and Himmler quickly asked Pohl to create a centralized office to coordinate the SS's Reich budget and, as part of this, the finances of the Inspectorate of Concentration Camps (IKL).[1] This otherwise minor administrative alteration put Pohl in a direct supervisory role over the camps, where slavery was already under way. The second development that drew the SS administrative chief into prison industries occurred as the Third Reich entered a new stage of economic development. In the early 1930s, almost every individual concentration camp had proposed forced-labor programs, but these were years of widespread unemployment. A broad coalition of Reich ministries and municipal government blocked the diversion of any meaningful employment to camp inmates. Fearing unwanted competition, private businessmen joined these protests as well. After 1936 genuine economic growth rendered such objections moot. Labor shortages began to appear, and whereas before 1936–38 it had seemed a disgrace to give "criminals" work that could otherwise employ civilians, Reich authorities began to demand that prisoners "earn their keep."

The economic upturn and a string of other achievements coincided with the increasing intervention of the state in the economy. Hitler's successful opposition to the Treaty of Versailles, sustained employment, and economic growth seemed to confirm even Hitler's inflated claims to genius. The feeling grew among Nazi activists that the time had come to initiate bolder programs, and the Reich experienced a recrudescence of fanaticism, now backed by the power to turn from agitation to policy. Hermann Göring quickly rose as the most influential force in the Nazi economy. He promoted strict policies of autarky and the hypertrophic development of war-related industries.[2] Göring's power continued to increase with the promulgation of the Four Year Plan, which attempted to plot out a course for the entire economy, fed a massive armaments drive, and accelerated development of raw-materials industries as well as huge investments in infrastructure.

Cooperation rather than power struggles characterized relations between the SS and the Four Year Plan up to this time. Himmler placed the Gestapo at Göring's service and worked with the German big banks to ferret out Jewish entrepreneurs as Göring initiated an "Aryanization" campaign. The SS also entered into close relationships with private industry for the strict surveillance of the German work force and founded new detention centers exclusively for the punishment of workers who cut their shifts or were suspected of sabotage or goldbricking. (These were the Labor Education Camps, or Arbeitserziehungslager, and should not be confused with concentration camps, which were centers for the long-term detention of political dissidents.) As the Four Year Plan soaked up Germany's reserves of civilian workers, various Reich authorities began to demand the contribution of prison labor, and the SS was quick and willing to comply with the demands. In each case the SS proceeded on the strength of broad-based consensus among Germany's economic leadership.

The German Earth and Stone Works

By 1937 the SS had generated plans for large-scale concentration camp industry. The historian who has studied this most closely, Paul Jaskot, points out that much initiative first came from regional officials in Thüringen. There a regional geological surveyor suggested the location of Buchenwald near suitable clay deposits, and the interior minister urged that "camp inmates should be occupied within the framework of the Four Year Plan with the production of bricks."[3] Perhaps at this suggestion, the SS settled on brick factories and slated several camps to employ thousands of prisoners each. The greatest ally of the SS "Labor Action" (as it came to be called) did not come from Thüringen or from within the Four Year Plan. That role was taken by Albert Speer, the future armaments minister of the Third Reich and favorite architect of Adolf Hitler. Since 1934 he had been preparing massive projects intended to stand as monuments to Hitler's "Thousand Year Reich," among them the Reich Chancellery, the Nuremberg Party Rally Grounds, and other so-called Führer buildings. By 1938 he was also relying upon Himmler's Gestapo to evict Jews from properties around Potsdamer Platz in Berlin.[4]

The general economic climate created a congenial setting for Speer's flirtation with the SS, for by 1938 even the pet projects of Hitler's most darling architect began to suffer from the scarcity of both workers and materials sucked into the Four Year Plan. Due to Hitler's personal favoritism Speer's projects continued, yet the high-prestige status of Hitler's architectural fantasies did not exempt Speer from all constraint. The thorniest problem of all was the Reich's growing scarcity of labor. Speer began placing huge orders for bricks, dressed granite, marble, and limestone, for he and Hitler preferred such "natural materials" for public buildings. But these kinds of raw materials also required labor-intensive production. Thus Speer and SS proponents of concentration camp industry both sensed that they could use each other to mutual benefit.

The entire German brick industry could conceivably fill only 18 percent of Speer's orders, even if its hands were not full with military orders. Likewise, Germany's largest granite quarries could cover only 4 percent even after pledging to double their output. The Führer buildings' sheer scale created a practically unlimited demand for these specific materials and nullified any conflict that the SS might otherwise have encountered from private industry. The Führer's known personal association with the vast projects, his insistence that the buildings would reinvigorate German culture, and, no less, his occasional puttering intervention on Speer's blueprints further anointed the Führer buildings as an exercise in Nazi devotion. Perhaps inevitably Himmler sought to play some role in this architectural adornment of Hitler's reign, for the SS posed as the servant and protector of his person, his ideals, and his state. It would now try to help build his buildings too.[5]

The strongest evidence suggests that Hitler personally brought Speer and Himmler together sometime between the end of 1937 and the beginning of 1938, when the Inspector of Concentration Camps was surveying sights for prison industries. Much evidence also suggests that Speer was eager for this meeting. He may even have initiated it. Over the next two years he would continually urge the SS to new industrial undertakings, and he made suggestions to improve factories at Sachsenhausen, the largest of the SS's brickworks and indeed the largest in all of Germany. After the war, Albert's older brother Hermann, perhaps jealous of his sibling's postwar celebrity status, remarked that his "little brother" had gotten involved in "that stupid anti-Semitism" at this time. Albert had cleared Jewish apartment buildings around Potsdamer Platz in preparation for the Führer buildings, and Hermann recalled that he let drop, "After all, the Jews were already making bricks under the Pharaohs," and therefore suggested to Himmler that prisoners might as well be put to work in the same way.[6] Whatever the origins of these dealings, it is clear that Speer hoped to benefit from the SS and sought out its cooperation.

The SS obliged and invested in quarries at Flossenbürg in the Oberpfalz as well as granite deposits by Mauthausen, and it planned brickworks for Buchenwald, Neuengamme, and Sachsenhausen. Characteristically, the SS justified these projects as "cultural" endeavors. Quarries associated with the camp Natzweiler in Elsaß are a case in point. Like all SS quarries they cut "natural stone" (*Naturstein*), a special architectural designation for high-grade materials dressed primarily for aesthetic properties. The Natzweiler operation was unprofitable, but the SS acquired it nonetheless by 1940 because Speer planned to use its rare, red granite liberally in the Reich Chancellery building. In a pattern that the SS would repeat over and over, the "cultural" service (in this case to Hitler's architecture) took precedence over immediate profit.

From the beginning Oswald Pohl assisted concentration camp industries in the acquisition of properties, but otherwise his role remained almost wholly advisory. His closest involvement at this time was with the SS Commercial Operations of Dachau, small workshops serving the incipient paramilitary units of an SS Troop Training Center also established at Dachau. They were mostly for garment making, shoemaking, or carpentry, and they were separate from the concentration camp, though contacts were close between Eicke's units and the center. These industries lay briefly under Pohl's and August Frank's jurisdiction as administrators of the SS's finances, but their control at this stage was indirect at best and brief in duration. They hardly warrant the name "industries" and seem to have been of little interest to either Pohl or Frank. By the fall of 1935 these "Commercial Operations" were placed directly under the Kommandant of the Troop Training Center.[7]

The guiding force behind SS industrial management in 1937 and 1938 seems to have been not Pohl but Arthur Ahrens, an SS officer about whom little information survives. Ahrens apparently owned a brickworks in eastern Germany and had also been managing the SS's publishing house, the Nordland Verlag, since 1935. He was likely Himmler's personal acquaintance, and, if he bore any resemblance to other early SS entrepreneurs like Anton Loibl or the Bauer brothers, he may have posed as a fast-dealing business "expert" within the Personal Staff. Yet unlike Loibl, a former chauffeur, Ahrens could boast some experience. His personnel file listed some vague apprenticeship in "banking," and he assured the SS, "I took the opportunity to educate myself in all branches of our modern economic apparatus."[8] He further claimed to be an experienced manager (*leitender Beamte*). In whatever way he insinuated himself into the top circles of the SS, he, and not Pohl, acted as chief executive officer (*Geschäftsführer*) and founder of the first large-scale SS concern, a building supply company named the German Earth and Stone Works (DESt, or Deutsche Erde- und Steinwerke), in April 1938.

Ahrens's enthusiasm for "modern economic apparatus" far outstripped his ability. As events soon demonstrated, one of the certainties about Ahrens was his consistent incompetence. He possessed only a basic secondary-school education and no university or vocational degree (unlike the officers Pohl would later recruit into SS industry). "The internal organization of the firm [DESt] is worse than wretched," stated a blunt report little more than a year after Ahrens had started the company. "Neither the brickworks in Weimar, Hamburg, and Oranienburg nor the stone quarries in Flossenbürg and Mauthausen were planned and organized according to any general fundamentals of competent business and salesmanship."[9] After 1938 Pohl had to bring his authority to bear on SS business enterprises in order to save them from this man's bungling.

And the DESt was hardly exceptional. Ahrens also founded an Experiment Station for Alimentation (Versuchsanstalt für Ernährung und Verpflegung), another indulgence of the SS's scientific pretensions and puerile appetite for technological wonder. The experiment station set out to research all-natural medicines, organic food products, and revolutionary means of horticulture as a research institute. This may seem at first a throwback to ancient remedies, but Ahrens placed the bulk of the firm's capital in a greenhouse for "modern" pepper production and invested in a supposed technological wonder, a newly patented pepper mill. The venture was every bit as absurd as Himmler's claim to advance "technical inventions of all kinds" by making bicycle pedal lights, only in this case the SS did not proceed alone. Its officers were deeply involved in the Reich Ministry for Agriculture and the Four Year Plan's push for independence in foodstuffs. They promised to promote novelty and autarky at the same time. "It is truly new territory," announced a publicity article in *Das Schwarze Korps*, the official magazine of the SS. Research would not only lead to breakthroughs in science but also enable the Nazi state to become "independent in spices," many of which were otherwise controlled by British colonies.[10] The fact that, in the larger scheme of things, pepper is hardly a strategic commodity did not diminish enthusiasm that the agricultural experiment stations were "of the greatest importance for the future of the German nation."[11] Behind the pretensions, however, Ahrens invested blindly, giving the pepper mill's "inventor" a carte blanche so that costs quickly overran sales accordingly.

The experiment stations also offered 3 million Reichsmarks for a new scientific cure for hoof-and-mouth disease by a Dutch "scientist" named

Johannes van den Berg, a man later exposed by Pohl's office as a mountebank. Van den Berg's "patent" was already owned by a company in receivership. Other investments included experimental pig stalls, sheep herds, and herb gardens—part of Himmler's faddish desire to promote "naturally grown medicinal herbs . . . systematically planted in great quantities."[12] Two consistent fantasies integrated these disparate projects: the wish to associate the SS with science and modern technology on one hand and prevailing ideals of agrarian utopias on the other. It was the same drive that Himmler had pursued in his *Artamanen* days to bring German citizens back into contact with "Blood and Soil," the same drive that animated the idyllic garden cities of the Cooperative Dwellings and Homesteads GmbH and the housing development of Berlin, Krumme Lanke. If absurd in practice, nothing appeared contradictory to the SS. The agricultural experiment stations hired "specialists and scientific collaborators" to carry out joint projects with industries and universities like any modern institute.[13] Himmler sought to make a show of scientific rigor in order to present his utopian desires as "modern" (hardly as a reaction against "modernity"), and thus it should come as no surprise to find a new group of emerging professionals involved in this project. These were academically credentialed landscape architects, striving to establish the legitimacy of their new degrees by arguing that the management of nature, in their hands, could and must serve to reinforce race and nation. Many, like Alwin Seifert, advised not only the SS but other public institutions such as the famous Reich Autobahn; others worked closely with the Reich Agricultural Ministry. The multifarious nature of National Socialist organizations allowed them much room to make contacts and work connections. That so often the SS or these new professional movements turned out to be laughable and pseudoscientific hardly diminishes their genuine enthusiasm for "modernity."[14]

Ahrens further indulged the SS's cultural pretensions by installing a sword smithy at Dachau in October 1938. A blacksmith named Müller convinced him that he could preserve the "art of the Damascus forge," and Himmler wished to award Müller's swords and daggers to members of the Personal Staff as gifts, much like the "cultural" figurines of the Allach Porcelain factory. Ahrens also invested in a bakery at Sachsenhausen to mass-produce 100,000 loaves a day, which he intended to sell to the concentration camp and the Waffen SS, making the SS both producer and consumer of its own product, a clever ruse to skim moneys from the Reich. In addition he added an experiment kitchen in Cölbe bei Marburg, where the SS indulged its scientific dandyism yet again by developing chemically treated wrapping paper. The paper was supposed to preserve bread for mobile field troops, although Pohl later exposed this too as worthless.[15]

Ahrens's most egregious decisions, however, concerned the largest and most important SS company, the German Earth and Stone Works. He left the entire planning and construction of its brickworks to a single, outside contractor, the Spengler Maschinenbau GmbH, which botched the job in almost every detail.[16] Ahrens had leapt into this deal with an enthusiasm for modern technology quite typical of SS industry, but this time the difference between the SS's pretensions and industrial reality quickly came a cropper. Because the DESt was supposed to supply prestigious national projects, the debacle quickly spurred more than a few embarrassing inquiries into all SS business ventures.

Given this ragtag ensemble of SS companies—from herb gardens to bakeries to porcelain works to brick factories, sand pits, and stone quarries—how could the SS defend its elite pretensions? More specifically in the case of DESt, what did "modern technology" actually mean to Ahrens, to the SS, or to Nazi Germany in general? Almost all Germans agreed that technological modernity meant mechanization, standardization, mass production, and the centralized direction of hierarchically managed organizations. Along with many other construction engineers, Fritz Todt, the celebrated builder of the Autobahnen, had even singled out the brick industry in this cause of modernization: "More than any other sector of the building economy, [it] demands energetic measures of rationalization and mechanization." He disparaged traditional producers, "who still work with manual craft processes that have not changed in operation for 30 to 40 years."[17] This backwardness, he pointed out, meant that annual output was expected to sink by 2 million, this at the very time that the Führer buildings proposed to expand the demand for bricks almost infinitely. The replacement of traditional, craft-oriented, small- batch producers with large mass-production firms using mechanized—that is, modern—works offered the only way to escape shortages. The SS swiftly joined its voice to this chorus by seeking out the latest and best machines available for DESt. In the effort, however, Ahrens (and later Pohl) soon discovered that modern technology and organization can be as multifarious in application as any other product of human ideology.

In particular modernization embraced a communitarian ideology in Nazi Germany. German engineers—the greatest enthusiasts of industrial modernization in Germany—advocated a vision of factory community along with mechanization. Their advocacy did not stop at the factory floor but widened to include all of society. From the most prestigious annals of engineering societies to the mundane trade journals of the stone and earth industries, all expressed cherished visions of a progressive future and lifted up favored technologies as the means to its achievement. Many histories dwell upon distinctions between Nazi fanaticism characterized by Hitler or the SS, on one hand, and the "rational pragmatism" of "normal" engineers or "technocrats" like Fritz Todt, on the other. But these distinctions can be maintained only if we consider machines themselves or modern management to have no social or cultural history, as if "pragmatic" choices between differing factory systems did not inherently involve choices among different visions of "community" and "society."

Artifacts commonly assumed to be "value neutral" never appeared as such to the SS. If SS managers inclined to dilettantism, experienced, successful managers took it all the more to heart that, as the historian of science Alan Beyerchen has noted, "technology is fundamentally about ways of organizing the world. . . . Organizing human beings and human affairs is itself a kind of techne."[18] Contemporary engineers would have agreed, as when one "father" of the building supply industries remarked, "Organization is also technology."[19] When Ahrens had to choose specific machines, any "pragmatism" necessarily involved judgment about reforms that, in the context of Nazi Germany, mingled racism, modernity, and class politics with machinery.

A political cartoon taken from an industrial tabloid illustrates what overarching goals the DESt sought to pursue. The cartoon was not written by SS officers, but it did express the attitude they adopted. On the left, the proletarian of yesterday—that is, the Weimar Republic—bends his back over an impossible load of bricks in front of an unmodernized, small-batch dome kiln, the kind of craft operation Todt characterized as backward. In the scenario on the right, intended to represent what Hitler had wrought, a skilled laborer works in a clean, ordered workshop after planned apprenticeship. Satisfied with his world, he strikes a manly pose before a factory, presumably automated. It is paneled by Bauhaus-style windows. The message for social change was clear: Germany had to modernize, to increase the scale and mechanization of factories in order to meliorate the troubled existence of low-wage, unskilled, proletarians with their depraved inclination toward communism. Machines were the crucial means to replace the disgruntled proletarian (standing by a pile of trash in the comic strip) with the self-confident, high-wage, skilled machine operator. In an academic venue, the SS engineer Hans Kammler had written that National Socialism was a necessary catalyst of modern organization in the building trades and vice versa. Here was a tabloid version of the same productivism. The clean, well-lighted factory behind the proud worker on the right did not just turn out mass-produced bricks; it manufactured new virtues of the German soul, a healthy body politic.[20]

The original ethos of concentration camp factories fit well with this goal, for they were founded at the onset of a new wave of arrests. In December 1937

Political cartoon "Earlier" (left) and "Today" (right) from the Fachliches Schulungsblatt der Deutschen Arbeitsfront Stein und Erden *(12 July 1938): 97. Captions, from "Earlier," top to bottom: unhygienic workplaces, hard and unhealthy labor, no systematic education, unworthy treatment, the result: the proletarian; from "Today," top to bottom: clean and healthy workplaces, exercise, paternalistic trust, apprentice workshops, the result: the self-conscious skilled worker. The general caption (not shown) reads: "A contrast between educational relationships in earlier times and now (see the picture) shows the direction for the measures to be taken."*

the Third Reich issued an "Edict Regarding Preventive Crime Fighting through the Police," further supplemented and expanded in January of the following year. The edict singled out new social groups for "criminal prevention." Previously the SS had usually sent prisoners into "protective custody" for political reasons; now the "asocials" or the "work-shy" made up the largest group of new inmates. (Jehovah's Witnesses also found themselves arrested at this time, not the least due to their pacifism.) The SS now arrested most individuals on the pretext that they had willfully withdrawn their contribution to the national economy. The SS initially took quite seriously the cynical phrase set in wrought iron above the gates of Auschwitz, "Work Sets You Free." This was part and parcel of productivist ideology: SS companies set out to produce a strong German nation no less than bricks and stones.

Ulrich Greifelt, Himmler's representative within the economic planning authorities of the Four Year Plan, expressed this mission in the following words:

> The chief of the SS administration . . . [has] established or [is] in the process of establishing production centers for costly building materials which are needed for the major construction enterprises of the Führer. . . . In view of the tight situation on the labor market, national labor discipline dictated that all persons who would not conform to the working life of the nation, and who were vegetating as work-shy and asocial, making the streets of our cities and countryside unsafe, had to be compulsorily registered and set to work.[21]

As Karin Orth has noted, the waves of arrests throughout 1938 quintupled the camps' population, bringing the total to 24,000 by November. Fully 70 percent of these counted as "asocials"; at most there were 2,000 women. A spike of arrests also followed the Reich Crystal Night pogrom of 9 November 1938, and the camps briefly detained an estimated 30,000 Jews. The Death's Head Units subjected them to the harshest brutality and tortured hundreds to death. But between six to eight weeks later, most "Crystal Night" Jews were released. Among long-term prisoners, Jews had been overrepresented, to be sure, but nevertheless made up no more than 10 percent of inmates.[22] The vast majority in the concentration camps, in 1937–38 as before, therefore remained non-Jewish, German men. They were the targets of Nazi policies against the "asocials" as well as the supposed beneficiaries of ongoing Nazi industrial reform. If modern factories could reform the proletarian pictured in the political comic strip, the SS hoped, at least initially, to "reform" prisoners in the same way; and if the camps could force what Himmler called "the scum of humanity" to turn out the raw materials of Hitler's monuments, this added poetic justice for the SS.[23]

Ideals of productivism were hardly unique to the SS, and they were unambiguously associated with modernization. Industrialists called for mechanized technology in order to preserve the strength of the economy, the sanctity of German culture, and the purity of race by introducing labor-saving machines. To many engineers, this was all of a piece. For example, an article in a mainstream building-supply journal blamed a "tragic" mongrelization of the white race in the United States on insufficient mechanization: "In America . . . Negros were first dragged in as cheap, agricultural machines. Now they present a difficult problem. If they continue to multiply at their current tempo, and if the birthrate of the white slave owners continues to sink, they will make America into a black continent in little over a century."[24] To preserve the purity of its own white race, Germany had to modernize its technology and abandon lingering nostalgia for craftwork:

> Craftwork is not only less economic; it is also poorer in quality than machine production. Let us burden our machines with labor and be satisfied as their creators. Let them work for us. We wish to use our intelligence for the invention and control of these machines—and only in this way— will we achieve what we need, the highest productivity [*Leistung*] of our people [*Volk*] and the highest culture.[25]

SS factories would be but a minute portion of the German economy, but their drive for modernization in the name of culture and race demonstrated one aspect of a broad-ranging consensus throughout the building trades, indeed throughout German industry.

Obviously the prisoners trapped in the concentration camps represented a completely different labor force than the idealized skilled worker in the cartoon, and they were quite a bit more demoralized than even the slouching proletarian. Did Ahrens therefore adjust the SS's vision systematically to accommodate forced labor? In fact, he did not. Instead, only the SS's enthusiasm for modern technology marked the first deals to equip the DESt. Ahrens entered foolish negotiations to get the latest automatic machines, regardless of their suitability. He ended up with the wrong technology precisely because of the prevalent blind enthusiasm for technological modernization within the SS. There is no other explanation.

What machines in particular captured Ahrens's attention? The newly invented—and therefore prestigious—automatic technology in which he placed all his hope was a "dry press" manufactured by the previously mentioned

Spengler Maschinenbau GmbH. All brick production, by whatever means, passes through five stages: the mining of clay, mixing, molding, curing, and firing. Spengler had played a leading role in developing a radically new system for pressing molds (instead of extruding them). Not without some technical elegance, the firm's "dry press" operated fully automatically, in continuous motion on a rotating table by filling and pressing molds much like filling and capping machines in bottling plants.[26] The main advantage to the Spengler system was the time it saved in curing bricks for the kiln. In contrast to the "wet process," which squeezed wet clay through stencil molds much like Play-Doh, dry presses handled raw material with half as much water. Yet the machine had corresponding disadvantages: clay is much more difficult to mold the drier it gets. It quickly becomes crumbly. Anyone attempting the dry process had to pay close attention to the composition of raw materials. Uniform preparation of the mix required special roller mills to ensure uniform granules, and success depended upon the strict regulation of water content in the clay that made up the mass.[27]

The Spengler machine was sensitive to misuse and small variations in preparations, it demanded a skilled tender, and it was expensive. "Wet" presses were much simpler and more rugged. Concentration camp prisoners usually possessed no skills specific to the brick industry, constant violence hardly stimulated their concentration, and, if anything, their true motivations inclined them to abuse the sensitive machines. It is hard to imagine anything less suitable for low-skilled, slave labor than the Spengler system. The only possible explanation for Ahrens's attraction to the process is Spengler's reputation as an inventive German firm and thus its aura of "modernity." Albert Speer wrote after the war that "someone turned up with a new system for manufacturing brick" and this nifty machine was the selling point.[28] There is no reason to doubt Speer's account in this instance. Himmler had jumped at every opportunity to promote German inventors and had previously backed firms much more suspect than Spengler. The SS was likely attracted to Spengler due to the belief that the "will" of German inventors needed every encouragement in an economy that was still too capitalistic for its taste. Thus the SS chose brickmaking technology in a conscious pattern of investment but did so neither to make effective, efficient use of prisoners nor to optimize profits. Ahrens bought Spengler's machines for their symbolic character as icons of modernism and, so doing, cast aside many criteria that are actually the essence of modern production: careful predictions of capital returns, evaluations of raw materials, or the consideration of labor, management, and technology as a systematic whole.

Like the dog that Sherlock Holmes listened for in the moors, the one that

The Spengler rotating "dry press." From the advertising pages of Johannes Fischer, Die Ziegelei. Anlage und Betrieb (Berlin: Paul Parey, 1955). As difficult as it is to believe now, such automated machines and their aura of modernity held as much attraction to the SS in the 1930s and 1940s as new computer technology and "cyberspace" can hold today.

did not bark, SS managers are often most conspicuous for what they did not say and what they did not do. It should also be noted, therefore, that the German Earth and Stone Works could have acted differently. Specific modern machines by no means precluded slave labor and could have suited it quite nicely. Take the modern kiln, for example, the core of any brick factory. Its routine operation offered plenty of opportunity for the camp SS to punish and demoralize inmates. The modern kiln used almost exclusively in Germany during the 1930s was the ring kiln, a model of the heat efficiency for which German industry was rightly famous. True to its name, it consisted of three concentric, doughnut-like chambers. Along its arc, the outside ring was divided into segments accessible through the kiln's outer wall. A small ventilation hole connected each of these chambers to a middle, exhaust chamber, which, in turn, connected to the smokestack, the innermost "doughnut hole." In the sealed segments, combustion proceeded around the arc and never went cold, except perhaps for seasonal repairs. In addition, the burn cycle set the pace of production, not the human hand of a craftsman. In this, the ring kiln resembled the assembly line despite other obvious differences. Likewise, its design was meant for large-volume, continuous, that is, modern, production. They were not fully mechanized, but advanced German brickworks of the 1930s and 1940s concentrated mechanization on fuel supply, the forming of molds, and the intensified use of energy for curing. (They often used the collectors' residual heat, for example, to drive the water from curing molds more quickly, again in a continuous process.)

The specific nature of this kiln demonstrates modern technology's openness to multiple interpretation and use. For one, the design left plenty of backbreaking work to be done by hand. Workers loaded each outer chamber with bricks and then sealed them consecutively. The outside portals had to be bricked up and broken into once a cycle. Bricks had to be stacked, unstacked, and restacked repeatedly, work usually done manually, even in advanced civilian factories.[29] Thus "modernity" scarcely eliminated the unloved, grimy nature of this toil, and Kommandanten could have easily used the kilns to great effect in order to gratify their predilection for driving inmates into the ground with demeaning tasks.

Instead of lengthy discussion of such options and advantages, Ahrens fixated on the dry press. He gave the Spengler Machinenbau GmbH, a firm highly specialized in one machine, a carte blanche to construct all aspects of every SS brick factory. The choice turned out to be catastrophic. Ahrens not only failed to hire specialists for systems that Spengler lacked experience to build; he also neglected to supervise what Spengler installed and the money it spent. Kilns installed at Hamburg, Oranienburg, and Weimar had to be de-

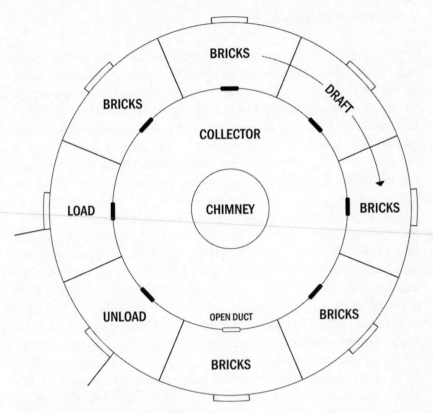

BRICKS

DRAFT

BRICKS

COLLECTOR

LOAD

CHIMNEY

BRICKS

UNLOAD

OPEN DUCT

BRICKS

BRICKS

Sketch of a ring kiln. Drawing by Steve Hsu.

molished and built anew little more than a year later, and in one case a fuel generator was so faulty it threatened to explode.[30] As already mentioned, the dry press demanded acute attention to the quality of clay. Yet no one in the SS bothered to evaluate the DESt's pits. Only after the fact did an engineer discover that the clay near Sachsenhausen was mediocre and impossible to use in the dry press.[31] The only conceivable objective the dry press served was the SS's desire to trick itself out in modern technology and champion innovation for its own sake, which, however, it did not bother to understand. If the SS was acting on an impulse to "modernize" that was mainstream in German industry, seasoned managers would have acted on their ideals of social reform *and* sound engineering knowledge of the material world and its technics. In the absence of competent industrial engineers who might have transformed machines, raw materials, labor, *and* social ideals into a coherent whole (clearly possible in this case given the nature of other available technology), DESt quickly blundered ahead into contradiction and incoherence.

Profits from Women's Work

The failure of DESt may come as no surprise to many, for DESt may seem to be an example of exactly what happens when "fanatics" put ideologies of social change first and neglect "business sense." Although, as already shown, civilian German managers, the very ones engineering the robust growth of Germany's war economy, never viewed social reform alien to their tasks, perhaps there is no better counterexample to the conception that business is somehow inherently "pragmatic" and immune to ideological extremism than one of the SS's own successful corporations from this same era. This book argues that the modern management of technological systems "works" only when both managerial consensus (based on ideology) and sound knowledge of the complex material realities of production can be brought together in a coherent system. DESt was a mass of contradiction, but a newly founded SS textile and garment works, the Textile and Leather Utilization GmbH (Textil- und Lederverwertung GmbH, or TexLed), earned profits where the DESt failed. Its management translated the same enthusiasm displayed by Ahrens into a viable factory system, and its history is an interesting reminder that there is no inherent contradiction between modern business organization, slavery, and barbaric ideology.

SS officers had planned the TexLed since late 1939 and booked a profit almost immediately. They owed their success to several factors. The first was the unusual presence of top technical management that identified with the SS, the very thing that DESt lacked. TexLed rested on an axis of cooperation formed by Fritz Lechler and Felix Krug, both of whom were exceptionally young, even among SS businessmen. Krug was born in 1908 and Lechler, his nominal boss, four years later. They were also independent from the very beginning of the influence of Arthur Ahrens and formed a clear, corporate hierarchy in line with the new style that Oswald Pohl had already imposed on SS financial administration. Likely common youth and experience cemented the bond between Krug and Lechler, and they had already formed a working relationship before their debut as SS managers. Despite their youth, the two were ideological "old fighters" (*alte Kämpfer*): Lechler joined the party and the SS in the first months of 1931; Krug had joined in 1930 and had joined the SS before the party.[32] Lechler further boasted a family heritage of committed Nazi managers and bragged about his father's activism: "In his capacity as a factory director at Siemens and Halske AG in Munich, he has demonstrated the purity of his Nat. Soc. world view [*Weltanschauung*] in the thoroughgoing social improvements he has undertaken for those under his charge. He is a member of the NSDAP."[33] By "social improvements" Lechler almost certainly meant the Nazi rationalization drive for a communal work force at Siemens, which Carola Sachse has documented in detail.[34]

In addition, Krug and Lechler had grown up around industry and commerce and had rounded out their experience with systematic education. Lechler learned the techniques of financial administration by attending one of Germany's commerce schools (*Handelsschule*), after which he worked in the payroll office of a machine tool factory. His career was successful, and he had become an office chief by 1935. He quit voluntarily to work full-time for the SS. It is reasonable to assume that conviction, not economic necessity, compelled him to do so.[35] Krug's father was a master tailor in Munich. Although nothing indicates the scale of his father's enterprise, Krug was undoubtedly exposed to the latest production practices when he attended a trade school for garment makers. He continued to work in the family's business until, like Lechler, he switched to full-time employment with the SS (in 1934).

Krug's vocational training did not count as elite in German society but proved of great advantage in Ravensbrück's factory halls. Krug evidently felt comfortable getting his hands dirty on the production line and provided the TexLed what almost all other SS companies lacked: a competent, technical supervisor who shared an ideological commitment to SS industry. From the beginning, Krug sought out a small staff of experienced foremen and added a quality assessment station, another hallmark of modern production control that afforded Krug continuous, reliable oversight of labor organization.[36] Combined with Lechler's skills in financial accounting, Krug's technical expertise made the TexLed the only SS company to employ the sophisticated methods of modern technical management in slave-labor supervision. As defined by Alfred Chandler and others, statistical measures were (and are) the lifeblood of modern management; their substitution for the direct surveillance of operations made impersonal corporate hierarchies possible and freed the middle manager from constant, personal, face-to-face intervention in production.[37] Rather than keeping simple records of transactions in bound ledgers, typical of traditional business enterprise, Krug wrought statistics out of the material world of production for internal command and control. He tracked them over time, evaluating unit costs, labor costs, and depreciation rates on machinery; and he combined aggregate statistics to yield calculations of production minutes per unit product and unit labor costs. Other SS corporations strove for such information management but lacked the production engineers to master it. In Krug's capable hands, production statistics became transferable: they meant, for instance, that Lechler or Pohl—neither of whom understood the details of technological production—could nevertheless comprehend the TexLed's operations at a glance down the balance sheets.

Coached by Krug, the supervisors of the TexLed maintained steady-flow production. They simplified standard patterns and routinized processes in order to cut training time among the unskilled women of Ravensbrück.[38] Once they had formed a team of workers, they moved for incremental increases in output. "Due to the excellent trade knowledge of the works managers employed [led by Krug] very good rates for piecework were achieved."[39] On the other hand, they adjusted some aspects of production to large fluctuations in the labor force. Lechler reported in 1941 that his work force at Ravensbrück varied daily by 100 to 200 women. This variable portion of the camp's workers was set to hand knitting that could be picked up and put down.[40] "Through rationalization," read one report, "the services of the technical management must be highlighted all the more."[41] Krug was then achieving productivity rates close to those of civilian workers.

The presence of a competent and dedicated engineer thus made much of the difference at TexLed, but, as seen in the case of Ahrens's disaster with the German Earth and Stone Works, much also depended upon what kind of machines Krug came to manage. Ironically, when choosing machines, TexLed had done exactly as Ahrens. "The installation of the workshops was effected in accordance with the most modern principles," Lechler wrote. Like Ahrens he had invested in the newest machines, "only the most modern Pfaff- und Dürrkopf fast sewing machines. Special machines (buttonhole machines, bolting machines, picoting machines, two needle columns and flat-knitting machines) were introduced as far as was possible. All machines have individual motors."[42] If the same blind exuberance for modern machines guided this factory setup, why then did TexLed succeed where DESt failed?

The nature of these machines differed from the "dry press" purchased by Ahrens, and the way in which they differed had everything to do with broader currents of Nazi modernization ideology. In particular, Nazi modernity foresaw different roles for men and women and, correspondingly, different technologies. As such, the garment-making sector presents another example of modern technics' multivalence. Pohl had located the TexLed at Ravensbrück because it was a woman's concentration camp, and he deemed garment making to be "woman's work." This had been Lechler's primary consideration: "The employment of female prisoners was possible mainly in the field which is most suitable for them."[43] In other words, sewing, knitting, weaving, and spinning were considered "natural" feminine activities.

Thus when Lechler and Krug set out to found a factory for female prisoners, they acted not only on their own values of gender, but on gender stereotypes that had long influenced technological systems in their sector. German textile engineers had shaped their practices according to the perceived special needs of German women and mothers. Nazi rhetoric of national community identified three distinct kinds of citizens: the German man of the cartoon depicted earlier, the German woman, and outcasts. Thus, even if Nazi ideology drew a clear dividing line between German mothers and women prisoners (whom they labeled as outcasts from the body politic), Nazi ideals of Aryan women, ironically, had everything to do with the eventual organization of slave labor at TexLed because these ideals had long shaped the "sweet" machines that Krug and Lechler subsequently tried to adapt to Ravensbrück.[44]

As shown already, the rationalization of German men's labor meant mechanization and aimed to create a high-wage, high-skilled working man. But this did not hold for women's industrial labor, where deskilled operations and transitory work were the norm. The drive to modernize men's skilled labor included endless pontificating about the cultural superiority of quality, capital goods produced with capital-intensive, labor-saving machines and long-term employment.[45] On the other hand, rationalization literature did not define the "cultural" role of a woman by the goods she made or the pay she received. She was destined for marriage, family life, and unpaid domestic labor. In the industrial work force, women's work was "rationalized" through deskilled, labor-intensive production—all conditions that prevailed in the garment-making industry. Thus, when Lechler and Krug installed the latest machines for women's work, they quickly tapped into a long-established technological tradition configured for short training time, low skill levels, low wages, and high turnover rates. Civilian labor in the garment industries cannot be flatly equated with concentration camp sweatshops, but both shared these basic characteristics.

Machine systems embodied such values. Unlike the newest machines in the brickmaking plants of DESt, for instance, much top-of-the-line technology in the garment industry increased the output of single operators rather than automating their labor. Rather than labor-saving, such machines might be called output-multiplying. It is important to understand the difference, for the SS had already faced analogous choices in its building supply industries. In textiles the automatic loom and the sewing machine underscore the contrast. Power looms replaced hand looms as labor-saving systems. One tender could monitor several looms set up rationally on the shop floor, thus working cloth in parallel because no single bolt demanded her full attention. The machine substituted for the necessary presence of the human hand and for much of the oversight of human eyes. Like the power loom, the sewing machine also made prodigious advances by substituting mechanical action for the human hand: its needle and undercarriage replicated and far surpassed the fingers of the tailor, and the electric drives that came to stand as the sine qua non of modern garment making in the 1920s further replaced and improved upon the treadle. Due to these progressive innovations, the garment industry had experienced eleven-fold gains in productivity around the time that Krug was learning his skills in Munich. Nevertheless, mechanized sewing had not (then or today) eliminated the requirement that a single operator interact directly with clothing turned out on the line. The sewing machine *multiplies* the output of a single operator but was never designed to *eliminate* her labor.

The difference is also one of capital investment. To return a profit, output-multiplying devices cannot represent an extraordinary investment per worker. While the purchase price for sewing machines can be considerable, it is small as a fraction of production costs and attention must focus on reducing wages. The labor-saving machines, on the other hand, usually represent a large capital investment next to which payroll costs are often marginal, even for a highly skilled operator. In fact, the operator *must* be skilled to ensure the proper, maximal utilization of the machine. Every ineffective use or, worse, any damage represents a loss of return on the expensive technology, exactly Ahrens's problem with the Spengler dry press. The very opposite holds for output-multiplying devices: the bulk of costs remain tied up in wages. This is why output-multiplying machines operate best when wielded by low-skilled and marginalized workers. They are the lowest paid, and this is why "sweatshops" still flourish in the garment trades.[46]

Steady-flow production that relies on output-multiplying machinery was (and remains) particularly suited for forced labor. Had the main motivating force behind SS concentration camp industry been a "rational" profit maximization based on efficient use of prison labor and modern technology, all SS companies that employed prisoners would have turned toward output-multiplying technology. Yet SS companies were not organized according to profit. They were dedicated to hazy, inconsistent maximization of the SS's "cultural" authority, "modernization" at any cost, the Führer principle, racial supremacy—in short, Nazi fundamentalism. Only in the case of TexLed, Lechler's ideals of the most modern machines happened to converge with a state of the art suited to the labor force at hand, and this was due to the way ideologies of gender had shaped the technology. Engineers in Germany (and elsewhere) had long designed machines for textile and garment making to take advantage of an unmotivated work force due to preconceptions of women's work; or, conversely, they had recruited women to operate machinery deemed unsuitable for men.[47] Inmates at Ravensbrück, like all concentration camp inmates, possessed irregular levels of skill. Their population also fluctuated drastically. Output-multiplying machines had been designed for just such a worker. More by accident than design, the TexLed happened to be situated in a sector in which "modernization" directed the SS toward machines that would work well in the camps. In addition, TexLed possessed a unified, competent technical management capable of orchestrating the whole in a complex system of production.

This production system further demonstrated that certain kinds of modern production could accommodate brutality *and* efficiency at the same time. Krug's system rested on violence and the threat of violence no less than sound technical management. Beatings awaited anyone who failed to reach piecework quotas. As one prisoner described, "We led a life of fright from morning till night."[48] Broken machines were considered sabotage, for which prisoners also faced whippings. Prisoner memoirs report that Krug's floor managers showed zeal in meting out such punishment, suggesting that they identified with the use of force and willingly dedicated their creativity and initiative in using it. There were no outstanding orders demanding beatings on the production line and even some directly from Himmler which sought to forbid and curtail it. Himmler had actually visited the camp shortly after it opened in 1940 and gave strict orders during his visit to regulate whippings.[49] Yet Krug's technical overseer for the highest-priority orders, Gustav Binder, routinely bashed women's heads against their worktables when he noticed them asleep or, indeed, suspected them of being sleepy. He also beat several women to death. These are further examples of local initiative, which mattered much more than bureaucratic proscriptions within the camps and SS industry alike. The French survivor Germaine Tillion, who augmented her own experience by interviewing former cell mates, described Binder as a bullnecked Bavarian, "whose workshop was the daily scene of brutalization of thirty or so of the women, over whom the reign of terror took every conceivable form."[50] The language of the women she interviewed, written down twenty years after the experience, still resonates with anxious dread.

Of no lesser importance than direct violence was the threat of even deadlier conditions outside the TexLed industrial court, which represented a camp within a camp. Even when facing the brutality of Binder, TexLed workers found shelter from the elements in the factory barracks. Everyone in the camp knew other women who had died, and they knew all too well that prisoners outside the industrial court worked in freezing weather without shoes. Although a whipping awaited anyone who failed to meet Krug's quotas, by fulfilling them women could also earn crucial extra rations that enabled many to survive. Women often worked hard at their benches to avoid more dangerous work and more deadly brutality outside, and thus visceral fear gave them powerful incentive to do Krug's bidding.[51]

Some may object that the TexLed's success did not have to do with the

The Textil- und Lederverwertung's sewing machine line as seen from the supervisor's room, 1940. Sammlung der Mahn und Gedenkstätte Ravensbrück, Stiftung Brandenburgische Gedänkstätten, Fürstenburg.

nature of its management and machines but rather was due to a labor force it did not have to pay. Indeed private industry noted the TexLed's 18 percent profit margins in 1940–41 with some alarm. Fearing that the SS might usurp private markets, Reich officials originally set limits on the corporation's output and permitted the TexLed to make only a single, standardized garment: prison uniforms.[52] Yet even if Lechler and Krug did benefit from such structural advantages, these alone by no means account for the company's profits. Firms like DESt had emerged under even more propitious conditions with guaranteed state contracts but failed miserably. Therefore success at Ravensbrück must be sought not in its structural advantages but in the nature of its management, its machines, and, no less, its consensus of purpose.

Opportunistic Idealists and the Shady Legality of SS Industry

Pohl brought both Lechler and Krug into SS industry after a concerted effort to root out the influence of Ahrens. Although the success of TexLed was fortuitous, it demonstrates what might have come to pass had Pohl located like-minded industrial engineers such as Krug for all SS companies. Pohl did recruit new SS business managers as he attempted to transform himself from a navy paymaster into a modern CEO, and did so immediately after one of his most talented officers, Walter Salpeter, began to uncover the mismanagement of DESt and the amateurish nature of Ahrens's business in late 1939. Salpeter worked hard to bring the idiosyncratic mismanagement of earlier years to an end, and his engagement marked a change toward management directed through impersonal hierarchies and controlled through the statistical transfer of information. His career therefore illuminates how ideological visions could animate aspects of modern bureaucracy that are otherwise as dead and tedious as corporate law and accounting: to Salpeter modern corporate structure possessed no lesser aura than the Spengler "dry press" did to Ahrens.

Salpeter's pathway into SS management was also typical of the handful of young managers drawn to the SS companies from 1939 to 1940. This cadre came from the internal ranks of an existing, dedicated officer corps that had not been recruited initially to manage SS corporations. Salpeter had started SS service in September 1937 to handle property law in Pohl's Legal Department (within the RFSS Personal Staff). First joining the SS only in February 1933, he did not belong to the old faithful, but this should not obscure his ardent convictions. In fact, if the SS had a business ideologue, Salpeter was it. He was anything but a fellow traveler, as his boosterism of the Führer principle in modern management quickly showed. When Salpeter came to work for Pohl, he was in his early thirties and, like so many capable, ambitious, young SS men, had never been unemployed. He already had knowledge of modern financial and legal administration and had even published a book on corporate taxes.[53]

In its quest for modern technology the SS had acted as a *Doppelgänger*. Modeling itself according to prevailing values of modernity that promised to lead toward a progressive future, it then tried to radicalize them in the name of Nazi cultural values. Salpeter's pursuit of modern organization was no different. Many historians and sociologists speculate that "ordinary Germans" like Walter Salpeter, banal and obscure functionaries, approached National Socialism as opportunists. William Brustein, for instance, argues that the rank and file joined out of nothing more than "material self-interest" based on "rational choice." In contrast to Hannah Arendt, who accused Eichmann of acting out of an "inability to think," Brustein urges us to believe that the typical Nazi acted out of an inability to think about anything but venal personal interest.[54] This theory has the advantage of freeing us from the misconception that Nazi bureaucrats were constrained and could not have acted otherwise, but only by postulating that they acted exclusively to line their pockets.

Let there be no doubt: Salpeter was an opportunist, as were countless other ambitious SS officers. But he was unlikely to have sought out SS employment for the salary alone. He noted that SS men earned less than those in private business, and to him idealism was adequate compensation: "Those who wish to work together in the VuWHA at a responsible post must do without some comforts." Possibly as a jibe at Ahrens, he added that statement about SS men already quoted in chapter 1, "They are more valuable than those who simply wish to put on the uniform of the Black Corps and strive after an influential job in the Main Offices of the Reichsführer SS—when possible immediately—out of material interests."[55] Nevertheless, his opportunism showed when he seized upon the chance to reform and expand SS corporations while he was supposed to be preoccupying himself with property rights. Yet characterizing this as "opportunistic" tells us nothing unless we ask to what ends Salpeter bent his opportunities. The young lawyer did promote his own career, but if Salpeter or Nazis like him were opportunists, they were "idealistic" opportunists: they selected what occasions presented themselves in order to transform their ideals into deeds. Like any *Doppelgänger*, at times this course of action meant adapting flexibly to circumstances, but it also meant that individuals like Salpeter would try to bend those circumstances to their values as well.

The first issue at stake was scarcely more inspiring than brickmaking machines: corporate tax law. As a specialist in tax regulations, it is no accident that the first problem Salpeter uncovered was that SS industries had actually made Himmler, as RFSS and nominal "Führer" of all enterprises, personally liable not only to income tax but to enormous capital gains and profit taxes as well. Salpeter warned Pohl of the fiasco and the fact that the Reich Finance Ministry currently demanded payment. At the same time, Salpeter slowly began to realize the utter chaos Ahrens had wrought. The DESt's books had been kept in shambles, if recorded at all. Salpeter quickly guessed that the brickworks of DESt were running at a loss, but no one could say how large that loss was. He believed the stone quarries might be running in the black, but again, no one could say how much money they had made. Sudden alarm further arose because Reich law barred the SS, as an organ of the Nazi Party, from diverting public funds for market speculation. The SS had long broken this stricture without attracting notice, but now the DESt was a nationwide concern. The NSDAP treasurer (Hans Xaver Schwarz) called upon Pohl to justify its existence.[56]

Salpeter very quickly used his expertise to gain increasing control over the SS companies. Had mere career advancement been enough for him, no doubt his ascension into top management would have sufficed, but in the process Salpeter labored to develop the basis for the SS's future corporate philosophy, both its consistencies and inconsistencies. From the beginning Pohl had intended the entire structure of his organization, of which the SS companies were now becoming an increasingly important piece, to manifest ideals of a "national community" (*Volksgemeinschaft*). Salpeter shared this view and meant to put the SS companies at the service of Nazi cultural unity. As he declared in exemplary bureaucratic prose, "The possible achievement of capital gains is not the task and goal but at most the result of activity that unfolds in fulfillment of our communal goal."[57] Just as the NSDAP claimed simultaneously to manifest and serve the will of the nation, the SS claimed to be one with the NSDAP, a unified communal entity in will and deed. By extension, just as the NSDAP claimed to lead a renaissance of German culture, so too the SS conceived its industry as part of that crusade. As "a political community called to the highest tasks," Salpeter maintained, the SS must use its resources and power for "cultural and ideological ends."[58]

Although according to popular conceptions the typical Nazi was a rigid, "mechanical" bureaucrat, Salpeter sought to differentiate the DESt and other SS companies from what he called "static elements" by declaring the SS dynamic, organic, and the defender of "community." Its managerial leadership was made up of men "who generally bring more discipline and the unconditional, necessary dedication to the work of the SS."[59] He argued that SS companies wished to fashion products to serve Nazi ideals and to fashion their managers to serve as ideal Nazis. At the same time, he viewed slave labor as equally essential to this communal service, for prisoners, in his eyes, had sinned against the National Socialist "community."

Salpeter, in his quest for modern corporate organization, developed a theory of the SS's mission as slave driver. He referred to concentration camp inmates as "fallow work power" (*brachliegende Arbeitskräfte*), whose unused if not useless economic potential the SS was duty-bound to put to work forcibly for "communal" purposes.[60] In the Nazi lexicon "fallow" was a relatively mild phrase but still derived from substantive judgments about the nature of prisoners as degenerates, whom the SS as well as German economic planners considered to be a danger, not a contributor, to national output. Such individuals were considered "burdensome lives," "life unworthy of life," "useless eaters," and a host of other, harsher terms that defined individual worth in terms of productivism and use value to the body politic.[61] Salpeter claimed for the SS the task of laying hands on such "ballast existences" and driving them to useful production. One of his undated memos to Pohl explicitly combined these sentiments:

SS industries [*Unternehmen*] have the task . . . to organize a more businesslike (more productive) execution of punishment and adjust it to the overall development of the Reich. . . . The economic results from this accomplished task will in all likelihood prepare a "Revolution in Prisons" [*Stafvollzug*], which could never be expected, at least in any time in the near future, from the Justice Ministry as the state's "static element."[62]

In keeping with enthusiasm for modernization, Salpeter further demanded that the SS forge "modern institutions for the execution of punishments,"[63] by which he clearly meant the application of modern management to prison industries whose productivity could "be solved only by adopting the private business point of view."[64] By "businesslike," however, he explicitly meant the techniques of modern management, not necessarily profit taking. He argued repeatedly for service to the national community regardless of profitability.

To realize his vision of corporations for "community" rather than capital, Salpeter urged Pohl to ground the legality of SS enterprise in a special structure. The Nazis had pushed through several measures regarding state industry during their first years in office, including special status for enterprises that helped force political dissidents to work in public building projects. Such projects were labeled *gemeinnützig*, that is, dedicated to communal goals under the special aegis of the state. The sophistry involved here was a general phenomenon of Nazi vocabulary politics. Once National Socialist Germany was defined as the National Community, all functions of the state automatically became "communal" or "cooperative," that is, *gemeinnützig*. This definition included nationalized, state-run corporations as well as consumer cooperatives, many of which had an anticapitalistic bent that no doubt appealed to the SS. In fact, the SS had already begun with the Cooperative Dwellings and Homesteads GmbH a few years before. The Hermann Göring Works and the Volkswagen Works were also justified as communal enterprises. Not inconveniently, communal corporations qualified for special tax exemptions.

In these matters Pohl, at Salpeter's urging, did not fail to seek "the instructions of the Führer," and alluded to "a meeting between the Führer, the Reichsführer SS, and the architect Speer."[65] This work culminated in a drafted letter from Pohl to the Reich Ministry of Finance:

Companies of the Schutzstaffel are operated in order to fulfill the task of the Reichsführer SS to bring prisoners in the [concentration] camps once more to work that is worthy of men. . . . The SS pursues its enterprise exclusively to fulfill discrete tasks that are completely cultural and communal in nature. The SS fundamentally avoids business endeavor for the sole purpose of earning money. . . . The very fact of our cultural goals leads our companies down certain paths that a purely private businessman would never dare, and this causes losses from time to time. . . . It is the will of the Reichsführer SS that profits from lucrative corporations be diverted to cover the losses of others that must labor under the constraints of their noncapitalistic [*nicht privatwirtschaftliche*] end goals. At times these goals damn our corporations to years of future losses.[66]

The letter pleaded for mundane tax exemptions yet reads like a manifesto. "Communal" enterprise offered a legal form for doing business with a social mission. In such context attempts to analyze "material interest" as something neatly divided from ideology is nonsense, for, in fact, Salpeter and Pohl's ideals of what constituted "communal service" led them to define the SS's "pragmatic" interests in the first place.

At the end of March 1939, as Salpeter's revelations made the reorganization of DESt urgent, Himmler placed all industrial administration directly and exclusively under Pohl's guidance.[67] Himmler also petitioned for and received special permission for his continuing entrepreneurial adventures through Hitler. Meanwhile Salpeter and Pohl had begun to lend the SS companies an ideological raison d'être.[68] This reorganization came as Himmler undertook to reshuffle the entire SS, and, effective on 4 April 1939, Himmler raised Pohl's former Office IV–Administration to the stature of an independent Main Office, the Administration and Business Main Office (Verwaltung- und Wirtschaftshauptamt, or VuWHA). As Main Office Chief, Pohl was now second to the Reichsführer SS alone, a considerable advancement. He also received a special appointment to the Reich Ministry of the Interior (officially titled Ministerial Director of the Chief of German Police) as sole SS representative in all Reich budgetary matters. From now on, he exercised his influence over all SS finances directly instead of tacitly from an advisory position in Himmler's Personal Staff.

Pohl first divided his new Main Office into three subdivisions, two of which encompassed traditional civil service functions: Office I worked on budgets and payroll for all SS organizations. An Office II controlled SS construction and marked a moderate expansion of Pohl's authority. Now, ever so slowly, he began to exert more direct control over all SS building, especially in the concentration camps. Yet the SS companies presented the most urgent tasks to the new VuWHA, and Pohl organized an Office III–W, which he christened with a special letter to signify its unusual functions: "The Office III is concerned with industrial and business tasks as the name 'W' announces [for *Wirtschaft*]."[69] Here a brief note on translation is perhaps necessary in order to capture the radical departure Pohl and Salpeter were taking with the Office

III–W. *Wirtschaft* is untranslatable directly into English, for its equivalent depends upon context. The business pages of any German newspaper are the *Wirtschaftsseiten*. Working in business is working in *Wirtschaft*. But the word also means economy; thus national economy in German is Volks*wirtschaft*. It also means, loosely, "commerce." Adding to the confusion, military supply is called Truppen*wirtschaft*, and Pohl's main duties of previous years had indeed focused on troop administration and annual budgets for the SS. Yet his office at that time was simply called *Verwaltung* (civil administration). What *Wirtschaft* meant in the new Office III–W is clear in its exclusive dedication to corporate management; it consisted of a staff of company executives, in other words businessmen, and they differentiated their work clearly from traditional troop administration or economic planning.

These differences had consequences. The Office III–W faced Pohl with an undertaking unlike any he had ever previously dealt with. It demanded the management of machines, labor, raw materials, *and* finances, that is, the orchestration of industrial systems. Especially given Pohl's desire for ever new and challenging tasks, it is no accident that the SS companies captured the lion's share of his attention from this point onward, but these efforts proved difficult. Losses at DESt were no longer secret, even if Pohl and Salpeter were stalling to avoid public audit. The only remaining secret was the magnitude of Ahrens's disastrous investments. Pohl testified after the war that the Reich Ministry of the Interior raised objections to the Office III–W because it feared, not without reason, that the SS would funnel state revenues handled in Office I and Office II into dubious industrial ventures. Himmler and Pohl quickly responded by dividing off the SS companies into a freestanding Main Office. The VuWHA now concentrated solely on the corporations of Office III–W, while a new and hastily organized Main Office for Budgets and Building (HAHB, or Hauptamt Haushalt und Bauten) took care of Pohl's civil service duties (the Offices I and II). Needless to say, the division of tasks was never as tidy as external ministries wished. Pohl's personnel flowed back and forth. Nevertheless, the new structure solidified from April onward, and Pohl's own activities demonstrated the increasing bifurcation. True to his promised "fury to work," he began to dedicate most of his attention to the modernization of the SS companies and left traditional civil service duties to subordinates. He had to divest the DESt of the personalized, eccentric management characterized by Ahrens and sweep out other would-be entrepreneurs in Himmler's Personal Staff. For this task, Pohl needed to recruit new men for an impersonal hierarchy of modern managers just as he had recruited officers to build up modern financial administration within the SS from 1934 to 1936.[70]

Pohl took major strides to accomplish this by the end of 1940 and seemed to be repeating his successes with past SS affairs. He quickly organized an impersonal corporate hierarchy. With Salpeter's help, he also began to pull together a cadre of SS businessmen from within the ranks of existing SS officers. Many also differed from the civil servants Pohl had sought out in the mid-1930s. They had experience in corporate law, corporate accounting, sales, and advertising; and some few were even industrial engineers like Felix Krug. At first both Salpeter and Pohl believed that only committed SS men, as manifest representatives of the Nazi will, could serve the SS's communal goals; therefore, Pohl first pulled his new managers from among officers like himself, those who had come to the SS to tackle jobs unrelated to industry but who now greeted the opportunity to experiment with the SS's "communal" enterprises. They saw in the unusual tasks of the VuWHA a chance to advance both themselves and the Nazi movement.

In early 1938, another officer within Salpeter's legal staff, Karl Mummenthey, began to dedicate increasing time to the DESt's books. Like Salpeter, he had training in corporate law and had also studied at a business school (Höhere Handelslehranstalt Chemnitz). In the course of 1939, Gerhard Maurer, the future top manager of all SS slave labor, also entered full-time work for DESt. He had a previous career in midlevel corporate management but had discontinued this to join SS administration at the Troop Training Center near Dachau (separate from the concentration camp). Karl Möckel, none other than the officer Eicke had once accused of incompetence and nervous collapse, also became an executive of several SS companies at this time. Like Maurer, Lechler, or Krug he had been working to audit Dachau's Troop Training Center. The only officer who had just joined the SS—moving laterally from a career in merchandising and sales—was Heinz Schwarz, whom Pohl quickly diverted into the SS industries. Viewed collectively, these men were extremely young, even younger than Salpeter. All except one were still in their twenties; all had joined the SS after successful careers; and all had experience in some aspect of modern management, either by education or by vocation.[71] They formed a small group of dedicated, intellectually spry, ideologically motivated, and hardworking young men. And who could deny that they quickly produced results? The new managers began to root out the mistakes of the past and reasonably expected to master them in the future. From 1939 and into 1941 they continually exposed the sources of loss due to the Spengler debacle. Over half a million Reichsmarks had disappeared in 1939 alone.[72]

However, despite the ideological cohesion and skills this group possessed, with the exception of Krug they lacked the knowledge and experience to wed sophisticated financial and legal administration to the intimate technical organization required for modern production. Equally important, there were simply not enough of them. In order to steer a modern corporate hierarchy, the SS needed a large staff of top-, middle-, and lower-level managers. In the mid-1930s the SS had effectively drawn many professional groups into its officer corps, but entrepreneurs and top managers of corporations were scarce among them.[73] No matter how much Pohl extolled the "will" of the SS man, he was forced to look outside for the specialized skills of professional managers. When he did, he often found talented individuals and tried to induct them into the existing cadre of his SS organization men. But increasingly, especially as the Third Reich converted to a total-war economy, newcomers would clash with his old guard.

The first outsider to enter SS corporate service was Erduin Schondorff, a Diploma Engineer Pohl recruited to take over the technical management of DESt. The appalling state of DESt had resulted mostly from misplaced investment in one unsuitable technology and the firm, Spengler, that made it. Pohl and Salpeter both recognized that only competent engineers could remedy this predicament. Schondorff was eager for the task. He was a trained engineer from Germany's polytechnic universities who had been seeking all his life to reform the brick industry. In academic publications he had attacked exactly those mistakes typical of Ahrens's mismanagement. In one harsh critique published in 1931 he wrote:

> The brick industry presents an odd spectacle when factories install new equipment or address production failures or judge the quality of their raw materials; the brickworks owners or foremen frequently depend on the so-called "expert" because they themselves have not restlessly [an adjective that would become standard Nazi jargon after 1933] enlightened themselves about the chemical properties of clay or about thermodynamic processes in factory operations. . . . The result is that one . . . allows equipment to be damaged.[74]

Nearly a decade later, DESt offered him the unprecedented opportunity to transform these ideas of modernization into practice at the largest brickworks in Germany. The progress Schondorff made initially also allowed the VuWHA to apply convincingly for enormous new loans needed to refinance its debt and undo past blunders.[75]

It is often assumed that men such as Schondorff opportunistically served the SS because they were politically naive or wholly ignorant of its purposes. "The technocrats," as John Ralston Saul characterizes such functionaries in modern society, "are highly sophisticated grease jockeys trained to make the engine of government and business run but unsuited . . . to have any idea where it could be steered if events were somehow to put them behind the wheel."[76] Or as Albert Speer avowed to a gullible audience, "The task I [had] to fulfill [was] an apolitical one. . . . my person and my work were evaluated solely by the standard of practical accomplishments."[77] Once again the history of technology might provide a more realistic portrayal of the context repressed by Speer and wholly misunderstood by Saul. Schondorff's publications and his actions placed him firmly within an ideological and overtly political tradition within the history of German engineering. He did not come to the SS as an opportunist blinkered by political naiveté; rather he came with political ideals of his own and saw the DESt as an opportunity to realize them.

First of all, like Salpeter or Ahrens for that matter, his ideological drive also focused on modernization. He had come of professional age during the turmoil of the Weimar Republic. Those years had not dimmed his faith in technological progress; instead, frustration made his enthusiasm for the modernization of industry all the more poignant.[78] In the midst of the world depression—whose impact was far greater on the building sector than elsewhere—he announced, "Naturally, brickworks schools must . . . expand their curriculum from old-fashioned processes. They must bring their students to the habit of independent work and make them into men who think constantly about progress."[79] Schondorff was an academic engineer engaged in research, doubtless aware of Fritz Todt's condemnation of the backwardness of German brick factories and, at the same time, eager for the unprecedented demand that the Führer buildings had created. Schondorff's own speciality lay in mechanized brick production. He held several patents in artificial drying technology, and he felt frustrated at his industry's slow pace in adopting large-scale, steady-flow production. In other words, his publications and actions show that he envisioned the brick industry's future through aggressive technological modernization, a program that National Socialism robustly promoted.[80]

Schondorff also sensed potential for another ideological program with a long history in the engineering profession: the advancement of rigorous scientific training as a means of reinforcing a division of labor between factory workers and professional technical management in favor of the latter. Elite academic engineers like Schondorff tended to promote this movement.[81] His Diploma Engineer was not quite the equivalent of a Ph.D. but did signify the completion of a scientific research thesis (the *Diplomarbeit*). Although he had not gone on to study for a doctorate but had become a docent (lecturer) in the Technikum Lage in Lippe, a vocational technical school, his position there situated him firmly behind the push for engineering professionalization, which Monte Calvert has labeled the "school culture" within the American context.[82] Academics like Schondorff stressed scientific knowledge over factory experience. "Essentially the brickworks engineer must pursue a more intensive education," he advocated; "a brickworks engineer must have a thoroughgoing knowledge of thermodynamics, of chemistry, and of machine building."[83] The movement criticized the traditional brick master (*Ziegelmeister*), an old-fashioned guild title, and sought to free the "engineer" from his association with tinkering foremen and factory supervisors. It criticized experience gained on the factory floor and those who came from "shop culture." Schondorff's articles attacked factory owners who resorted to "cut and try" methods or relied on their "knack" (*Spitzengefühl*) to solve problems.[84] Far from merely a debate about "objective," "pure" science, or "rational" industrial development, this doctrine included an inescapable advocacy of social reform. It does not fit with our traditional conceptions of ideology, polarized between "left" and "right," communism, liberalism, or fascism, nor did it reach the global proportions of Nazi fundamentalism, but it was nonetheless an argument about who, in German society, should gain the privilege to control the shop floor, namely engineers over foremen and workers. Schondorff pushed for scientific, university education over apprenticeship and shop experience as a road toward that reform.

Schondorff had already begun consulting with the SS in mid-1939 and Pohl subsequently urged him to join full-time. The offer must have seemed like the chance of a lifetime. The SS, as expressed in its own florid rhetoric, was one of the dynamic, progressive organs of Nazi state service, and it is very likely that Schondorff initially heeded its call because he believed the DESt would provide an opportunity to advance his version of engineering professionalization. Before his employment with the SS, he had also done work for another industrial combine founded by the Nazi state, the Hermann Göring Works, which at this time exerted a tremendous draw on the technical talent of private industry.[85] Pohl made clear that no expense would be spared to equip the DESt with the most modern machines available. In addition, Schondorff could erect, from the ground up, the largest brickworks in Germany.[86] He was officially named to the board of directors in October 1939 and soon became one of its executive officers.[87] Pohl placed complete faith in him. "We possess in Schondorff perhaps the best practical and theoretical expert in the brick industry; a man who has rare abilities," he wrote to Himmler during the war.[88]

On the surface, Schondorff was just what the SS needed. He brought students with him from Lippe and infused the VuWHA with a tightly knit community of men with engineering competence. They reorganized the brickworks according to Schondorff's conception of technological modernization (later they would help organize cement factories as well). Schondorff himself analyzed core samples from the various SS clay pits, oversaw the progress made by subcontractors as they installed new equipment, and recommended that Pohl erect new construction bureaus within the concentration camps for this purpose.[89] But Schondorff's ideology of engineering professionalization had consequences, and backing him as a "theoretical expert," as Pohl put it, meant once again placing faith in scientific dandyism. Schondorff had prepared his pupils to strive for technological modernization but not to master the factory floor. Here their "school culture" showed its weaknesses in the continuing failures of DESt. "Shop culture" stressed the management of technology at the site of production and included the orchestration of labor. This was its strength over academic training. Labor management had never been part of Schondorff's courses. An ideal curriculum that he published, for example, contained not a word about work site supervision but dwelt mostly on the complexities of engineering science and labor-saving technology as isolated machines rather than holistic factory systems.[90]

Nothing could have been more inappropriate for DESt. The SS's concentration camp industries presented the most knotted factory problems at exactly the juncture of machines *and* labor passed over by Schondorff's "school culture." Had he or his students really been "technocrats" driven only by efficiency, they might have adapted themselves uncritically to solve the DESt's problems by integrating slave labor and modern machinery in a coherent system—as was clearly possible. But Schondorff and his students acted out their own visions of the engineer's tasks, including a rigid separation between expert management and factory floor supervision, and they saw the SS primarily as an institution to realize those interests. This is not to say that they were repulsed by the concentration camps. However, especially in the "school culture" represented by Schondorff, modernization encouraged an almost exclusive fixation upon "sweet machines" while neglecting to integrate them with the factory floor. Had Schondorff been dedicated, like Salpeter, to erecting "modern institutions for the execution of punishments," he might have provided the SS with detailed plans for the integration of low-skilled, manual labor into modern technological production—as did engineers in Pohl's future corps of civil engineers, for example. Technologies like the ring kiln were suited for this. Had he applied exclusively "instrumental reason" to production, he might have even proposed measures to preserve the lives of prisoners, if only as a kind of industrial input. But Schondorff's reports concentrated only on automated machines.

Erduin Schondorff's counterpart in the VuWHA's top financial administration was Dr. Hans Hohberg, a certified public accountant and a specialist in modern financial management. One historian has gone so far as to call Hohberg the "gray eminence" of SS industry, which may capture the influence he gained over the Office III–W; but, like almost all those who came to build SS institutions, he was young, only thirty-four when Pohl first sought him out. In 1939 the accountant was just establishing himself in a rapidly ascending career. By the mid-1930s he had climbed the ranks, first within a Hamburg accounting firm, and next within the Berlin Public Accountants Firm (Berliner Wirtschaftsprüfergesellschaft). By 1939 he was directing a branch in Königsberg. Pohl came across Hohberg just as the public accountant was auditing the Reich Organization for Peoples Care and Settlement Aid. Although this organization was not associated with the SS, the meeting was probably no accident. The SS's involvement in settlement policy was becoming one of Himmler's main concerns and, as such, an area of expansion for Pohl's industries.[91]

Perhaps Pohl believed he had found a man for the SS's grand crusades, for Hohberg shared much in common with the managerial community of the VuWHA. He was ambitious, highly educated, and possessed specialized knowledge of modern administration. In the late 1920s he had also worked as a director of a cooperative, a "communal" enterprise. But one thing Hohberg did not share was the SS's cultural philosophy of productivism, and no evidence suggests his dedication to the Nazi Party or its organizations. He never applied for SS membership or joined the NSDAP.

Like Schondorff before him, Hohberg's ambitions seem to have been sparked primarily by the promise of gaining a free hand to build an organization from the ground up. One of his friends testified to this effect after the war: "Dr. Hohberg accepted the position as a certified public accountant for the SS enterprises [because] his first audits aroused in him a great desire to reorganize the complexity of the economic activity of the SS."[92] The complexity of modern organizations and the challenge they posed to his personal talents drew him in. Once again, he might seem to be the classic "pragmatist," "technocrat," or "opportunist." Commenting on SS executives like Hohberg, one historian remarks, "These men can neither be described as ideologically motivated nor as misfits of the depression but simply as very ruthless entrepreneurs who, quite clear-eyed, saw opportunities for profit."[93] But this misses the spirit of revolutionary reform associated with the techniques of modern management in this era of its birth. As with things so utterly lacking in glamour as brick molds, so too corporate bureaucracies and accounting sheets could glow in such men's minds with the aura of futurism.

As the organization men responsible for the new institutions of production and distribution, men like Hohberg pointed to the burgeoning wealth of the second industrial revolution in full realization that more could be accomplished under their control than ever before in human history. They rightly felt responsible, even privileged. They measured the progress and fitness of their nations by the extent to which men like themselves could implement the revolutionary techniques of modern management. The aspirations of the future chief of SS civil engineers, Hans Kammler, had expressed no less (e.g., in the book he published discussed in chapter 1). Thus if they were "opportunistic," they looked to opportunities that seemed pregnant with the potential to remake society in their own image. As unlikely as it seems today, an aura of excitement then surrounded professions that now attract scant interest and even derision. The certified public accountants were only one group among various emerging professionals who had newly fought for their legal status since the turn of the century, and German universities had just established doctorates for CPAs. One of Hohberg's contemporaries compared their calling to that of heroic doctors whose patient was the German economy. To play the role, he assured his fellows, the certified public accountant must do more than a good job; he must be a paragon of human character, a sage, an exemplary citizen: "Everything is at stake. . . . This requires a complete man. Service to the customer means service to the economy, to the fatherland. Indeed, the accountant with a pure nature, in command of a healthy human understanding, he's the one we need."[94] By claiming "everything is at stake," the author urged his fellows to imagine their role as greater than themselves, greater than just a "pragmatic" business task or career advancement. He laid claim to a superior, spiritual role through the mastery of modern administration. Of course it would be wrong to believe that all accountants viewed their life's work in such terms; but it would be equally wrong to assume that none believed in the virtue of their profession and settled only for the "iron cage" of bureaucracy, Herbert Marcuse's "one-dimensional man," or venal greed.

The ideas of National Socialism lent such hortatory appeals a specific twist and a specific promise. The Führer principle aggrandized the spiritual role of the manager by idealizing him as the embodiment of the Germanic "will" and encouraged him to believe that control and leadership stemmed from his soulful ability to manifest the interests of his employees. The manager claimed to be more than the leader of a team in the pursuit of material goals; he claimed to be their conscience; he claimed to know their minds better than they themselves did. There is every reason to believe that Hohberg saw the SS as an opportunity to realize such ideals even as he dissented from other SS aims. As he once formulated for Pohl, "The strict operation of the Führer principle has been rigorously introduced into the SS companies."[95] He also brought to his work a confidence in his own abilities that surpassed the point of vanity and no doubt believed he embodied the "spirit" or "will" of those under his aegis. During the course of his career with the SS he began signing his correspondence as "Chief of the Business Staff" (*Chef Stab W*), although officially only Pohl should have been allowed to use this title.[96]

To Hohberg the Führer principle provided a pathway toward the realization of large corporate hierarchies that operated through the statistical transfer of fungible information. One of his first acts was to distribute a clipping from a managerial self-help book.[97] It urged SS administrators to display a new esprit de corps: "Managers have a weapon which can be used to dig themselves out of every problem, a magic word, namely '*Stürmisch*' [vehement or strong, literally, "stormy"]. *Stürmisch*! What does this word mean? *Stürmisch* means rigorous. It means no more, no less than to attack problems from top to bottom and do everything anew instead of ad hoc or bungling."[98] The call to be "stormy" echoed the call of Pohl and Salpeter, which demanded constant, "restless" action as the foundation of their organization. This was unmistakably part of the vocabulary that accompanied the Führer principle: the decisive swift deed would put an end to tedious reflection about problems. The word *Sturm* had further, military inflections, meaning to "take by storm," and no doubt appealed to the SS's glorification of crisp, martial command. *Stürmisch* also meant "passionate" and thus simultaneously stressed both uncompromising action and unmediated "will." In other words, "pragmatic" techniques of modern managerial control were indistinguishable in Hohberg's reorganization of the VuWHA from meditations upon the ideology of leadership. Like Schondorff, he would eventually clash with SS "Old Fighters," but he would not do so as a "rational pragmatist" opposed to the "fanatics." He would come to loggerheads with Pohl precisely at the point that the SS disturbed his ideological conceptions of his own leadership role and corporate community.

The "Organic Corporation"

Working closely with Walter Salpeter, the changes wrought over the next two years in the VuWHA were mostly Hohberg's doing. He suggested a holding corporation to Pohl, one that contained the Führer principle in its very structure: a recent innovation in German business law, the "Organic Corporation." Normally holding companies maintain control over subsidiaries through mutual share ownership and interlocking directorates (in German, *Schachtelung*); they are tied less by legal strictures than by relationships among directors and boards. Subsidiaries usually maintain nominal independence. But Hohberg's "organic" holding company held its affiliates unconditionally by contract and assumed liability as a judicial person (or, by Nazi metaphor, a "Führer entity"). The word "filial" is therefore more appropriate than "branch" or "subsidiary" for the subordinate SS companies, for the SS's holding company was supposed to stand in relationship to them as the head of a family. As such it was supposed to unify the "will" of all filials. Legally, filials had the status of employees (*Angestellten*) rather than independent companies. Philosophically, Hohberg's "organic" company supposedly constituted the totality of its filials, just as Führer managers supposedly manifested the will of all subordinates: "An organic corporation legally exists in the sense of German legal and financial vocabulary when one company (filial) is financially, economically, and organizationally so structured in relation to a suzerain company (the holding company) that one must see it as a dependent entity, as a limb [*Gliedorgan*]. Independent negotiations at a firm's own risk are precluded. A filial must receive its instructions from the parent company in all fundamental business transactions."[99]

Hohberg argued that the holding company would allow the SS to achieve economies of scope in top management, to impose strict centralized control, and to facilitate the transfer of profit and loss uniformly throughout SS enterprise. He could have easily chosen less strict, more "pragmatic" structures. One of the advantages of holding companies loosely related to their filials is that they *do not bind themselves* to the management of "all fundamental business transactions" so as to shed the burdens of capital costs, inventories, or losses. In a sense, they let filials do all the work while they get all the credit. In such schemes, holding companies can combine joint-stock companies (the German AG) with limited-liability companies (the German GmbH) as well as others. The "organic corporation," however, required strict homogeneity. The needless stringency of Hohberg's scheme can be explained only by his own preference for and faith in the Führer principle, which he believed would unify the whole in one indivisible "will." The allusion to "organic" operations was an additional rhetorical coup, for it associated the holding company with primal community, a motherhood concept in the Nazi phantasmagoria.

As pleasing as the "organic corporation" sounded to the ears of Pohl and Salpeter, other, finer points of ideological commitment slowly began to divide them from Hohberg. Salpeter picked up Hohberg's ideas and ran with them, arguing all the while for the SS's communal and anticapitalistic service to Nazi fundamentalism. Hohberg had placed his hope in the Führer principle because, to him, it was coupled to managerial modernization and supposedly lent meaning and added efficiency to large, impersonal bureaucracies. Salpeter was committed to the panoply of causes that had germinated within the VuWHA Law Department for over a year. Hohberg, on the other hand, seems to have remained indifferent to the SS's proclaimed renaissance of German culture and its trumpeted opposition to capitalism. He therefore saw no need to seek financial concessions through direct state mandate on idealistic grounds and urged that the "Organic Corporation" be registered as a private business. But the VuWHA and all its filials were, in Salpeter's eyes, dedicated to the service of German culture and the national community instead of, even in opposition to, private business:[100]

> The employment of concentration camp prisoners in the building supply sector has been ordered by [the Reichsführer SS] in his capacity as Chief of German Police, under whom the concentration camps are subordinated, as a *political directive of the state*. The privileged state status of this organ of the Reichsführer SS is also proof of the communal nature [*Gemeinnützigkeit*] which this organ follows. . . . The motivating force for the Reichsführer SS in carrying out these tasks is solely the promotion of the public weal.[101]

In Salpeter's vision, therefore, the Reich Finance Ministry should mandate the SS's industrial enterprises as special state companies essential to the public weal, itself defined by National Socialism. Salpeter and Pohl wanted the SS to take a role in the forefront of the Nazi revolution through public, communal industry; they wished to eradicate the cultural scourge of capitalism; and they loudly proclaimed the SS's mission to force concentration camp prisoners to serve the national community. In fact, Hohberg's lukewarm interest in such issues later began to irk Salpeter's subordinates, another example of the divisions that could occur within National Socialist institutions precisely because of the multivalence of ideological motivations.[102]

Adding to the sheer complexity of the "organic" holding company, the VuWHA continued a horizontal expansion into new ventures. Pohl had inherited an overextended and disjointed organization from Arthur Ahrens, but without pause he began to acquire mineral water and vitamin drink companies. Here again was another cultural crusade. In a country renowned for its beer, Himmler feared that the pervasiveness of alcohol threatened to sap the vitality of Germandom. Identifying profiteering capitalism as the culprit, Himmler ordered Pohl to secure a monopoly position so that the SS could force prices down, making fruit juices, vitamin drinks, and mineral water "cheaper than beer."[103] The SS had acquired springs near Marienbad and already founded a new company by the end of 1938 (Sudetenquell GmbH). Pohl seems to have taken the initiative in the acquisition of additional mineral springs after March 1939 following the annexation of Czechoslovakia. Here authorities were quickly stripping Czech Jews and the opponents of National Socialism of their property and transferring it to German entrepreneurs. Pohl displayed his naive business sense in these endeavors by seeking an absolute monopoly, a market position well beyond that needed to leverage prices. By the spring of 1941 the SS further tried unsuccessfully to confiscate the property of a British firm active in Bad Neuenahr, the Apollinaris Springs, a subsidiary of the Gordon Hotel Company, which Karl Möckel declared "an exploitative object of capitalism of the English variety."[104] The SS eventually managed to acquire about 75 percent of the market, and even when some plants ran heavy losses, the VuWHA defended them vehemently.[105] Nothing better demonstrates that Himmler and, following his lead, Pohl and Salpeter did not think about the practicalities of monopolies, but were driven far more by their self-proclaimed mission to mold a proper, right-thinking German citizenship.

In May 1941 the VuWHA also founded the German Equipment Works (Deutsche Ausrüstungswerke GmbH, or DAW), which set about the thankless task "to slowly transform into the new structure [of the VuWHA] the special business enterprises which represent a special capital formation of the General SS in connection with the concentration camps."[106] These were the scattered workshops for graft maintained ad hoc by the Kommandanten since the earliest days of Eicke's Dachau. Thus the years leading up to 1940 were filled with "restless" action indeed. Even before the consolidation of existing shops had been completed, the DAW added a glass factory in the Czech Protektorat. In addition Pohl had laid plans for Krug and Lechler's textile factory at the new women's camp at Ravensbrück. All the while Salpeter and Hohberg set about homogenizing the legal structure of all SS companies in order to recast them as "limbs" of an "organic corporation." Hohberg also began to require regularized audits and distributed standardized forms to gather statistical information as he and Salpeter began disentangling SS filials from the personal ownership of SS officers. By the summer of 1940, Pohl asked all SS companies to submit complete development plans for the coming decade, and on 26 July he founded the long-promised holding company, named the German Commercial Operations (Deutsche Wirtschaftsbetriebe, or DWB). It registered carrying a loss of over 100,000 Reichsmarks.[107]

Compared with the extent of SS operations, however, this figure cannot be considered large, especially next to the losses heaped upon loss by Arthur Ahrens only a year before. Ahrens had believed himself to be a "modern" man and apparently even succeeded in selling himself as such to Himmler. But instead of erecting a large, impersonal hierarchy of professional managers working with statistical means of surveillance, negligence and ignorance had marked his entire effort. Rather than consciously adopting machines that could merge labor, raw materials, and organization into a systematic expression of SS intentions, he had invested blindly in Spengler's dry press for little more than the "gee-whiz" value it held as a "modern" invention.[108]

Now Pohl made significant strides to "rationalize" SS industry. On the way, however, he was forced to recruit managers from outside the close-knit cadre of SS men that had gathered around him by 1939. The new managers, such as Schondorff and Hohberg, came with ideals of their own and sought to bend SS industries to those purposes. At first all shared a common enthusiasm for modernization, whose roots ran deep in mainstream German industry and had always been anchored in concepts of social reform. In particular "modernization" meant the conversion of proletarians into skilled, proud, and good Germans, and it accorded pride of place in management to academically trained professionals. SS fundamentalists like Salpeter tried to radicalize these reforms according to their peculiar fantasies of "communal" industries and their anticapitalist productivism. With Pohl—but also initially with Hohberg—he set out to make the SS's Führer corporation as much a manufactory of German values as of bricks and stone. Yet exactly here rifts began to emerge. Schondorff and Hohberg began to split subtly from important core values that the SS espoused (although they championed others). Ideals mattered in these supposedly most technocratic and banal of professions, and they mattered vitally, precisely because those who built the SS's slave-labor industries had to move within a broad current of ideological choices—even when these conflicted.

Some have argued that the concentration camps represented an ultimate expression of a capitalist rationality which supposedly seeks to maximize the exploitation of labor, even unto death. Gerhard Armanski is more accurate when he condemns the concentration camps as a "political economy of misery."[109] We can condemn the SS's industries with precision only if we acknowledge that the political economy within which the SS worked included decisions about values proper to Nazi society that colored its managers' calculations right down to the factory floor. The process of including such decisions is neither foreign nor unusual in any political economy; the SS was unusual only in the content of its radicalism: an admixture of mainstream ideals, racial lunacy, and anticapitalist productivism. The very machines and industrial organization the SS deployed—the Spengler dry press, the Pfaff and Dürrkopf fast sewing machines, and the "organic" holding company—displayed choices about what suited Hitler's Germany. To view them otherwise would mean pretending that technology or organization somehow lacked its own history, that its "internal" logic is always and everywhere the same, and that such logic binds managers to act in only one supposedly "rational" way.

CHAPTER 3
MANUFACTURING
A NEW ORDER

After the conquest of Poland in September 1939, the direction of SS industry changed fundamentally. Himmler asked for and received a new, grandiloquent title, Reichskommissar for the Reinforcement of Germandom (Reichskommissar für die Festigung deutschen Volkstums, or RKF). As such he secured Hitler's direct blessing to proceed with "the agricultural march of conquest into the East with the plow and the sword," which he had envisioned since his days in the Artamanen Bund.[1] The SS's initiatives were manifold and involved both the pacification of indigenous peoples, their extirpation, and the transplantation of "Aryan" settlers as the vanguard of German imperialism in the territories of the East. SS Action Groups (Einsatzgruppen) began the systematic murder of Polish intellectuals, clergy, and prominent politicians to obliterate indigenous community leaders. Other SS offices screened natives in a "dragnet for German blood" and developed questionnaires and medical examinations to test for "ethnic Germanity" according to specious standards of genetic and cultural purity. Ominously, although the Holocaust was still two years away, Adolf Eichmann began organizing the forced transport of Jews and Gypsies out of German territory, first in Vienna. The zeal with which Himmler's henchmen carried through such measures was only one indication of a high optimism that swept over Nazi fundamentalists at the time. Rather than a sober estimation of grim, looming world conflict, great expectations spurred the SS onward.[2]

Himmler's program culminated in the promulgation of a "New Order," a blueprint for racial imperialism, and it incorporated ideals that were by no means particular to the SS. A welter of overlapping organizations emerged, each eager to take charge of occupation and settlement policy of Nazi-occupied eastern Europe. The SS's unique position lay in its executive power base. Rolf-Dieter Müller suggests that Hitler appointed Himmler as Reichskommissar in order to reward him for the success of the Action Groups, which had "cleansed" Poland's cultural and political leaders. This is entirely plausible, and Hitler probably also considered the SS's impressive capacity for effective modern organization, of which the Action Groups were one, particularly horrific manifestation. By 1939 Himmler's forces were well seasoned. They had helped consolidate German hegemony in Austria and Czechoslovakia and now stood at the ready for further action. Within the RKF, Himmler set up planning offices to work out the details of model settlement communities, labeled SS and Police Strongholds (Stützpunkte). The name associated them with defensive military bases, but the RKF intended much more. Himmler looked to the East as a "fertile field of German blood," in which strongholds would serve as greenhouses, cultivating Aryan population growth.[3] He also once referred to Germandom as "cultural fertilizer for foreign peoples," a comment whose comic irony was likely lost on the humorless policeman.[4]

Throughout the war this "New Order" for Germandom remained a fixation of Himmler and the SS. Even as late as November 1942—his optimism undimmed as the Red Army closed upon Stalingrad—the Reichsführer still could not contain himself: "Today colonies, tomorrow settlement regions, the day after that, it all becomes part of the Reich!"[5] RKF architects and geographers plotted cities, towns, and homesteads, including standardized houses, rationalized workplaces, and the minutia of the settler's private life. The inculcation of a new National Socialist community was the expressed goal of all aspects of design, and the blueprint for this grandiose German utopia, the General Plan–East, eventually projected a massive building schedule of over 66 billion Reichsmarks, near 700 billion in today's Deutschmarks.[6]

Racial supremacy, cultural productivism, militarism, and agrarian romanticism all found their expressions in the RKF:

> The New Order of property rights and the constitution of towns in the eastern territories will be freed absolutely from the economic principles of private industry. The dominant concept must remain a single commandment: to restlessly reconstruct the East as German and especially to occupy the countryside with racially worthy and capable German families that will yield many children.[7]

The SS's "blood and soil" ideology is often taken to epitomize Nazi antimodernity, yet Himmler's expert for regional planning adamantly opposed any return to traditional society: "The organization of new farmsteads will exclude from the very beginning any romantic or farfetched fantasies of 'bucolic forms.'" Instead he spoke of the New Order as progressive, not least because of its drive for technological modernization: "German agriculture has just begun to overcome the backwardness of the past decades. The technical and economic criteria for optimal production are still in a great deal of flux and represent a complex problem."[8] Architects designed the home, the workshop, and the farmhouse for labor efficiency and included new gas or electric appliances as well as the latest agricultural equipment. They emphasized scientific planning and design that integrated settlements into a modern industrial economy. The RKF included economic planners and worked closely with Hermann Göring, who had meanwhile organized a new office to "Germanize" the eastern economy, the Main Trustee–East. Göring ceded responsibility for agricultural policy in the East to Himmler's RKF and also helped the SS acquire support industries like sawmills, dairies, slaughterhouses, or granaries, all of them industries that could contribute directly to the strongholds.[9]

The desire to base the New Order on Nazi principles of a modern economy offered a new role to Oswald Pohl's Office III–W (Business), which had been striving to position its "communal" corporations in the vanguard of economic reform. In addition, Pohl's companies had already invested deeply in industries that produced the very building materials called for in great quantities by these new settlement plans. Thus on 18 November, only a month after Himmler's appointment as Reichskommissar, Pohl received new duties as "General Trustee for Building Materials Production in the East."[10] The RKF continued to plan what to build, but Himmler entrusted Pohl with securing the means of construction.

To realize the RKF's goals, Himmler necessarily began to apportion different tasks among his Main Office chiefs. In an otherwise excellent essay, Mechtild Rössler remarks that, on the founding of the RKF, "there must have been some conflicts over jurisdiction [because] a separation was made here between individual aspects such as technical development, infrastructure, railroads, highways and communication lines, and a total 'environmental design.' "[11] But this overlooks the fact that divvying up tasks is an unavoidable part of any modern bureaucratic endeavor, not merely or even necessarily evidence of power struggle. Any truly far-reaching institution—not just those unique to National Socialism—must be "polycratic" to some extent; it must bring multiple competencies to bear in order to master any complex task. Himmler's SS was no exception. The RKF was a planning office staffed by geographers, demographers, and design architects, not business executives, accountants, or civil engineers. Himmler naturally looked elsewhere to meet the logistics of construction, and Pohl eagerly promised that the German Commercial Operations (DWB) would "prepare 100 percent for the supply of the SS projects."[12] No evidence suggests friction between the VuWHA and RKF. They were always meant to be complementary.

The New Order also provided an overarching, unifying goal otherwise lacking up to this point. SS companies had already participated in idealized settlement developments like those of the Cooperative Dwellings and Home-steads GmbH. Various SS companies had also initiated settlements for their civilian employees. The tasks that lay ahead made these seem like finger exercises. Now the RKF set out to lay the foundation stones of the new millennium: "Every great epoch leaves its mark in the construction of its cities that reflect the order and form of that epoch," Himmler once mused. "The sweeping power of our times will express itself in the service of the communal order of life under National Socialism."[13] If the plexus of ideology within the SS was broad and often contradictory, the dream of making the East German promised such grand vistas of destiny; its mission was so capacious that it could subsume all other objectives, even contradictory ones. The first large-scale SS industries had owed their existence to the Führer buildings. As massive as these monuments had been, they paled in comparison to the construction of a racial utopia across the entire subcontinent of eastern Europe.

The New Order quickly inspired further expansion. The IKL planned to open concentration camps at Auschwitz, Lublin (Majdanek), and Stutthof. Unlike in 1937–39, when Pohl had served mostly in an advisory position, he intervened directly to make these new camps crucial components of a vertically integrated construction and building supply enterprise. The VuWHA intended to use prisoners for the bulk of its labor at settlement building sites; at the same time slave-labor factories were to produce materials. In each new camp the DAW (German Equipment Works GmbH) claimed exclusive control over prison workshops and began to orient them toward outfitting SS homesteads. At the first strongholds near Lublin, the VuWHA also planned to build a natural gas plant to provide German settlers and town buildings with fuel. In addition, other filials of the German Commercial Operations that did not rely on forced labor expanded. The Allach Porcelain Manufacture confiscated the Victoria AG from Jewish owners in Bohemia and Moravia and began firing household tableware as well as other practical items for everyday use to augment its former specialization in SS memorabilia and kitsch.[14]

The pull of the New Order was evident even in preexisting industries that had little to do with building supply. The Textile and Leather Utilization GmbH (Textil- und Lederverwertung GmbH, or TexLed) is an interesting case in point. Its garment manufactory had proceeded before the RKF catalyzed the corporate mission of the SS's "organic corporation." Like executives at other SS companies TexLed managers announced their intention to contribute to German culture, and in small ways they did seek to produce directly for the New Order by turning out straw mats "for the purposes of shading residential gardens" as well as matting for plaster work and facades.[15]

Yet most of its products—prison uniforms, socks, watch caps, and infantry jackets—fit only after great efforts of sophistry into the rhetoric of cultural crusades.

The SS's mission was, however, no less essential to the TexLed. Long before the RKF had established its tasks in the East, Walter Salpeter had argued that the winnings of profitable firms could be transferred to cover money-losing operations that were indispensable to Nazi rejuvenation. The TexLed was somewhat unique because it did generate fat profits; and perhaps because this was a slight embarrassment within the anticapitalist culture of the SS, after one audit an SS officer wrote, "Profit is not the central goal. . . . eventual surpluses will be put at the disposal of general cultural and social goals."[16] Some speculate that the SS founded its industrial empire to finance the Waffen SS and even to become independent in armaments. Marginal evidence can perhaps be construed this way. Corporations like TexLed or DAW sold rudimentary products to SS troops (jackets, socks, caps in the case of the former; desks, bunks, chests, and snow skis in the case of the latter). They also provided the same goods to the Wehrmacht. None of this amounted to self-supply. Himmler also intended to reward veteran front soldiers with homesteads and positions in what German Commercial Operations came to call "settler industries." It is a stretch, but this could be conceived as a bid for the self-supply of paramilitary units, for Himmler did fancy Aryan settlers as "warrior farmers." But the SS actually planned the first strongholds for the Order Police (regular civilian police) under Kurt Daluege, not the Waffen SS.[17] Far from plotting monomaniacally for the financial independence of the Waffen SS, SS companies did the opposite. That is, companies like TexLed or DAW tried to cash in on state purchases placed by the SS's military formations (and other state organs) so that the German Commercial Operations could funnel the profits to support Himmler's dreams of racial strongholds.

In fact, Himmler constantly reiterated this point long after total war should have taught him otherwise. As he lectured to his Main Office Chiefs and SS Chapter leaders in the summer of 1942:

> The war will have no meaning when, 20 years hence, we have not undertaken a totally German settlement of the occupied territories. . . . If we do not provide the bricks here, if we do not fill our camps full with slaves—in this room I say the thing very clearly and unambiguously—with work slaves, who, without regard to whatever loss, [are to] build our cities, our towns, our farmsteads, we will not have the money after the long years of war in order to furnish the settlements in such a fashion that truly German men can live in them and can take root in the first generation.[18]

Thus, if it is true that Himmler sought some measure of financial independence through the SS companies, the more pertinent question is, Independence toward what end? Rather than making the Waffen SS "independent," as if only the expansion of institutional authority were at stake, Himmler hoped to make good on his cultural fantasies. In such a scheme of things, SS managers could construe production, even of socks or wooden writing desks, as contributions to German national identity. The New Order thus provided an outlet for the SS's cultural productivism on a much grander scale than ever before.

Besides reorienting existing companies, the DWB also founded new ones specially conceived to manufacture the New Order, in particular the East German Building Supply Works (Ost-Deutsche Baustoffwerke GmbH, or ODBS) and German Noble Furniture (Deutsche Edelmöbel GmbH). Pohl started up East German Building Supply by acquiring factories confiscated from Jews and Poles. Here the SS availed itself of collegial contacts to a member of the Dresdner Bank's board of directors, Dr. Emil Meyer. Meyer, who enjoyed membership in the General SS and often visited Himmler, extended his bank's cooperation through informal networks that mixed friendship and business. The SS was by no means the Dresdner Bank's biggest or only customer, yet the two had developed a keen interest in mutual endeavors after the bank had made "Aryanization" one of its specialties. The Dresdner Bank collected 2 percent of every transaction in takeovers and relied upon the executive muscle of the SS to enforce them. Long before Pohl's and Meyer's dealings, the two institutions had established commodious mutual interest, a common commitment to driving Jews out of legitimate businesses, and a joint enthusiasm for the Germanification of the East.[19]

Almost immediately after the Germans overran Poland, the Dresdner Bank published a booklet, *People and Business in the Former Poland. Only for Confidential Use!*—as the subtitle implied, it could be acquired only through personal contacts to bank officials such as Meyer. Shortly after the New Year, Meyer sent a copy to Salpeter. "German *Lebensraum* [living space] in Europe has expanded . . . by almost 820,000 square kilometers through the New Order of politics in the East,"[20] the introduction announced enthusiastically and further proposed that, in a very short time, this region could be coordinated as a unified industrial region. Long tables listed all the companies that Göring's trusteeship was putting on the auction block for German entrepreneurs. The Dresdner Bank attested, in its own words, that these endeavors were more than *just business*; men like Meyer considered Aryanization to be a just national cause.

Pohl and Salpeter quickly availed themselves of Meyer's advice. Pohl assigned two SS financial administrators to organize the ODBS, Leo Volk and Hanns Bobermin, who quickly drew up a report listing 413 Polish brickworks that the SS wished to acquire. The report's preamble shows the kind of messianic drive that filled these men with purpose, even in something so outwardly banal as brick production:

> The results of the Treaty of Versailles have been pushed aside. The Great German Reich at last has the chance to bring back Germans who were formerly forced to live among foreign peoples. . . . One of the Reichsführer's main duties is the creation of German settlement areas that will bind the returning Germans to their homeland. In order to fulfill this task, above all the Reichsführer needs to secure the production of building materials.

The report then continued with an ethnic slur: "Under Polish rule, housing developments worthy of human beings were not built whatsoever"; "the leadership in the sector of building materials production must therefore be transferred into the hands of Germans [i.e., the SS]."[21] No job requirements, enforced censorship, or threat of being fired coerced statements such as these from Bobermin or Volk, and there is no reason to doubt their sincerity of expression.

The two men were outstanding examples of the competent financial administrators who had recently joined Pohl and Salpeter after the overhaul of the SS companies in 1939. Under their ambitious guidance, the ODBS consolidated brickworks, cement factories, and various gravel and clay pits by the end of 1940. Both understood modern, statistical business accounting and brought deep commitment to the SS's ideals. These ideals and business techniques were, in fact, inseparable, for the ODBS intended to take up its self-proclaimed "leadership" role through modern management and technology.

Volk's own background and career reveal deep dedication to the myriad goals of the SS's state-run, communal industry, its cultural productivism, and its new utopian settlement drive. Even before he had finished his law exams, he had given up free time to help organize the Krumme Lanke settlement for SS officers in Berlin. Late in the war, he even wrote a kind of manifesto to help indoctrinate a new generation of managerial cadets at an SS officer school.[22] Remaking the occupied East, predicated as it was on the "concrete" and "lasting" values of German culture instead of "greed" and "abstract" preoccupations with profit, gave him an outlet for altruism. It was time at last, as he wrote, to "stand by the point, the state commands the economy; the state is not there for the economy, but the economy is there for the state."[23] The ODBS offered the chance to take such projects to a new and unprecedented level.

Bobermin's career had paralleled Volk's. He also had elite academic training (a doctorate in economics). Although both men had joined the SS only in 1933, once they did join they fully dedicated themselves, and before coming to Pohl's office Bobermin had sacrificed free time to the SS press service. He had become a press officer by 1937. In this position he actively formulated and packaged SS ideology, a higher degree of engagement than simple membership or perusal of an occasional political flyer. He was also successful in private business, becoming a top manager (*Direktor*) by 1938 in an advertising firm. The VuWHA offered him the opportunity to apply his organizational talents in the service of his political ideas. When he became the ODBS's chief executive officer and, simultaneously, head of an independent department of the VuWHA, he and Volk implemented the organizational innovations that had been initiated a year before in the DWB. Bobermin established headquarters in Posen, which monitored decentralized work groups clustered according to industrial operations—cement, bricks, or gravel pits—and linked these to regional distribution headquarters.[24]

Bobermin and Volk asked Pohl to invest in new machinery, sought out additional works with top-of-the-line technology, and urged the VuWHA to consider technological modernization as one of its main duties. Göring's Main Trustee–East originally left many backward and unprofitable factories to the ODBS. Many of them verged on bankruptcy. Bobermin and Volk saw this "disadvantage" as a challenge to modernize. In their eyes, the run-down works represented an opportunity. Pohl, Bobermin, and Volk made clear that the ODBS would not let meager profitability hold them back, for the SS had pledged itself to tread where private businessmen would not dare:

> The insufficient price for brick offers too little profit incentive, so the private entrepreneur is interested only in high-priced products (roofing tiles, clinkers), not in the many bricks needed to build walls. . . . The great difficulties that must be overcome to acquire the necessary machinery and replacement parts today; likewise the plethora of bureaucratic stumbling blocks can be overcome best by an enterprise inspired by the SS's will to build and its power to get things done—we are an enterprise that is able to succeed because of our greatness and purpose.[25]

At this very time, Fritz Todt also warned of shortages of just such "bricks needed to build walls" (*Hintermauersteine*).[26] Leaders in the building trades were constantly calling for mechanization, and the SS proposed to leap into the breach now that the great cultural dreams of the RKF required it. The VuWHA argued that it could be trusted to lead the way to modernity, whereas capitalist entrepreneurs, who "adjust production only to meet economic demand," who neither "expand their works nor produce a large inventory in preparation for the coming projects [i.e., of the New Order]," would simply fail to do so.[27] Modernizing even otherwise inglorious industries was thus central to SS executives. They needed to modernize in order to believe in themselves as a vanguard of a New Order of industry no less than a New Order of Nazi racial imperialism. "The expansion and foundational work of the East German Building Materials Works serves not only the good of the Reichskommissar but the entire building economy of the East," crowed Pohl.[28]

The German Commercial Operations had expanded into building supply, where the SS already owned substantial holdings. The second new enterprise born of the RKF's settlement crusade, German Noble Furniture, represented a new industrial departure for Pohl's VuWHA. In the summer of 1940, the German Equipment Works prepared to bid on the Emil Gerstel AG of Prague, whose Jewish owner of that name was being forced to sell. Here the SS followed its enthusiasm for technological modernization yet again, but in this case, the attempted takeover brought the DWB into a fracas with the civil administration of occupied Czechoslovakia. The acquisition of German Noble Furniture therefore presents an example of what was really at stake when true struggles for power with the SS did arise.

In the fall, Pohl had submitted a bid to the Reichs *Protektor*, Constantin Freiherr von Neurath, the regional governor of all former Czech lands that the Reich had not directly annexed (the Protektorat, encompassing Bohemia and Moravia). Neurath opposed the SS's goal of technological modernization because he feared the ensuing standardization of goods might lead to a "limitation of production to fixed models, which threaten free artistic expression."[29] Thus Neurath was opposing the modern nature of SS business, namely, its preference for mass production and large-scale, centralized organization. The brief conflict is therefore instructive.

First of all, Neurath was hardly representative of Hitler's movement, much less of its activists. He was a baron of the German aristocracy and an old-fashioned conservative. "This man has nothing in common with us; he belongs to an entirely different world," wrote Joseph Goebbels.[30] In the last days of the Weimar Republic Neurath had served Chancellor Franz von Papen as foreign minister. When Hitler became Chancellor in January 1933, Neurath remained in his ministerial post. In fact, he remained there until February 1938, when Hitler replaced him with Joachim von Ribbentrop. Nevertheless, Neurath had left the government before the round of annexations that began with Austria, and he was not closely associated with either the NSDAP or Hitler's expansionist policies. At times the foreign press even speculated that Neurath was "anti-Nazi" due to his aristocratic, "cultured" background. Hitler seems to have appointed him Reichs *Protektor* so that Neurath, as an "outsider" to the Nazi Party, might elicit some measure of cooperation from Czech citizens. From the start, however, much more dedicated Nazis among the Sudeten German nationalist movement challenged his authority, as did others within his own administration.

Because of Neurath's marginal and weak position, his efforts to champion a craft-based economy cannot be considered typical of Nazi policy in the Sudetenland, certainly not of National Socialist "antimodernism." It is also important to note that Neurath lost, as antimoderns repeatedly did in the Third Reich. To characterize the entire Third Reich as antimodern is out of all proportion to the weight such voices carried. In addition, when antimoderns raised objections, they never actually claimed to be against modernity; unlike those Nazis who championed modernity, people like Neurath avoided the word. One searches in vain for quotations in which they cry out, "We will destroy modernity," or "I hate modernism!" Instead Neurath based his objections on typically smug hypocrisy and cultural pretension to what rightly counted as "German" design and "artistic expression," the very rhetoric mobilized by the Third Reich's modern productivism.

The SS countered that its unified management within Hitler's National Community could ensure that standardized production would ring true. "If the entire field of design down to the level of the last detail of utility is seen as a unity and focused on man," explained an engineer who designed SS furniture at this time, "then the senseless contrast between 'soulless industrial wares' and the exclusive, spiritual quality of craftsmanship falls away."[31] The SS, by its own admission, strove to enforce a homogeneity of "cultural" taste, but declared that this very homogeneity expressed the German will. Neurath remained unconvinced and continued to frustrate the merger throughout 1940 and into 1941. So began a case of true SS industrial espionage. In Prague the SS solicited the help of a well-positioned officer in private industry, Dr. Kurt May, the son of the furniture manufacturer, A. May of Stuttgart. In 1932, at twenty-two years of age, he had taken over the firm upon his father's death.[32] By the late 1930s he had appeared in occupied Czechoslovakia seeking new investments in the crusading spirit of an industrial modernizer. As such he shared some key elements of the VuWHA's corporate vision. As described by a friend, "By means of special production of individual parts by contracting firms, planned mass production, and modern wood-saving methods, a price level was to be achieved [by May] within the means of the majority of buyers but which, at the same time, guaranteed sound and tasteful work."[33] According to his contemporaries, "Dr. May was . . . concerned with the creation of culturally valuable household articles" and his vision included rhetoric of production norms for German culture.[34] His father had founded the Industrial Society for German Cultural Living (Verband Deutscher Wohnkultur), and May had himself founded a German Home Design Society.[35]

May transferred his German Home Design Society to the SS (an *eingetragener Verein*, or nonprofit society) while the SS founded a GmbH (a limited liability corporation) of the same name with a high-profile storefront in Berlin's Potsdamer Straße. Meanwhile, May, as a civilian entrepreneur, bid on the coveted Gerstel factory. In Brünn, he also purchased a 75 percent interest in another Jewish firm, the D. Drucker AG. Several of his partners from Stuttgart simultaneously founded a German Masterpiece GmbH (Deutsche Meisterwerk GmbH) to subsume these industrial properties.[36] These dealings had the appearance of private business, beyond the SS's direct involvement. Pohl quickly began negotiations with industrial authorities in the Protektorat of Bohemia and Moravia to subsume German Masterpiece as a subsidiary. Because German Masterpiece was nominally German-owned and run by private citizens, the transfer remained hidden from Neurath's direct scrutiny. The closure of all transactions took place in 1941, and thus Pohl had deftly manipulated modern organization—that is, a national corporate bureaucracy—to overcome the merely regional scope of the Reich *Protektor*'s authority. The SS created a new filial, German Noble Furniture GmbH (Deutsche Edelmöbel GmbH), to consolidate all furniture manufacturing, and May entered as CEO of "woodworking industries."[37] This poised the SS not only to deliver the building supplies needed to put up settlements but to furnish their interiors as well.

The "Final Form" of the German Commercial Operations

Amid the excitement over the RKF's grand designs, the VuWHA's long-drawn-out effort to place its "organic corporation" on sound footings reached its zenith. Although initiated in 1939, the holding company did not officially register until the summer of 1940, a delay that only hinted at the difficulties Pohl's administrators faced. Finally, in the spring of 1941, Hans Hohberg distributed standardized contracts that bound all SS business officers to the "organic" corporation, christened the German Commercial Operations GmbH. Simultaneously SS managers were bound by rank within the quasi-military organization of the VuWHA. It took the bulk of 1941 to hammer out technicalities, but the resulting homogeneous structure, tailor-made by Hans Hohberg, now facilitated statistical surveillance and routine transactions by a small, centralized staff in Berlin.[38]

At the beginning of January 1941, Pohl had been forced to excuse his sluggish progress to Himmler, but by 4 September he could proudly call together the DWB's chief executive officers to announce the final organizational form of all SS industrial enterprise. This was the culmination of two years' effort to build the Führer principle into a militarized, modern management. Pohl told his men that the new structure would yield unity: "According to this principle, I have built up this staff in all the various aspects of labor law, social benefits, and structural organization."[39] As expounded on numerous occasions, Pohl meant the unitary "German will," personified by Hitler, in which he expected his managers to extinguish their own personal identities and act as one. Pohl also stressed the "fulfillment of tasks that fall to the Reichsführer SS as the Chief of German Police, such as the concentration camp industries," including "leadership of corporations that promote the National Socialist world view [Weltanschauung]" and "tasks ordered by the Reichskommissar for the Reinforcement of Germandom."[40]

Just a few months previously, in the spring of 1941, the last spring of Nazi Germany's unblemished optimism, a public exhibit in Berlin had also capped the VuWHA's efforts to manufacture the New Order. Lightning war against the Soviet Union was then in its planning stage, and, despite the failed air war against Great Britain, Axis forces still reigned supreme and unassailable on the continent. The SS was sanguinely unconcerned with the war economy and already referred to the New Order as the "Peace Program," whose realization would be postponed only as long as it took to achieve total victory. All expected that hour to be soon. The exhibit therefore offers a glimpse of the utopia the SS intended to construct in the full blush of confidence. DESt displayed its wares for future settlement construction. Salpeter arranged a model "Colonial House" with furniture and china as well as heating stoves, all "from our own companies."[41] Even the beverage and bottling plants gave away "Trink-o," an energy drink to promote healthy living. Pohl also asked the civil engineers of the concentration camps to prepare exhibits of camp architecture and officer settlements, although it is not clear if these actually appeared in Berlin.

Kurt May helped organize displays of the kind of interiors that the SS's "German Home Design" intended to impose on the East. An engineer from the Industrial Arts School of May's native Stuttgart, Diploma Engineer Hermann Gretsch, helped. In a litany of cultural productivism that had attracted so many of the young business executives of the DWB, Gretsch criticized the tastelessness of past ages and the slavish imitation of foreign styles, especially those motivated by a mindless liberal capitalism: "If we find so many houses so ugly today, it is because they were made in the period of beloved mammon!" He then suggested that German values and race had to be crafted into material artifacts: "Race, heritage, tradition, and life-style are important, but designers completely forgot them. They have forgotten that they must also satisfy cultural needs."[42]

Gretsch called his own style "agrarian objectivity," a "timeless" aesthetic he located in an imagined epoch before capitalist spoliation. He consciously coined the term as a direct attack on left-leaning artistic movements associated with "new objectivity," whose consummate representative was the Bauhaus of Dessau. The Bauhaus had sought to create homes and living spaces as "machines for living" and made ostentatious use of mass-produced materials like steel, concrete, and glass. Its designers strove to strip away flourishes that did not serve practical, functional needs, a style perhaps best represented by Marcel Breuer's famous chair made of bent tube steel and flat-black leather.[43]

To those who would define "modernism" by the international style of architecture and design promoted by the Bauhaus, Gretsch and the SS's German Home Design undoubtedly seem "antimodern." Were they not simply reactionaries who longed to turn back the clock to an imaginary past? (Gretsch acknowledged his own nostalgia for the epoch of Biedermeier.) Gretsch tried to derive racially pure "German" design from what he considered the eternal virtues of peasants, whose "life-style was more objective than the so-called new objectivity." By contrast, "the farmer took it for granted that everything he designed had to be practical as well as beautiful. He obeyed the eternal laws of nature."[44] In Gretsch's living spaces, the woman's place was unmistakably in the kitchen, preferably with many children, and Gretsch took care to place a baby carriage in the SS "parents' room." As decor, scenes of peasant life hung on the walls of his displays. Again, this style would seem to confirm the view of those who saw the Nazis as hopelessly backward-thinking retrogrades. In a survey conducted in 1929, for instance, Erich Fromm claimed to have discovered an overweening preference for conventional aesthetics among Nazis, including wall pictures of dictators and generals as opposed to progressive, original works of art.[45] Gretsch's taste could not have proved Fromm wrong.

On the other hand, as Mechtild Rössler and Sabina Schleiermacher aptly note, "the history of the modern is not only the history of international developments in philosophy, architecture, art, and aesthetics, is not only the Vienna Circle, the German Werkbund, and the Bauhaus. . . . [M]odern also means a new culture of technology and science."[46] In fact, when it came to technology and industry, Gretsch shared the same impulses with Bauhaus design, like the attempt to plan domestic spaces as "machines for living." The SS's "German design" confirms Rössler and Schleiermacher no less than

The Marcel Breuer chair. Bauhaus Archives, Berlin.

Fromm. Gretsch did not object to the new objectivity's spirited idealization of technology; he simply found Bauhaus designs ugly, pretentious, degenerate, and injurious to Germandom. He also called upon the SS to hammer out norms for functional dwellings: "The useful is always the beautiful and lends simple, unified principles of form."[47] Furthermore, simple design aided mass production: "We should not aim for short-term fashion 'hits'; we should aim for standardized designs that are universally viable. Only this will reduce the serial production expenses over long-term production and thus make designer furniture profitable."[48] Unlike the Bauhaus, however, Gretsch believed that traditional materials like wood suited the "German spirit" better than tube steel, and Kurt May, whose firms specialized in woodworking, intended to adapt modern production to the values advocated by Gretsch.

Their efforts suggest that Nazi modernity had much less to do with fine distinctions of intellectual history and more to do with technology, organiza-

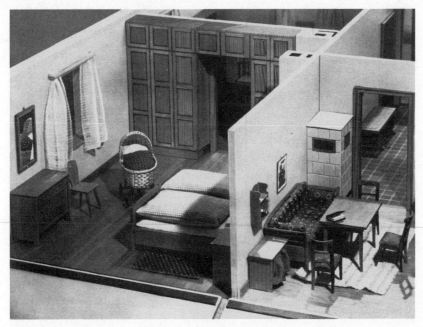

Hermann Gretsch's SS interior design, from Planung und Aufbau im Osten.
Erläuterungen und Skizzen zum ländlichen Aufbau in den neuen Ostgebieten *(Berlin:*
Deutsche Landbuchhandlung, 1941). Gretsch's caption below this photo reads, "Sleeping
room of the parents (Wall Closet!) and room for the cottage farmer." He was especially
proud of the expandable wall closet, which could be extended to meet the needs of families
"rich in children."

tion, and their fusion in production. As Frank Trommler has pointed out, the aesthetics of the Werkbund and the Bauhaus do not necessarily define modernity; rather they often betray the attempts (which had failed by 1933) of high artists and architects to claim a leading role in the mastery of the ascendant technology and organization of the twentieth century.[49] In this light, the Bauhaus's effort to control the cultural meaning of modernity was no different *in kind*, only *in content*, from that of the SS. "The multiplication of various styles afflicted Germany like a disease," stated Gretsch, and he argued that German cultural identity was better served by the imposition of style by a centralized, hierarchical organization.[50] What was totalitarian about the RKF and SS settlements, but no less a part of Nazi modernity, was not the SS's aesthetic but a sinister drive to subject all aspects of life to design and mass duplication and, in addition, the ready willingness to police and enforce that homogeneous design (through modern administration). The VuWHA promised to manufacture that world and to build it, even as other SS Main Offices set out to eradicate all that polluted the Nazi dream of uniform German culture.

The Venality of Evil: Modern *Mis*management of Slave Labor

As the ODBS and German Noble Furniture demonstrated, manufacturing the New Order demanded another round of expansion, and, to accomplish it, Pohl had to add personnel yet again. Of necessity this posed challenges to the SS's corporate community and managerial culture. Pohl was aware of this, and, beyond hiring new officers (as he had Hohberg and Schondorff), he also started a special curriculum for midlevel managers at the SS officer school Bad Tölz. The SS had founded a "Führer" School in 1934 to train cadets, first at Bad Tölz, then a second at Braunschweig followed in wartime by schools at Klagenfurt and Prague. Himmler had also rechristened them *Junker* Schools in 1937, after the *Junker* aristocracy that had dominated the Prussian army. Hearkening to this name, most cadets were destined to lead Waffen SS regiments in combat. However, as early as 1936 Pohl initiated special courses for SS paymasters and quartermasters. These had never caught on, and Pohl considered eliminating them, for the reorganization of SS civil service administration at that time had proved successful without cadets. Yet the advent of the Reichskommissar for the Reinforcement of Germandom made Pohl reconsider, and he directed Bad Tölz to start up administrative courses once again. The Economics Ministry also encouraged the development. Unlike a few years earlier, however, now Pohl intended to overhaul the curriculum in order to provide a steady stream of recruits to the mid- and lower-level management of SS industry. In keeping with the SS's general adulation of martial organization, the young officers were still to receive military training, but Pohl now wished them to receive lessons in the mechanics of corporations as well. As soldiers *and* competent businessmen, he expected to shape them as better organized and more disciplined than their counterparts in private industry.[51]

Despite Pohl's hopes, however, the new courses incorporated all the weaknesses of SS enterprise. The man responsible for them, Hans Baier, had no engineering training and would only exacerbate the DWB's already catastrophic management. Baier also later reappeared in a key role in DWB top management and is of further interest because he was one of Pohl's longstanding navy buddies from World War I. Both had served as paymasters. Baier's character therefore reveals much about how Pohl chose men to fill important positions, how the VuWHA set out to shape its own "organization men," and what consequences this had for the VuWHA. In addition, if anyone in the VuWHA qualified as the best representative of the banality of evil in modern bureaucracy, Baier was that man. He was venal and petty. But, of note, he was also astoundingly incompetent. This was perhaps no mere coincidence. Throughout the history of SS industrial enterprise, those individuals who acted most out of "material self-interest" or mere cupidity were consistently the most ineffectual.

There is one more reason for dwelling briefly on this petty bureaucrat. Baier also left behind an extensive personal correspondence. It affords rare, candid insight into his most intimate convictions. First, his letters show him to be anything but a "cog" trapped in an iron cage of bureaucracy. Second, despite his venality, Baier seems to have felt a genuine need to *believe* that his motives were altruistic and his actions competent. Thus he acted out his crass opportunism on a stage set by the values and ideals he (and his friend Pohl) considered legitimate; he interpreted his life as nothing less than a grand national cause beyond himself. In this context, "self-interest" and outright greed prove slippery categories of historical analysis indeed.

More so than most SS "opportunists" Baier originally came to Pohl in order to advance his career. In 1938, still serving as a paymaster, he wrote to ask if his talents might be useful to the SS, and Pohl pulled strings to secure an appointment as the instructor of the administrative curriculum at Bad Tölz. Baier's exuberant gratitude recalls Pohl's own excitement upon appointment to the SS four years earlier:

> I have given notice to the German navy not because I was displeased with my position there; rather because I have now found even bigger and better tasks with the SS. Celebrations accompanied the opening of the administrative course . . . whose designated leader I am. You can well believe that the first days of my teaching duties were full of work and still are.[52]

In reality, however, Baier was singular only in his laziness, at which he excelled even Richard Glücks, the Inspector of Concentration Camps. Baier devoted most of his energy to the pursuit of perquisites. "In one of my bedrooms the curtains have become unusable," he once wrote during wartime shortages. "I therefore request reimbursement for two, 6-meter curtains." Then he stated in the same letter, "I want no special treatment."[53] But even in such petty affairs he was spectacularly incompetent. He got caught trading on the black market in Prague, cavorting in uniform with two mistresses at once, and using a car at staff expense to taxi his daughter to school and transport his mistresses around Munich; he even used his status as "professor" to harangue his daughter's schoolteachers, whom he believed were too hard on her.

Baier was a kind of incompetent twin of Oswald Pohl. He was one year Pohl's junior and, like Pohl, had come from a humble but respectable background. Right after demobilization at the end of World War I, Baier drifted through some truncated university studies in economics, first at the University of Kiel with his friend Pohl and later at Göttingen. He eventually found work as a tax specialist without finishing any formal degree. Then, in 1935, he returned to the navy as a paymaster and climbed to the rank of lieutenant captain. But a few years later his career had turned sharply downhill. "The navy gives me neither a pension nor good wishes to send me on my way," he wrote in a paranoid letter to Pohl. "Be sure: here I not only stand isolated; no, all stand against me beginning from above."[54] Contrary to his festive testament upon entering as a "leader" at Bad Tölz, Baier had never "given notice" to the navy; the navy had pushed him out with good riddance. When Baier announced his switch to the SS, his superior officer immediately asked him to take all remaining vacation leave and clear out his office early.

Why did Pohl, himself an extremely energetic and competent military organizer, choose to rely on this man for anything? Baier's entire career spoke of unfinished initiatives, failure, and greed. On paper he was a highly educated man; in practice, a disaster. The answer seems to be that Pohl saw in Baier a like-minded man, one whom he could work with and trust. In this, Pohl had not deviated from the Führer principle, which encouraged leaders to believe that they embodied the "will" of their subordinates. He seems to have based his evaluation of Baier, as of other men, not on concrete estimation of his friend's capabilities but on criteria of personal feeling. In some past cases, as with Walter Salpeter, Pohl's judgment led him to place his trust in men who possessed deep conviction *and* administrative skills. In other cases, as with Hans Hohberg or Erduin Schondorff, Pohl had fingered men who had considerable knowledge but misjudged their wholehearted commitment to the broad plexus of the SS's cause. Pohl's mode of operations led to total disaster when he chose men like Baier who were committed but incompetent.

There was no mistaking, however, that Baier's shallow self-interest coincided with unmistakable, deep political feeling and a self-image as an upright man. He counted himself among his nation's best sons, and therefore in the rising star of his personal comfort he could divine national destiny or the blessings of progress itself. Thus his honestly dumbfounded reaction when the navy refused him a pension: "In any case," he whined to Pohl, "I ask advice in the name of the Reich and the German people, what is to be done?"[55] He was not the first or the last officer to use the VuWHA to promote his own advancement, but because men like Baier identified their individual interest so strongly with those of "the German people" or other grand entities beyond themselves, they readily developed genuine attachments to the ideals of those organizations which promoted their careers. Their beliefs had consequences, and institutionalized success had consequences for their beliefs. The two mutually reinforced each other.

Baier took pains to explain that his work with the SS "must give me the chance to dedicate my energy to the renewal of Germany in more important and distinguished service than is currently allowed here in my position with the navy."[56] Upon first glance, it would be easy to write this comment off as deeply veined cynicism. He joined the party only during the grand hurrah of the seizure of power and never bothered to join the SS until offered the position at Bad Tölz.[57] Was not Baier merely using his old friend? Yet such a caricature overlooks the amiable cohabitation of opportunism and genuine idealism in the same small soul. Counting material interest as evidence that ideological motives did not or could not play a role is simply a grotesque false dilemma, for Baier's ideological conviction ran just as deep as his venality. In one instance, he and Pohl joined their voices to proselytize for the cause of National Socialism. Consider a letter to their mutual friend: "I tried to discourage Wolf [Lubbe] from his pessimistic view of our times," Baier confided to Pohl,

> although I believe I will never succeed in doing so in this life. Nevertheless I will try again and again. When the conversation turns to party or political matters, I will try to convince him—I say, I wish to free him from his tragic burden of pessimism. Then we can bring him into the right mood for struggle [*Kampfstimmung*] in our movement.[58]

"My dear Hans," his friend eventually replied,

> This war for the existence of our fatherland has proved anew that soldierly comradeship is the highest form of life! . . . It has proved that the Führer has established himself as a symbol at the head of the army and the Waffen SS. In exactly this same way we must all stand together after the war, in exactly the same way that we do as soldiers. Only the One remains.[59]

The "One" in the closing line expresses the unity these men felt, and their sense of individual importance as part of that greater unity under Hitler, "the highest form of life." Meanwhile, Baier carried on, embroiled in the meanest trivialities of petty gain. When he received his friend's letter, while his nation, "the One," was descending into ruinous total war, Baier was just then applying for a two-month holiday at a sauna resort.[60]

Baier's actual curriculum for modern management mirrored the man. He sonorously titled his lecture series "First Seminar for Administrative Knowledge." The lessons were repetitive and often overlapped, for example, "Money, Bank, and Stock Market Accounting," "Financial and Budgetary Accounting for the SS," "Requisitioning and Administrative Accounting," and, finally, "Accounting." Then came one course each on computing travel costs, typewriting, shorthand, and, to show cultural flair, great books in German literature.[61] Baier and Pohl assured Himmler that these lectures would form "the kind of man that did not just join the SS to be part of some kind of educated elite or social club, but the kind of man who has been well seasoned. By this I mean an SS Führer pure and simple."[62] The circular rhetoric and banal sentimentalities were typical of Baier. What he meant, "pure and simple," by developing a training course for men that were supposed to be already "well seasoned"—that is, with no need for book learning—remained hopelessly vague. Nor did he specify what he meant by an educational program for men who were not supposed to belong to an "educated elite." Nor did he detail the exact content of his twenty seminar plans. In a movement that conflated commitment and competence, activity and organization, Baier did not have to explain. Apparently, Pohl expected that his friend would carry through without monitoring his achievements, for Baier spoke the language of Nazi fundamentalism. Undoubtedly Pohl could see only the latter Baier, not the one who begged for new curtains and spa vacations.

If Pohl intended the managerial course at Bad Tölz to serve as a kind of vocational training for midlevel managers—the kind of education routinely arranged between modern firms and the new local commerce or technical schools (*Handelsschulen* or *Technikum*)—Baier's curriculum proved a sensational flop. Students shirked and revolted openly in class. By the end of 1939, when the SS began to need them for the new tasks of the RKF, they were skipping Baier's class entirely. He suffered personally from his students' rejection and asked Pohl to release him from the *Junker* School, but, like all bad teachers, he blamed his pupils, complaining that they spent their time drinking and carousing around Dachau. Open insubordination of the administrative cadets caused Pohl to consider eliminating the course altogether, yet in light of the new expansion of SS industry for the New Order in 1939, Pohl kept it going. He never sought out anyone with industrial experience to teach it. Baier's curriculum was allowed to stand, and Bad Tölz continued to neglect technical knowledge, although it was supposed to train industrial managers. Professional ignorance was partly to blame: Pohl and Baier were financial administrators and civil servants and had never confronted the complexities of industrial management. Yet given the astounding breadth of Baier's incompetence, the disastrous administrative curriculum had also resulted from Pohl's style, based as it was in the Führer principle. Pohl had complete faith in himself to judge the "will" of men and extended that trust to Baier as a true National Socialist. He and Baier both believed that inspired altruism would strike the right tone, but Pohl never monitored the results. Bad Tölz therefore evinced the blundering style of the DWB in general.[63]

As Pohl's industrial responsibilities grew, so did the catastrophic *mis*management of slave labor. To such daily tasks *Junker* School administrative graduates were supposed to turn their skills. Just down the road from Bad Tölz, the German Equipment Works was then in the process of incorporating all the concentration camp workshops at Dachau (as well as other camps). Labor management had already become a trouble spot, with its roots in the vexed expectations and ideological identities that SS business executives and the Inspectorate of Concentration Camps brought to their joint projects. Eicke had pledged full cooperation with the SS industries at the end of 1938 yet did little to enforce this pledge. In practice, the camp SS continued to act upon the primacy of policing: they either hindered production by brutalizing prisoners or, at best, remained indifferent to factory operations.

A top SS manager complained (about SS clay pits near Sachsenhausen): "In the interests of business, prisoners must not be left exposed to the elements for an entire day, which has happened just recently. In such cases, the damage to our enterprises cannot be ignored. Prison industries must run continuously."[64] This report, from the summer of 1939, showed the difficulty SS business executives encountered when confronting concentration camp staff. First, the camp SS continued to use labor as a means of brutality, even when this endangered useful work. Here prisoners were purposefully left at the mercy of the weather regardless of the consequences for industrial output. Second, watchmen showed a careless disregard for the continuity of labor. Kommandanten broke off work in midstream, or they changed prisoners' assignments day after day, destroying the investment of time necessary to teach prisoners a specific job. In addition, the report went on to complain that Kommandanten continued to think of prison labor as an in-house service industry. These points of conflict are no surprise considering the different heritages brought to bear on camp labor from the Death's Head Units and the VuWHA. According to the Death's Head way of life, it was almost a duty to mistreat prisoners. As managers, they ignored or failed to grasp the nuances of production and also accepted poor organization as the norm.

Beyond seeking to train manager-cadets in the *Junker* School Bad Tölz, Pohl also attempted to create a new post in his top management to alleviate such problems within the Main Office for Budgets and Building. This was the Office I/5. Its sole duty was to smooth out the IKL's labor allocations to the German Commercial Operations in the VuWHA. There is no explanation for why Pohl did not place it within the VuWHA, but the divisions between the VuWHA and the HAHB as well as their personnel had always been extraordinarily fluid. Perhaps he positioned I/5 within the HAHB because this office controlled the camp's budgets. This convoluted structure only foreshadowed the mismanagement that would arise from I/5 over the next two years. From the very start its headquarters were located at the IKL in Oranienburg instead of with the HAHB in Berlin. Later, in September 1941, Pohl actually detached the Office I/5 and allowed its full integration as a subordinate office of the IKL. This almost ensured that the I/5's efforts to promote SS industry would fail. Like Baier's failure, however, the history of this debacle sheds light on SS ideals and individual agency. Conflict arose quickly between the IKL, with its primacy of policing, on one hand, and, on the other hand, the German Commercial Operations, whose managers espoused productivism.[65]

Pohl had specially recruited a new manager, Wilhelm Burböck, to lead Office I/5. Burböck also received a special title, Deputy for the Labor Action (*Beauftragter für den Arbeitseinsatz*). His own career wholly matched neither the norm for managers in Pohl's companies nor that of the Death's Head Units, befitting his assignment as a go-between. Like many officers under Pohl, Burböck had started a secure, successful career that he abandoned for the movement (as a technical-administrative expert for the Austrian Post). This job had introduced him to modern bureaucracy, and yet Burböck's experience (like Pohl's own) was limited to civil service administration. He was undoubtedly familiar with information flow in large, bureaucratic hierarchies, but the world of the firm was alien to him. His former job never required him to link abstract statistical observations to the daily material conditions of large-scale production or factory engineering, and his technical ignorance, like that endemic to SS top management in general, would cost the SS companies.

He shared with both the IKL and VuWHA a deep devotion to the Nazi cause. He joined the SS before the Nazi Party and had belonged to German ultranationalist clubs since his adolescence. The Austrian civil service cashiered him in 1933 as a direct result of his political agitation. His membership in the National Socialist German Labor Party dates to just weeks before the Nazi seizure of power. On the surface this is not especially early, yet Burböck was Austrian, and membership in Nazi organizations was banned shortly after he joined. He continued undeterred even after the SS staged the assassination of the Austrian prime minister, and he made contact at this time with high-level SS subordinates who later promoted his career. Burböck also shared a long-standing association with SS settlement projects, even before the grandiose plans for the New Order were put to paper. From 1936 to 1937, he served as the adjutant of Herbert Backe, a high-ranking SS officer responsible for SS settlements before the creation of the RKF (Backe eventually became minister of agriculture and many of his officers transferred directly to the RKF). Burböck organized various settlement projects for Backe, including the "SS Comrades House" in Berlin, Krumme Lanke.[66]

Yet Burböck's social background was perhaps more typical of the Death's Head Units, from whose administrative ranks Backe had originally recruited him. Burböck's father was a blacksmith, placing him between the old middle class and the ranks of skilled laborers. He had received only a rudimentary education without going on to a university and had become an administrative expert on the job. His duties with the Austrian postal service were neither demanding nor especially prestigious (*fachtechnischer Beamte*), and without the SS he stood only a slim chance of ascending to the head of an entire office staff. In fact, the German Commercial Operations recruited him not for his administrative abilities but for his connections. As one top manager wrote: "[Burböck] is especially valuable to me for his personal connection to State Secretary Major General Dr. [Ernst] Kaltenbrunner, with whom I am in constant contact concerning the region around Linz (projected prisoner action for the regulation of the Donau, road construction, etc. with the camp Mauthausen and Gusen)."[67] Although Burböck had never managed industrial production of any kind, it was hoped that his good contacts would facilitate the allocation of labor. Managerial skills were only a secondary consideration.

On entering office, Burböck developed numerous initiatives in a burst of energy. He created a clear organizational hierarchy with well-defined tasks and channels of communication. He split his Labor Action Office into two divisions. One, in his central office near Berlin, gathered and collated information; the other, composed of officers he sent into the field, managed slave labor on-site. Within this second branch of his office, Burböck also established the Labor Action Führer of the detention camp (*Arbeitseinsatzführer*, later called the *Schutzhaftlagerführer "Einsatz"*), a new post within each Kommandant's staff, and made their main responsibility a quintessentially bureaucratic task: they had to preside over a cross-referenced card file.[68] As mundane as this seems, these banks of cards are of special importance, for they formed the core information system for all subsequent slave-labor management. The Labor Action Führer was supposed to list working inmates by name, number, skills, and history of labor experience within the camp system. If maintained diligently, a Labor Action Führer could tally at a glance through his cards the exact number of prisoners available, where they were, their skills, and their work history.

Burböck also laid down rules for the allocation of prisoners. Ostensibly, his

Labor Action Führer were to intervene in production management as liaisons to factories (almost exclusively SS subsidiaries before 1941). On one hand, the Labor Action Führer had to negotiate with industrial supervisors regarding the duration of labor, the feasibility of operations, and wages; on the other hand, he had to hammer out the selection of prisoners and security conditions with camp Kommandanten.[69] In theory, this system would allow the IKL to coordinate all camps with only a minimal staff in Oranienburg to review the Labor Action Führer's reports.

However, Burböck did not, at first, create the kind of statistical shorthand necessary for modern control: the reports he requested came embedded in prose instead of statistical codes, which had consequences when the SS began allocating large numbers of inmates. Burböck had introduced a new information system with quantifiable standards (wages, population, guard personnel), but he did not extend statistical surveillance to the sophisticated management of labor organization. Because the SS charged only a pittance to its own cost accounts for leasing prisoners, perhaps Burböck did not believe such controls were important, but, in contrast, low rates for slave labor never stopped German engineers in private industry from calculating such figures.[70] Labor still entailed an expense even when no wages went to the prisoners: for example, firms often had to carry the costs of provisions and security, among other things, and usually had to reckon *decreased* productivity and *increased* depreciation of machinery into their books. Burböck, by contrast, did not make reference to factory production at all; he sought instead to measure only crude numbers of prisoners' bodies with only a cursory acknowledgment of their skills, which he proposed to flag with color labels. In theory, the SS Labor Action set out to place dependable data at its managers' fingertips; in practice, it quickly reneged on its charge to manage production.

This failure is aptly demonstrated by the meager efforts Burböck did make toward supervising work sites. Burböck asked his Labor Action Führer to take snapshots of work details and send the pictures to I/5a in Oranienburg. As Anson Rabinbach has amply demonstrated, time-motion photography at work sites had become, by the early twentieth century, a fad in modern managerial circles in America and in Europe. In theory, Burböck would study these photos in order to determine more efficient guidelines for supervision.[71] Yet nothing is more typical of Burböck's modern *mis*management than this appeal from afar for photos. Although Burböck designed a system to control the flow of laborers and the supervision of work, the Office I/5 neglected statistical evaluation of productivity as a function of labor hours, cost, raw materials, throughput, or output. He also remained in Oranienburg and allowed individual camps to proceed without his direct intervention. He expected modern organization to run automatically once he had dictated it to his subordinates or drawn it upon a chart.

If modern institutions really constrained people to act within an "iron cage," no doubt it would have worked, but Burböck actually stressed the freedom of Labor Action Führer to act on their own initiative: "Because the duties of the Labor Action Führer cannot be ordered in its complete details from my office, it is pertinent and necessary that the Labor Action Führer themselves take initiative to recognize problems in their particular context."[72] In practical terms, this meant that Burböck avoided direct personal supervision and even responsibility for production management. It also meant that the brutality at work sites was more likely to result from conscious local initiative rather than rule-bound bureaucracy. On site, the Labor Action Führer used their freedom to delegate responsibility on down the camps' hierarchy. The actual oversight of work details was passed on to mere SS guards or Kapos, regardless of Burböck's orders to the contrary. Following the military organization of the camps, Burböck and Glücks also subordinated the Labor Action officers to the very same Kommandanten who had long mistreated prisoners without concern for production. Even when the Labor Action Führer were loyal to the cause of production (not all were), this nullified efforts to change work practices.

This can be seen in the efforts of Phillipp Grimm, recruited to Office I/5 at the same time as Burböck. Grimm fit in well with Pohl's managerial officers. Like many SS business executives, he had attended one of Germany's modern business schools (the Nuremberger Handelsschule) founded to teach the economics and accounting methods necessary for white-collar jobs in the commercial sector.[73] Grimm assumed the title of Labor Action Führer at Buchenwald with every intention of orienting the camp to production. As an officer of I/5, however, he quickly got caught in a collision of heritages: the productivist identity of the SS companies on one hand and the primacy of policing among the "political soldiers" of the IKL on the other. At Buchenwald he met with the immediate disapproval of the Kommandant, Karl-Otto Koch, who was notoriously corrupt and cruel even by SS standards. Embezzlement as well as his and his wife's outrageous sexual affairs eventually led to Koch's arrest and execution by the SS itself. Typical of the prevailing practices of Eicke's "Dachau School," he saw forced labor primarily as a means of torture, or, when the occasion presented itself, Koch and his wife also exploited prisoners for their personal gain. Perhaps the most gruesome example, one that fuses corruption, brutality, and petty industry in one, can be found in his wife's desire to have the tattooed skin of dead prisoners tanned and fashioned into decorative articles in the camp's workshops.[74]

When Grimm first arrived for duty, Koch did not even allow him into the camp's offices, and Grimm finally gained access only through Pohl's direct intervention. When he submitted labor evaluations to Koch, Koch refused to forward them to the Office I/5, and the entire Buchenwald staff remained intransigent. In one instance Grimm reported angrily, "The concentration camp's internal businesses and subordinate troop units still refuse to pay anything for the prisoners' services."[75] Resistance extended well beyond Buchenwald, as Grimm found out when he tried to manage the transfer of 520 prisoners from Dachau. "The Labor Action officer of the supplying camp is instructed to select the prisoners personally in the event of a transfer,"[76] he wrote to the Labor Action Führer at Dachau. Grimm's conflict with a fellow Labor Action Führer is likely no accident, for Grimm came as an outsider. As Karin Orth has shown, the HAHB I/5, due to either negligence or convenience, often simply assigned Labor Action Führer from within the hardened ranks of the IKL detention camp officers and the cronies of Kommandanten.[77]

Had the I/5 officer at Dachau really operated like a cog in bureaucratic wheels, this entire operation might have proceeded on the strength of directives alone. However, Dachau did not comply, showing how dependent modern managerial structures are on the willing complicity of human actors and how dependent that cooperation is on a shared consensus of purpose. Grimm eventually traveled personally to Bavaria to choose laborers himself. Even then, Dachau's Kommandant snubbed him by sending only a deputy. When the Dachau Kommandant did release the actual transport, a trainload of sickly prisoners whom Grimm had not selected arrived at Buchenwald. The trip to Munich had been a waste of time: Dachau had merely used the opportunity to purge unwanted inmates (a sign that this mismanagement always had the most dire consequences for prisoners). Phillipp Grimm complained bitterly: this was a flagrant violation of official guidelines. Yet in accord with those same official guidelines, he reported directly to his Kommandant, Koch, who stifled Grimm's initiative. Characteristic of the Death's Head Units, Koch and the Kommandant of Dachau united in their disdain for efficient industrial production, not to mention the barest modicum of bodily preservation of the prisoners.[78]

Burböck eventually recognized that his system was not working, but for his part he remained in Oranienburg, and, in fact, no evidence suggests that he ever intervened at a single work site. As early as November 1940, he called all Labor Action officers to a conference at Oranienburg; five months later he called another at Dachau; yet at all these meetings he only urged his subordinates to "overcome difficulties" and "foster more cooperation" with each other and their Kommandanten. He did not propose concrete measures of reform, nor did he recognize the chaos wrought by Kommandanten upon his organization. Instead he encouraged individual Labor Action officers to talk out their problems, as if Office I/5 had become a kind of self-help discussion group. Burböck's entire operation acquired the character of a sham. While his office became paralyzed and his Labor Action Führer stopped heeding his powerless directives, he threatened "sharpest disapproval."[79] Nothing ever came of the threats. Having erected no statistical surveillance of production, Burböck could not hold them accountable from afar. In February 1942, the Inspector of Concentration Camps, Richard Glücks, finally liquidated I/5b, comprised of the Labor Action Führer at each camp. Glücks simultaneously ordered Kommandanten to redouble their efforts to cooperate with labor management but did not stir a finger to ensure that this would actually be done. The I/5 filing system and records were to be turned over to the camp Kommandanten, which guaranteed that, for the time being, nothing would change. The I/5 had been a conscious pretense at modern management and looked convincing on paper, but it was riven by divergent administrative styles and conflicting purposes within the IKL. Here the multiplicity embedded within National Socialist institutions and differing shades of ideological commitments did cause dysfunction in exactly the way that so many histories of "polycracy" maintain.

One exception bears mention, however, one in which Burböck followed through with modern techniques of statistical managerial control. One simple Office I/5 innovation enabled the SS to manage the catastrophic attrition of the camps' labor force in conjunction with the genocide: the inclusion of "unfit to work" on standard reports. This last example of the I/5's operation also shows, first, that the DWB's "modernizing" spirit was in no way humane and, second, how modern managerial techniques began to increase the scale of murder. In addition, by the time Burböck left I/5, the IKL had become accustomed to operating within the prescribed channels of authority and communication that he had defined, and his successor, Gerhard Maurer, had no choice but to follow in his footsteps. Therefore it is important to understand the tools that Burböck left behind, for they shaped the suffering of working inmates for the duration of the war.

Burböck introduced the statistical surveillance of the "unfit" in the late summer of 1941. That prisoners were already sick and dying says much of the poor organization of supplies, the atrocious conditions of shelter, and the brutality of daily life in the camps. Epidemics and death had been endemic to the camps since the quantum leap in their population had started in 1936. By the summer of 1941, epidemics and the attrition of inmates were causing a labor shortage within the DWB, despite the relatively small scale of operations at this time (estimated at a maximum 60,000 total working prisoners—later the SS Labor Action encompassed over ten times this amount). Evidence that IKL officers or SS business executives first recognized the health of inmates as a production problem can be seen initially within the Office I/5. As shown earlier, Grimm complained of receiving sickly and dying inmates from Dachau at the end of 1940. Burböck's original managerial descriptions had also defined the removal of sickly workers as a chief duty of the Labor Action Führer, and he repeatedly warned his officers to stop allocating weakened prisoners to SS industrial enterprises. Burböck subsequently ordered all I/5 officers to add the new category *Arbeitsunfähig* (unfit to work) to their filing systems.[80]

Ideally, he intended this statistical innovation to help overcome the kind of problems that had plagued Grimm. The Labor Action Führer would, theoretically, be able to distinguish between healthy and incapacitated workers at a glance through orderly card files and thus make the Office I/5 more capable of mustering the work force that the SS needed. Yet the statistics proved most effective at facilitating a new method of murder, the Operation 14 f 13, in which concentration camps first culled prisoners systematically for extermination. It is an open question whether Burböck worked with those who began implementing the Operation 14 f 13, a project originally conceived to purge the IKL network of the mentally ill, the physically handicapped, and recidivist criminals (also included because their social deviance was considered a medical trait). Kommandant staffs had to conduct preliminary selections, and then special commissions of doctors arrived to pass ultimate "scientific" judgment on those who would be eliminated.

According to one doctor's testimony, categories for extermination were broadened over the course of 1941. As another testified, "In the autumn of 1941 an investigation was conducted on all Jews by the camp physician [at Buchenwald]. Those that were *unfit for labor* were sorted out" (emphasis added).[81] This was the period directly after Burböck stepped up his complaints about sickly prisoners in the Labor Action, and he remained in constant contact with Glücks's adjutant who was supervising organized selections for extermination. Jews were the largest single group of prisoners killed, but selections included Poles, Czechs, "asocials," and "inveterate criminals." (In one preselection of 293 prisoners "eligible for transport," 119 were Jews; the rest fell in other categories.)[82] The most consistent category for selection, however, one that all had in common, was "unfit to work." At this time, "special selections" on the basis of these same categories also began at Auschwitz. Starting with the Operation 14 f 13, statistical compilations of the "unfit" became a standard tool for the systematic liquidation of prisoners.

Yet even here, the IKL's style of mismanagement emerges more clearly than rhetoric of a well-oiled "machinery of extermination" or the "technocrats of death" might imply. For instance, on 19 and 20 January 1942 the doctor Fritz Mennecke selected 214 inmates out of a preselected group of 293 for extermination at Gross-Rosen.[83] The Kommandant's staff had chosen the initial 293 upon receiving a quota: Gross-Rosen had to cull 250 prisoners arbitrarily from the camp population. "The requested number . . . was exceeded by 43 in order to have the necessary elbowroom for possible losses";[84] that is, the camp staff saw the action as a chance to demonstrate initiative and ambitiously provided 293 instead. The Kommandant of Gross-Rosen sent in a concluding report in March:

> 214 inmates were selected [i.e., by the doctor, from the 293]. From this number 70 were transferred on 3/17/42, and 57 inmates on 3/18/42. Between 1/20 and 3/17/42, 36 selected inmates died. The remainder of 51 inmates consists of 42 Jews who are able to work and 10 other inmates, who have regained their strength owing to a temporary cessation of work (camp closed between 1/17 and 2/17/42) and who will therefore not be transferred.[85]

The tone of the report was one of pride. The Kommandant's staff had enthusiastically culled prisoners for extermination who actually recovered, which demonstrates their weak commitment to productivity as a value in selection for death. Of note, the Kommandant's numbers did not always add up (10 plus 42 does not equal 51), a further demonstration that the modern control that the Labor Action Office strove to impose was more sham than reality. Camp staff saw the Operation 14 f 13 as a convenient excuse to purge unwanted inmates; the desire to excise "disease" from the body politic overrode any concern for factory management; and doctors helped in the task, even when the possibility existed that the sick might recuperate for meaningful work.

Arthur Liebehenschel, adjutant to the Inspector of Concentration Camps, rebuked the mismanagement at Gross-Rosen:

> According to the report . . . 42 of the 51 inmates selected for special treatment 14 f 13 became "fit to work again" which made their transfer for special treatment unnecessary. This shows that the selection of these inmates is not being effected in compliance with the rules laid down. Only those inmates who correspond to the conditions laid down, and this is the most important thing, who are no longer *fit to work*, are to be brought before the examining commission.[86] (emphasis added)

This protest did not halt or alter the implementation of Operation 14 f 13. So ready and willing were concentration camp personnel to participate in the killing that some SS medical orderlies began giving lethal injections before the official commissions could do the work. Between 10,000 and 20,000 prisoners fell victim to this operation, although it was stopped by mid-1943.[87]

Despite the friction, conflict was not the norm. By inserting "unfit to work" on an information sheet Burböck had given Kommandanten and medical personnel a category with which to organize the sick and injured in files that they began to take advantage of. Kommandanten used Burböck's technique to expand the brutality that their calling had long demanded. The destruction of prisoners could now proceed on a new scale in part because new tools of information manipulation lay ready to hand. As an organizational innovation, the 14 f 13 program also began to draw the expertise of doctors into a system for managing murder *and* slave labor in tandem.

The doctor active at Gross-Rosen's special selection, Mennecke, even wrote extensive letters home expressing his near glee with the entire operation. Once again, this was no civil servant whose moral sensibilities were dulled by weary days totting up figures. "I am here collecting massive new experiences in this work," he shared with his wife, "and it is essentially good that Dr. St. [Theodor Steinmayer] and I do this alone!"[88] Absolutely nothing indicates that working with statistics in an extensive, organized division of labor helped distance him from moral considerations. To the contrary, as Ian Hacking notes regarding statistics covering an almost equally disquieting subject, child abuse, "what you count depends upon your theory about what you are counting."[89] And Mennecke made clear that moral judgments about race adhered within his statistics: "All [Jews] do not need to be 'examined,' " he wrote home, "but . . . it is sufficient to take the reasons for their arrest from the files (often very voluminous!) and to transfer them to the reports. Therefore it is merely a theoretical work." He also declared concentration camp inmates "parasites on the nation," especially Jews.[90] Typical of Nazi racial and productivist ideology, he conflated categories of health, criminality, race, and industrial efficiency. In the process the eugenic presumption that productivity was a function of biology found concrete expression in numbers. Rather than condemn prisoners in racial and criminal terms, however, Mennecke's reports now used Burböck's industrial term, "unfit to work."[91]

The Operation 14 f 13 was only one facet of the creeping rationalization of genocide from 1940 onward that would eventually culminate in the industrial killing camps. Over the past decade historians of science have argued that the Enlightenment gave issue to an "avalanche of numbers," an "orgy of rationalization," and, much in the tradition of Michel Foucault's *Discipline and Punish*, several recent essays on the culture of precision allude to the effort undertaken since the Enlightenment to alter entire populations through the use of statistical surveillance.[92] The Holocaust was undeniably one such experiment in the manipulation of demography on a vast, terrifying scale. It was precisely the desire, immanent within the Operation 14 f 13, to mold the clay of humanity into a "Thousand Year Reich," which Zygmunt Bauman has declared totalitarian and condemned as the root of the Nazis' contempt for human life. Yet to Bauman modern techniques of control and the systems they were intended to sustain often appear as purely instrumental: the alienated tools of morally stunted citizens. As he writes, a "meticulous functional division of labor" in modern states has led to the "substitution of technical for a moral responsibility."[93]

But the Operation 14 f 13 demonstrates quite the opposite, namely the persistence of moral initiative in modern, rational organizations even in the midst of an extensive division of labor and the information management new to the twentieth century. Active engagement occurred in the Office I/5 both when conflict rendered its operations defunct as well as when its officers found common ground with other IKL personnel. In the first case demonstrated here, the case of the Labor Action Führer, conflicting commitments to production and the primacy of policing derailed the labor management that Pohl hoped to impose. Burböck could not overcome them. The conflicts originated, however, not in organizational structure alone but in the professional identity and values of competing institutions. Individuals mobilized bureaucracies to get what they wanted; there can be little doubt of that, but these institutions alone did not generate the dynamism of the conflicts involved.

By no means did everything end in paralysis, however, and the Operation 14 f 13 shows that organizational innovations could mobilize several interests at once. Burböck had originally established the category "unfit to work" in order to safeguard production, yet Death's Head Kommandanten began to use Burböck's files for a purpose consistent with their identity as "political soldiers" rather than meaningful labor. Otherwise contemptuous of modern administration, they seized upon a tool that they could use to ramp up the deadly retribution meted out to their "political" foes. Burböck had, in fact, provided even more than mere managerial technique: he gave them a conceptual bridge. Now Kommandanten could believe that they too were actively safeguarding productivity by parsing camp populations into "fit" and "unfit to work." They simply concentrated most enthusiastically on dispensing with the "unfit." Nevertheless, as will become clear in chapter 5, the constant culling of the "unfit" was instrumental in the VuWHA's management of the "fit" as well. The statistics on absolute numbers of healthy, available workers became all the more reliable because the IKL so zealously eliminated the "unfit." This proved a perfect syncretic solution in a political economy of misery.

CHAPTER 4

ENGINEERING
A NEW ORDER

Historians of the Nazi genocide have increasingly emphasized the exceptional atmosphere created by war, especially the brutality of combat in the East, as a prime mover that radicalized Nazi racial policy into the full-blown Holocaust. Before the invasion of the Soviet Union, Hitler declared to his generals that Germany would be waging a new kind of war: "one of ideologies and racial differences [that] will have to be conducted with un-precedented, unmerciful, and unrelenting harshness."[1] At the same time, the material capacity of the German economy was strained to its utmost, making it clear from the beginning that Germany would sacrifice civilian populations in the name of the Third Reich's war machine. "In this year 20 to 30 million people will starve to death in Russia," stated Hermann Göring. "Perhaps that is good, because certain people must be decimated."[2] He had already an-nounced in May, even before the invasion of the Soviet Union, "Any attempt to save the population there from death by starvation . . . would reduce Germany's staying power in the war and would undermine Germany's and Europe's power to resist the blockade."[3] Simultaneously he was negotiating the "Final Solution to the Jewish Question" with Reinhard Heydrich of the SS Reich Security Main Office. Such evidence suggests that a desperate attitude of "all or nothing" drove the constant radicalization of plans for murder in the East, a theme emphasized by Sebastian Haffner and Arno Mayer.[4]

There can be no question that the conditions of war created an unprece-dented callousness and brutality, yet it is important to remember that war, in and of itself, never created the drive to make the East into a clean slate for designs of racial supremacy. War may have catalyzed but never compelled genocide. Well past 1942 optimism—not desperation—was still the order of the day among devout Nazis like Heinrich Himmler. True, the United States had entered the war, but early 1942 was the nadir of the Allied war effort. U-boats were preying on U.S. and British shipping with impunity; Rommel was chasing the allies out of Africa; the Japanese were driving the Americans out of the Pacific; and tiny Britain remained very much hemmed in on its island. The only bright spot was the Soviet Union's dubious success in avoiding total collapse by holding its core industrial regions in the winter of 1941–42. Surveying the European continent stretched beneath Germany's feet in the dead of that winter, Himmler proclaimed his wish to "uplift the political, economic, and cultural existence of the new eastern territories."[5] It is impossible to overstate the extent to which this "uplift" made occupation policy there different from that of France, Norway, or the Benelux countries. In the West the Nazis settled for puppet governments, while in the East various organizations competed to reconstruct the entire fabric of the conquered lands. It was here that the Reichskommissar for the Reinforcement of Germandom (RKF), deeming the eastern "races" to be base and worthless, attained almost a free hand. The Nazis by no means spared western Europe, where the SS began to round up Jews and send them to their death. Nevertheless barbarity progressed in the East on a more harrowing scale. The western Jews were delivered here to mass graves and the crematoria. The Third Reich set out to uproot indigenous peoples and to cultivate an ideal German society in their stead, and when Nazi leaders declared that "umpteen" millions of people in the East must die, they were talking about plans for destruction and reconstruction in equal measure.

Himmler had positioned himself at the summit of cultural policy in the East by virtue of the acumen and ruthlessness with which he converted Nazi visions into executive policy, above all genocidal murder. As late as 1941, however, the SS still lacked the capacity to implement the next crucial phase of Germany's racial utopia: the construction of settlements from which a new society of Aryans was supposed to grow. Very soon RKF planners would call for 66 billion Reichsmarks of building in the East; entire towns and cities were to be obliterated in order to erect them anew. Himmler had ordered his most capable administrative officer, Oswald Pohl, to organize a corporate empire to manufacture the New Order. At the same time, the Main Office for Budgets and Building (HAHB, or Hauptamt Haushalt und Bauten, also called the Office II–Building) set out to construct the New Order as well.

A High Degree of Order?

The original impetus for extending Pohl's authority over SS building predated both the RKF and the war and seems to have come first through the expanding concentration camps. Pohl's mandate to centralize all SS construction within the Main Office for Budgets and Building came with the foundation of new camps throughout the Reich and Austria. The days were now past when Eicke had built up Dachau outside Munich in an old, abandoned powder factory. At that time, Eicke had had to plead for money to put up a few fences, but funding now flowed regularly from the Reich, whose Finance Ministry demanded to know where the money was going. Pohl was the only SS officer with the sufficient experience, recognized stature, and organization capable of handling the myriad agencies involved. Chief among his problems was the increasingly strict rationing of building supplies. Already at the end of 1938 Fritz Todt had issued the first of what would be a series of priority ratings and materials rationing. Todt was Germany's construction czar, the celebrated builder of the Autobahn. At the beginning of 1939, with war still eight months away, he cut off supplies to any projects—including those of the SS—that could not be directly justified in the name of national defense. When Todt began to restrict shipments of cement, steel, and lumber, Himmler objected that his camp system was an essential institution for national survival. Convincing Fritz Todt, however, was a different matter.[6]

The problem of winning over Todt—a man for the most part well disposed toward the SS—was twofold, both political and structural. Politically the SS sought to win sympathy by arguing that the SS had to stand at the ready to prevent political enemies and traitors from "stabbing Germany in the back" on the home front. Structurally the challenges to SS legitimacy were no less crucial, for the HAHB faced years of mismanagement bred under the "Dachau School." Competent men like Todt were loath to waste scarce materials on any organization that could not utilize them effectively. When Himmler put forward a case for his organization's vital contribution to national security, Reich ministries considered the political legitimacy of the SS's claims and its organizational ability to live up to its ambitions simultaneously. Todt, for instance, supported settlement ideals but wished to postpone their realization until after the war. He also supported weapons allocations to the Waffen SS. On the other hand, he remained unimpressed with the SS's petitions for construction materials, and, needless to say, at the end of 1939 he accorded the concentration camps the second-to-last priority rating.[7]

Thus in 1939, on the eve of Himmler's appointment as RKF and to the SS's embarrassment, Pohl could offer an effective plan for neither future construction of the Third Reich's Aryan utopia nor expenditures on much lesser projects. Shortcomings had become clear when Pohl tried to pass SS construction budgets through state ministries. The SS's representative to the Four Year Plan, Ulrich Greifelt (who would be named the chief of the Main Office of the RKF), urgently requested Pohl to establish the HAHB as an independent state building authority (*Bauhoheit*) endowed with special legal and financial privileges. In fact, once achieved, this status later proved instrumental at Auschwitz, where the SS used it continually to override local civil authorities in order to expand the camp.[8] The end of June in 1939, however, was a different matter. When Pohl met to discuss matters personally with the Reich Ministry of Finance, the results proved so disastrous that months later he still warned, "We should maintain distance from the further recall of the incident."[9] Reich auditors were then exposing mismanagement in the Construction Directorates maintained by the HAHB in the concentration camps and likely suspected graft. The accounts of some directorates were in such shambles that auditors could find no evidence either to prove or to disprove the misappropriation of funds. Invoices were kept in loose piles if kept at all, and Pohl's auditing division was understating the obvious when it warned SS construction officers that "the Reich demands a *high* degree of order" (emphasis in original).[10] The directorates had proved incapable of meeting it.

Again and again Pohl, his deputies, and, on occasion, Himmler himself had to reiterate orders consolidating power for all SS construction in the Main Office for Budgets and Building. Other SS departments were not necessarily wont to acknowledge that authority. Pohl had made several key appointments and could claim responsibility for infusing the IKL, the SS Command Troops, and other SS branches with a few capable civil engineers, but no central leadership cadre of an SS construction corps had yet emerged. The construction bureau in Pohl's "V" Office went through a string of chiefs. Sometime between the end of 1938 and the beginning of 1939, the concentration camp Construction Directorates were rechristened the New Construction Directorates (Neubauleitung), likely in response to Erduin Schondorff's demands, but curiously no document in either the complete archival collections of the directorate of Auschwitz or that of Flossenbürg announces this transformation or what it entailed. The very *lack* of evidence is a sign that the New Construction Directorates remained decentralized and loosely controlled from Berlin. By October 1940 Pohl had gathered only seven officers of rank into the HAHB. Some departments went entirely unfilled. Those officers who did lead the office pushed for "an improvement in the ready distribution of labor resources" but confronted a general disarray.[11] If some individual engineers from the Eicke years were indeed competent, as yet no integrated, national hierarchy existed to mobilize them in a concerted effort.

Meanwhile the founding of Buchenwald, Mauthausen, Flossenbürg, Ravensbrück, and Neuengamme had doubled the IKL's construction needs. More new camps were yet to come: in 1940 Auschwitz, in 1941 Majdanek (Lublin) and Stutthof. Increased complexity only compounded the increases in scale. The SS companies, in particular, now required factory halls and land for quarries and clay pits, industrial buildings with which the SS, as yet, had slight experience. Pohl began a concerted effort to transform SS construction into a nationwide hierarchy for competent civil engineering, but the HAHB could exert control only at a snail's pace.

While the HAHB faltered, war began to make engineering the New Order an ever greater, not a lesser, imperative. At the end of 1939, after the speedy conquest of Poland, Hitler bestowed upon Himmler the authority to begin German settlements throughout the eastern occupied territories as Reichskommissar for the Reinforcement of Germandom, a responsibility that Himmler had won as a direct result of the SS's responsive executive organization. The embarrassing lack of modern organization in construction now threatened the SS's ambitions to construct a Nazi utopia. Fumbled IKL budgets—typical throughout 1940 and 1941—not only risked exposure for waste and corruption; they now risked failure to fulfill the Führer's direct order. The predicament only increased the potential for disaster as the first giddy months of the Soviet campaign made quick victory appear to be a certainty.

Beginning on 22 June 1941, Germany's armored divisions flooded across the Soviet frontier along a front stretching from the Baltic to the Black Sea. The scenario of the Polish invasion seemed to repeat itself. Stalin was taken by surprise. The German air force caught Red Army supply lines, communication centers, and divisions out in the open. Its primary target, the Soviet air force, was destroyed on the ground, and in this the German air force proved so effective that it has become the model for all such preemptive strikes.[12] The Wehrmacht took hundreds of thousands and then millions of prisoners. The Chief of the General Staff smugly announced, "It is very likely not saying too much when I observe that the campaign against the Soviet Union has been won in less than fourteen days."[13] By mid-July Hitler announced that the army could be reduced to prepare for the peacetime reorientation of the economy.

RKF drafting boards became crowded with designs on a region stretching from the Oder to the Dnepr, from Prague to Riga. We know now in retrospect how bitter the Soviet campaign would prove to be, and how it bled Germany white, but this dawned on contemporaries only slowly. In 1941 winning the war did not so much preoccupy top SS circles as winning an imaginary peace. Himmler wondered how quickly the RKF could turn lofty settlement visions into the mundane bricks and mortar of housing for SS war veterans and Aryan settlers. Time seemed to be of the essence.

Regional officials in occupied Poland quickly developed more radical plans than did top Nazis in Berlin, even as Pohl haltingly tried to control SS construction. In fact, planning lagged implementation as SS men in the field vied with civil servants and the army to shuffle huge populations of Poles, Jews, and "ethnic" Germans (*Volksdeutschen*) across the breadth of Poland in a kind of racial supremacist shell game. As one RKF planner, Walter Geisler, announced in a published book, "Anyone who wants to build up must begin with cleansing, which means that everything that does not fit the new plan, or that opposes it, must be destroyed or removed."[14] Ironically he was not dictating policy but rather describing what had already become practice. Each district tried to expel unwanted populations and provide model communities for incoming Germans. RKF planners as well as the Reich Security Main Office had developed their first reports as early as the spring of 1940. With repeated embellishment over the next three years, these coalesced into the General Plan–East (Generalplan-Ost). But by 1941 no comprehensive, overarching blueprint existed; instead multiple, parallel projects went forward in the climate of "restless" action that Himmler constantly encouraged. Even though Pohl's administrative offices were frantically reorganizing to help construct the New Order, Himmler chose not to wait.[15]

Odilo Globocnik: Handcrafting the New Order

Perhaps because Pohl's organization lagged behind the seemingly unstoppable German army, Himmler moved ahead and entrusted the construction of the RKF's first idyllic settlements to an officer outside the HAHB in 1940 (even before the invasion of the Soviet Union). Odilo Globocnik, the SS and Police Führer (SSPF) of the Lublin district, had been actively preparing settlements almost as soon as he had arrived in Lublin at the end of November 1939. As SSPF, he was Himmler's empowered representative for all SS functions within his region, including those of the RKF.[16]

Why exactly Himmler decided upon him for the first settlements of the New Order is not clear, but Himmler had considerable affection for the man, fondly calling him "Globus." He was the kind of "action man" Himmler valued—energetic, a self-starter. Globocnik shared much in common with Theodor Eicke, another favorite of Himmler's. Globocnik often preferred to cover over his bumbling with bravado and sought to prove himself again and again in outbursts of energetic activity, the more radical the better. He continually skirted the law and engaged in savage acts that he always took care to commit, at least ostensibly, in the name of the movement. While serving as an illegal Gauleiter in Vienna when Austria had forbidden the NSDAP, he murdered a Jewish jeweler during a robbery and had to flee to Bavaria. Once there he continued to aid the Austrian SS and participated in further terrorist acts until he eventually fell under suspicion of embezzlement. Just as the Reichs-führer SS had shielded Eicke, so too he protected Globocnik from prosecution for crimes and incompetence.[17]

As SS and Police Führer in the district of Lublin, "Globus" displayed his immense energy by constructing both the first idyllic SS settlements and the first death camps. An SS man in Lublin recorded that Globocnik "holds the eventual cleansing of the entire General Government [i.e., occupied Poland] of Jews and also Poles as necessary. . . . An activist cultural and settlement policy [*Volkstums- und Siedlungspolitik*] with broad goals will be set against the passivity of ossified governmental bureaucracy."[18] As this statement implied, the SS's racial supremacy here extended not only to Jews but also to Poles. Far from secretive, Globocnik announced his plans at a Nazi Party meeting in Kraków; the city's German-language newspaper reported that he would "surround" the Polish population of the General Government with Aryan settlers in order to "eventually crush them to death economically and biologically."[19]

Despite a past history of embezzlement and contempt for bureaucrats (*Bonzen*) akin to that endemic to the IKL, Globocnik nevertheless had some qualifications besides bravado and brutal language. He possessed a modicum of technical training, and, within the SS, his admirers said that the day he stopped building would be the day he died. He had earned his living as a skilled mason and probably had worked as a foreman managing building sites. Yet his lack of formal polytechnic education would have consequences when the scale and scope of his operations demanded modern bureaucracy. He had made his name more as a model of Nazi inspiration than as an organizational workhorse like Pohl, and, unsurprisingly, Pohl would also encounter the same conflicts with Globocnik that he did with the IKL.[20]

Globocnik chose the town of Zamosc as the site of the first SS settlements for reasons that remain unclear. Richard Evans, Charles Maier, and others have argued that the Nazi genocide is a unique event in history precisely because "Nazi mass murder was an end in itself," whereas the collectivization campaigns of the Soviet Union, by contrast, were carried out in the name of "social reconstruction."[21] It is easy to believe that a man with such a monstrous appetite for murder like Globocnik really had no other motive than nihilistic hatred, but Globocnik would have strenuously disagreed. No doubt he boasted of the "uniqueness" of his genocidal policies, and he reportedly bragged to those who worked on the gas chambers, "Gentlemen, if ever a generation comes after us that is so cowardly and so weak that it cannot understand our good and so necessary work, all of National Socialism was for nothing. Quite the opposite, one must bury a bronze tablet [with the Jews] that it was us, us, that had the courage to carry this gigantic work to its end."[22] Yet rather than an orgy of murder as "an end in itself" Globocnik's racial supremacy encompassed the SS's vision of a socially engineered utopia. In the first weeks of October 1939 the SS had begun pushing Jews and Poles out of territories destined to be annexed to the "Greater Reich" (Eastern Upper Silesia, Warthegau, Danzig–West Prussia, and Southeastern Prussia). Simultaneously, the RKF planned to replace them with ethnic Germans from the Baltic countries and Soviet territory.

Lublin became the nexus of these experiments because it was chosen early on as the "reservation" for expelled populations. At the same time it served as a transit node for ethnic Germans heading westward toward "cleansed" territory. Its status as a "reservation" for unwanted populations lasted only a short time, but long enough so that by March 1940 transports of expellees (Aussiedler) already totaled over 200,000 souls. Most transports through 1941 consisted of Poles, among whom expelled Jewish citizens were a minority; they most likely numbered in the thousands rather than tens or hundreds of thousands. In February 1941 Adolf Eichmann also began to deport the Jews of Austria into the General Government after inventing an "assembly line" system for processing their identification cards, papers, and confiscated property. The Lublin district was thus a way station for the Nazis' demographic experiments. The number of Jews there seems to have fluctuated between 250,000 and 300,000. Some estimates by the Nazis themselves put the number as high as 448,000.[23]

The geographic location of the district only amplified its importance to RKF planners, for until the invasion of the Soviet Union Lublin rested on the Reich's easternmost defensive frontier. Himmler had long fantasized about an ideal society of warrior-farmers as the vessels of German cultural rejuvenation. The General Plan–East called for SS and Police Strongholds in such outlying regions both to defend "the ultimate ownership of the land conquered by the sword" with self-sufficient SS militias (a short-term strategic necessity) and to increase "German blood" (a long-term goal of demographic expansion).[24] Thus there exists no single reason for the Lublin district's selection as the first Nazi settlement outpost, but this is precisely because so many different currents of Nazi policy converged upon it.

Attempts to convert the region into a model habitat for SS "warrior-farmers" were well under way at the end of 1940. Globocnik energetically began to acquire farmland, where possible by confiscating it from Jews. Here he secured the cooperation of local civilian administrators, and, as one official testified after the war, "the discussions over the Jews were generally so, as when one spoke of the lowest garbage."[25] Hastily formed SS "militias" (Selbst-

schütze) of newly relocated "ethnic Germans" expelled Jews and Poles from their homes. These militias displayed much of the wanton brutality typical of the Death's Head Units. An initial task set for the Jews of Lublin was the excavation of massive antitank ditches in which Globocnik's thugs worked many of their victims to death. Meanwhile, to make SS settlements self-sufficient Globocnik anticipated planting a total of 1,543 hectares of new homestead land in the spring (not including pasture, forest, stock ponds, lakes, and existing homesteads that the SS also acquired).[26]

Many have seen Himmler's agrarian romanticism as evidence of a drive to "turn back the clock" to an antimodern world of bucolic bliss and invented traditions, and the rhetoric of warrior-farmers does lend credence to this view, as does the anachronistic barbarity of slavery. Even if settlements had the bureaucratic-military sounding title "SS and Police Strongholds," some SS documents also referred to them wistfully as "knight's estates" (*Rittergüter*). As Robert Jan van Pelt and Debórah Dwork have shown, however, nostalgic dreaming of "knights of yore" existed side by side with rational settlement layouts and plans for the exploitation of industrial resources. "New technological operations and far-reaching mechanization" would be used to "raise the labor productivity of the farming family."[27] Far from turning back the clock of world history to the Dark Ages, the SS believed itself to be the modern culmination of the past: "Historical forms in city construction can serve the present only poorly," announced one of Himmler's planning papers. "In a community order based on the Führer principle and erected on ideal values, communal buildings must take the leading position in the cityscape much like churches in the Middle Ages or the forum in the antique city."[28] Such statements and countless others clearly announced the SS's intention, not to return to an antimodern past, but to construct a particular Nazi vision of the future.

As noted, Himmler saw Germans as cultural fertilizer. As a kind of "gardener," then, Globocnik had two related tasks: first, to "weed out" races considered dangerous to Germandom and, second, to nurture the "growth" of German settlers. In realizing these goals, modern organization and technology mingled with ideological doctrine. Globocnik attributed the development of industry and agriculture to the essence of the Aryan race and, conversely, interpreted poor organization as a mark of racial inferiority. He described Poles as "unambiguous, easily satisfied, and inert," whose "misrule in business has run conditions for production completely to ruin." He further betrayed his preference for the large-scale organization new to the twentieth century by deriding Polish factories as "midget operations."[29] He intended to bring the SS strongholds' agricultural production up to the standards called for by the General Plan–East, whose author declared, "[Settlers] also have the obligation to be exemplary managers and pioneers in the agricultural, technical, and economic aspects of farming."[30] To eradicate "misrule in business," Globocnik planned to introduce tractors for plowing, planting, and harvest, fertilizers to raise yield, scientific animal husbandry, and processing industries for meat, fats, and dairy products, not to mention, of course, automobile workshops and modern roads. Far from backward-looking peasant homesteads, he promised progressive sanitation facilities and mechanized housework. In short, he developed a Nazi vision of economic modernization, albeit one based on productivism, in which modern organization and technology were supposed to fashion Germandom.

Some have criticized Götz Aly and Susanne Heim's claim that the National Socialists conceived mass murder as a rational instrument applied in order to modernize economically the occupied territories. On the other hand, their argument that Himmler's SS undeniably carried its own vision of a Nazi modernity into the East would seem to hold firm. I would argue only that, rather than "rational capitalism," the SS based its modernity on productivism and racial supremacy. The SSPF of Lublin even included a special RKF officer, a man named Gustav Hanelt, whose duties combined "the holistic planning of the SS and Police Strongholds, Jewish cleansing, and scientific action within the framework of the SS communal buildings." Hanelt also planned a "research institute for the East," later christened the "Research Institute for Eastern Shelter," to bring "active, spirited intellectuals" to Lublin in order to tackle settlement problems.[31] To Globocnik's staffers, and within the SS more generally, the social engineering of genocide was part and parcel of engineering the New Order.

Central to SS modernity was the common belief that well-functioning institutions and the newest, latest developments in technology—even articles that now seem mundane, like tractors or gas water heaters—distinguished the superiority of Germans as a technological race. As the flip side of this coin, RKF planners derided the "backward" nature of traditional Jewish shopkeepers and craftsmen and also attributed the primitive Polish countryside to the degenerate nature of the Slavic "subhuman." In the eyes of Globocnik and Himmler, the supposed low productivity and lack of modernity among Poles and Jews justified their subordination as slaves and serfs to German efficiency—or their outright elimination as "unnecessary eaters."[32] Productivist ideology merged with racial supremacy in judgments passed on habits of economy and industry, which goes a long way toward explaining the seeming inconsistency between the SS's simultaneous drive for modernity and slave labor. The German social historian Werner Conze summed up this racial supremacy more or less concisely when he wrote in the weeks just before war broke out with Poland, "The poverty of the towns in Poland . . . is not a crisis phenomenon but, to the contrary, a necessary consequence of degeneration."[33] One year previously he had also written that White Russian villages had failed to develop any noteworthy industry because Jews dominated their commerce and manufacture.

Despite such common sentiments, Globocnik's report on the Lublin district near summer's end in 1941 revealed a crowning irony. As he continued to oversee the transport of "ethnic" Germans into his district in conjunction with settlement building, the only thing at which he distinguished himself was genocide. By mid-1941 he was forced to admit that his ideal farms could not even find horses for plowing, let alone tractors. His negotiations for even minor support industries had snagged in Reich agencies, his antitank ditches were likewise a disaster, and he could not secure all the property that his network of strongholds called for. Globocnik had ridiculed the "misrule" of Polish organization yet had proved incapable of replacing the backward huts of his district with the clean, hygienic farmsteads that had been promised on drafting boards. The SS's administrative and technological means were both the tools of its executive power and symbols in its ideological phantasmagoria. Globocnik's bungled initiatives left SS settlers looking little different from the "unambiguous, easily satisfied, and inert" Poles or the Jews that he had pledged to eradicate. This marked the cue for the entry of Pohl's Main Office for Budgets and Building.[34]

Himmler brought Pohl to visit Globocnik at the town of Zamosc at the end of July 1941. Reich officials were already proclaiming victory over the Soviet Union, and with the Wehrmacht racing toward Moscow, the vista of German settlement expansion must have made the two men giddy with expectation. Himmler seems to have avoided criticizing Globocnik directly. He simply shifted responsibility for construction directly to Pohl's HAHB. Globocnik stayed on, nominally in control of the design and layout of settlements (though, de facto, Pohl's new chief of engineers began to take over this work as well). Globocnik also finished up some temporary construction already under way, but his influence diminished rapidly. In March of the coming year, Himmler removed him from the planning of strongholds completely (at a time when the RKF announced plans to triple their number). This must have been a bitter blow, for settlement policy carried high prestige within the SS well into late 1942, and Globocnik's relationship was sour with Pohl ever after.[35]

Himmler had made a choice between two different styles of organization. By backing Pohl, he chose systematic, impersonal—that is, modern—administrative hierarchy over Globocnik's personal initiative. Globocnik had essentially set about handcrafting the New Order. He had failed to master the modern means that SS rhetoric led him to admire, and, at best, he presided over the Lublin district more or less as a foreman would run a construction site. He had never juggled correspondence with national ministries and agencies that could accelerate large projects or cut them off completely in a hail of audits, property rights inquiries, raw materials vouchers, and permits. The RKF was talking about dispersing tens of billions of Reichsmarks, and as always when the scale of the SS's responsibilities multiplied, the demand for accountability and competence increased in step. Whereas Globocnik had focused on a small constellation of settlements surrounding Lublin, Pohl had in mind an organization that could purchase property in the East, order steel from the Hermann Göring Works in Salzgitter and bricks and stone from the DESt at Sachsenhausen or Mauthausen, hammer out priority ratings and permits with ministries in Berlin, and mobilize the SS's available machinery, labor, and civil engineering corps within a matrix of closely calculated time schedules.[36]

The role of concentration camp industry also began to change at this stage, not by taking in armaments production in response to the war effort, but by reorienting to serve the SS and Police Strongholds. At Lublin, Himmler decreed a "concentration camp for 25,000 to 50,000 prisoners for the deployment of workshops and construction on behalf of the SS and Police [i.e., the Strongholds]."[37] This became known as Majdanek. The HAHB issued the first construction order by mid-September, and the next month Globocnik released 2 million Reichsmarks from his SS garrison to cover the initial building. By 1 November, the HAHB ordered an expansion of Majdanek from 50,000 to 125,000, then, little more than a month later, to 150,000 prisoners. A similar expansion was proceeding at Auschwitz.[38]

From 1936 to 1940 the SS had sought to supply building materials for the monumental Führer buildings, founding camps like Sachsenhausen, Mauthausen, Buchenwald, Neuengamme, Ravensbrück, and Flossenbürg as integrated institutions for punishment and industrial labor. Now Himmler wanted to make construction of his new architectural fantasy—the settlements of the New Order—into a self-sufficient business enterprise. Several years earlier new camps had brought an increase in scale and scope to the IKL. So too the new "settlement" camps planned in the East dwarfed all previous developments. In 1939 the entire prison population of the IKL had hovered around 25,000, with a sudden, temporary spike of up to 54,000 after the "Crystal Night" pogrom of 9 November 1938. Auschwitz (in Upper Silesia), Stutthof (near Danzig), and Majdanek (Lublin) were supposed to hold not thousands, but tens of thousands, then hundreds of thousands. Stutthof alone, organized by the end of 1941 and the smallest of the three, would have doubled the IKL's previous captive labor force. It foresaw space for 25,000 prisoners, "with whom we can then complete the buildup of settlements in the Gau Danzig–West Prussia," as Himmler put it.[39] He also intended Auschwitz, founded in mid-1940, to serve as a model for the SS's settlement fantasies. Its Kommandant, Rudolf Höss, had received his appointment in part due to his experience in agrarian communes. He and his wife had lived in right-wing settlements, and he had some formal training in agricultural sciences and administration. Auschwitz planned to incorporate industries for settlement development, including the Experiment Station for Alimentation, led at Auschwitz by an agricultural scientist and SS officer, Dr. Joachim Cäsar. Höss and Cäsar were to organize "*the* agricultural experiment station for the East. . . . Great laboratories and plant breeding departments must come into existence" (emphasis in original).[40]

At Lublin, Himmler bid the German Equipment Works (DAW) to "develop a special workshop for simple, electric heaters" for SS households as well as an extensive central heating system for SS communal buildings in the city. The VuWHA planned a sanitation facility complete with installations to generate natural gas for German homes. The DAW also started additional workshops for simple items like screws, hinges, and window frames. It also added an automobile repair shop. Thus SS industry, the concentration camps, and a newly formed SS construction corps began to emerge at Lublin as an integrated undertaking. This was, in fact, Himmler's intention:

> We are not conceivable without the SS economic enterprises. . . . The settlement construction program, which is the precondition for a healthy and social foundation for the entire SS as for the entire leadership corps, is not conceivable if I do not get the money from some place or other. No one gives me the money as a present; it must be earned, and it is earned by putting the scum of humanity, the inmates, the habitual criminals, to work.[41]

The SS companies, which began as pet projects within the Reichsführer's Personal Staff, were now, in the midst of war, becoming workshops and supply industries for the new order that the SS planned for an imaginary peace.

Hans Kammler: Modern Engineering in the SS

Although Pohl was a competent financial administrator, he could have never mastered the technical details of building the New Order alone. At the end of 1940, he therefore courted Dr. Hans Kammler, an engineer from the German air force, who would quickly prove one of the most capable and dedicated men ever to make his career under Himmler. The chief of Himmler's Personal Staff requested Kammler's release as "general adviser for settlement in the staff of the Reichskommissar for the Reinforcement of Germandom" and did not hesitate to allude to the "German building program of the Führer," a clear sign that winning the peace and constructing the New Order, not wartime concerns, dominated the SS's intentions at the time.[42] Kammler went to work for the SS on 1 June 1941. Now the pace and scale of the New Order demanded a different organization, and, instead of Globocnik, a bricklayer and foreman, Pohl and Himmler turned to a modern technical manager.

In Kammler technological competence and extreme Nazi fanaticism coexisted in the same man, a historical warning against the facile belief that technological rationality is the "highest" form of reason. For his intensity, his mastery of engineering, his organizational genius, and his passion for National Socialism, SS men esteemed Kammler as a paragon. The afterglow of this awe lingered well after the war, as Pohl testified:

Not only was he a wonderful construction engineer but also a son of an old Pomeranian officer's family. And as a cavalry officer [in the First World War] he was an excellent soldier and was also an honorary professor at the technical school at Charlottenburg [the Polytechnic University in Berlin] in scientific matters. Physically, he was tall, slim, with a haughty nose; he was elastic just like a bow which is eternally strong. Quite often I could not understand how this man coped with the tasks to which he was assigned in as short a time as one day. He was everywhere! Apparently he slept only during official trips in his car at night. And neither did he float over the surface of matters but somehow managed to know all tasks to the very bottom. . . . He was possessed of an agility which could not be explained. He called his men . . . to his office at any time of the day or night regardless of whether it was noon or at three o'clock in the morning. . . . In his manners he was very ambitious—but in a good sense because he never aspired to become important.[43]

Albert Speer later came to fear the chief SS engineer and numbered him among the calm, cool, and inscrutable experts who served Himmler: "Nobody would have dreamed that some day he would be one of Himmler's most brutal and most ruthless henchmen."[44] Most who worked with Kammler directly, however, including Speer's own subordinates, did not fear but admired him. "He thought his mind could encompass and master all problems," said one SS building engineer. "Far from regarding the enormous load of assignments which gradually came his way as a burden, he rather felt them to be an adequate appreciation of his personality."[45]

Kammler came from a privileged if not elite background. Pohl stated that Kammler's father was an officer and Kammler a cavalryman, almost certainly a romanticization. Kammler's father had been a gendarme of Stettin. Kammler himself must have felt a personal identification with the military either through his father's profession, patriotism, or both. He was born in 1901 into that generation of World War I latecomers that included Heinrich Himmler. Kammler never saw action, but he did claim to have mobilized with a cavalry regiment in May 1919. The German military effort had long collapsed. Nevertheless, Kammler participated in the Free Corps Rossbach, one of the right-wing paramilitary units full of older, disgruntled soldiers and nostalgic young men like himself. He eventually dropped out of the paramilitary scene to finish his studies, and, after passing through several humanistic high schools (*Gymnasien*), he completed a Diploma Engineer in architecture in 1923 at the Polytechnic University of Danzig. Even at this early date, he sought out practical experience in Danzig's Municipal Settlement Office and dedicated his thesis to techniques for surveying, cost-calculation, and housing development. His studies concentrated on the modern management of construction—not the aesthetics of form—and testified to the convergence of architectural and engineering practice at this time. Kammler also made himself familiar with industrial processes. During one semester break, he acted as a supervisor for a sugar factory in Hanover, where he managed transport work at the river docks. After graduation, he began work as a civil servant drafting and reviewing blueprints for the municipal government of Berlin and immediately began managing settlement projects in Zehlendorf.[46]

In 1928 Kammler passed his civil service exams. Up till then he had worked as an assistant in state offices; now he was qualified to lead one. As the scope of his professional tasks grew, the ideological character of his work came increasingly to the fore. Kammler's Diploma thesis of 1923 had dispassionately discussed left-leaning architects like Walther Gropius.[47] After 1928, his interests in settlement building converged with the particular advocacy of the garden city movement celebrated among right-wing architects who wished to unite German families with the "German soil." By 1930 Kammler was organizing settlements for agrarian communities, and in 1932 he joined the Nazi Party. Shortly after the seizure of power, he began work in the Reich Agriculture and Food Ministry (Reichs Ernährungsministerium), led by Richard Walther Darré, who had framed the rhetoric of "Blood and Soil" adopted by Himmler himself. Kammler rose to lead his own Garden Settlement Division by 1935 "at the expressed wish of the Reich Leader Darré," whom he had obviously impressed.[48] Kammler first donated his time to the SS in 1933 as an adviser to the Race and Settlement Office, also led by Darré, and three years later sought to serve his country as a military engineer for the air force, which, like the SS, owed its existence to National Socialism. In all these endeavors he rehearsed the organizational abilities that he later put to use under Pohl.

Not unlike Pohl's business experts, Hans Hohberg or Erduin Schondorff, Kammler shared a lust for complex organization. Pohl likely sensed this, admired it, and recognized in Kammler a kindred spirit, a man above the petty pursuit of authority who nevertheless reveled in the mastery of organizational dynamism. Kammler's tale is the tale of a man who mobilized the creativity of others outside himself, and for that reason the SS civil engineering corps depended on a shared consensus of purpose. A well-defined mission was essential in recruiting many of the talented officers upon whom Kammler relied. By and large, his top officers signed up because they believed in what they were doing. The mythology created by Albert Speer and historians after the war that technical work distanced such men from moral considerations does not bear out in scrutiny of their daily tasks, and Kammler would have strenuously rejected the label of "technocrat" applied to him after the war.[49] To him, not unlike Globocnik, modern technology, organization, and ideologies of German supremacy were one and the same, but unlike Globocnik, Kammler knew how to mobilize them in unison. The engineers under Kammler gave vent to their contempt for "typically Polish" misrule in matters of technical and organizational proficiency, just as the SSPF staff of Lublin associated poor organization with degeneracy.[50] Likewise, in the slang of engineers at Flossenbürg, inferior building stones were referred to as "Jews."[51] Given the glorification of the "German inventive spirit," it is not too much to assume that many of them believed that the "German will" was the wellspring of technological progress and sound organization and, conversely, that efficient institutions and machines were evidence of the restless action of the German will.

Nothing expressed Kammler's own ambitions in these matters more clearly than the book that he brought out jointly in 1934 with a docent at the Technical University in Berlin (Edgar Hotz). The authors concentrated on technical information, including guides to Tayloristic time-motion studies, modern bureaucratic hierarchies, and statistical methods of surveillance. But Kammler and his coauthor also raised their voice in a litany of productivism and racial supremacy that should be familiar by now. They wished to produce national culture instead of profits. They cried down the Weimar Republic, for "the man and the soil did not stand at the center . . . but rather materialism, bureaucratic technicalities, legalities, and the salesman's point of view."[52]

Kammler, like many engineers, believed that the methods of modern management and organization were part and parcel of the new society he wished to construct. Romantic belief that contact with the German soil was necessary to make good Germans existed side by side with this enthusiasm. Even banal bookkeeping, in this light, became an arena of political rejuvenation: "We are certain," states his introduction, "that a precondition for the unification of cost and price accounting between firms and customers is a complete change from liberal capitalism to a National Socialist economic order."[53] Whereas, in their eyes, the Weimar Republic had stifled the best methods of organization, the authors complimented the Nazi regime for committing itself to the unfettered release of those institutions and technological developments new to the twentieth century. Modernity and National Socialism mutually reinforced each other.

Many during and after the Second World War expected engineers to have been natural resisters of Nazi rule due to their "rational" profession. Not even Franz Neumann was free of this romanticization of technical knowledge as an inherently moral force: the engineer, he wrote, "will later constitute . . . the most serious break in the regime. The engineer exercises the most rational vocation and he knows what beneficent powers the productive machinery can wield. Every day sees how this machinery becomes an instrument of destruction rather than of welfare."[54] Nevertheless, as difficult as it is to believe, many young, intelligent, idealistic engineers, especially the most educated and elite among them, looked to National Socialism to fulfill their visions of progress. At the Polytechnic University of Munich, for example, the Nazis carried fifteen of thirty seats in November 1932. These were the last elections for student body representatives before Hitler's seizure of power. The voter turnout reached 90 percent, and a further ten seats were carried by other radical right representatives. Liberals or socialists carried not a single one. Nine years earlier, Polytechnic students had held rallies just days after the "Beer Hall Putsch" of November 1923, and fully 70 percent of those present had spoken out in favor of Hitler and demanded leniency.[55]

Since the turn of the century, professional engineering education taught students to draw a direct causal line between industrial growth and social progress, between the steady promotion of technical innovation and the betterment of humankind. Viewed abstractly, this drive is perhaps laudable, but the moral crux lies in *what kind of society* lay at the center of their vision. Increasingly in the late 1920s the extreme right succeeded in mobilizing the activism of Germany's elite engineering students. It is small wonder that engineers generally loathed the Weimar Republic, for they identified democracy and free markets with the wreck of German industry and a stunted economy at the very time when advances in modern production and management techniques opened unprecedented opportunities to increase distribution and production. The prostration of the Weimar economy not only blighted the career chances of engineers; it withered their vision of what a progressive Germany should be and do.[56]

Of course, economic disaster did not automatically turn engineers into goose-stepping Nazis. But it is significant that the support of engineers for Hitler's regime increased over time as the Nazis succeeded in sustaining industrial growth. From the vantage point of the present, historians now know that the Third Reich's boom years of 1933–36 rested on high deficit spending and choked-off investment in the consumer sector. Historians like Richard Overy are correct to warn that the sometime lingering admiration for the Nazis' "economic efficiency" (preserved in statements like "they built roads and got people working again") is hollow, ignorant, and facile, but the engineers of the 1930s were contemporary observers without the luxury of well-informed historical research. Who could deny that they were finding jobs again? No less, these jobs were exciting. While the Western democracies continued to flounder through the Great Depression, the Nazis' message of productivism and their promotion of technological prowess as a wellspring of identity made the engineer seem like a poet: the smithy of the German soul.[57]

Kammler quickly set out to construct the New Order by applying the panoply of modern managerial methods that had enjoyed their first widespread popularization in Germany in the 1920s. Here again his academic writings shared much in common with the work of Erduin Schondorff, the SS's expert in brick manufacture. Both ridiculed old-fashioned building masters (a guild title) who based management on "rules of thumb" instead of systematic bookkeeping and proven, scientific methods.[58] And yet Schondorff had concentrated on theoretical engineering science and neglected the factory floor—the arena of direct control over the material world. By contrast, Kammler and his coauthor Hotz had pointed out that no automatic machinery or scientific knowledge could substitute for clear-eyed foremen and keen managers who kept statistics on what workers were doing, when raw materials were coming in, or how their building sites progressed. The greatest demon of the work site was in all cases time, whose waste was always the guilty party whenever anything disrupted the steady confluence of construction or prevented the utilization of all resources at full capacity. Idle machinery, laborers, or materials, in the eyes of Kammler, wasted the potential, so crucial to the productivist aesthetic, to push the pace and yield of industry to its outer limits. They proposed managerial hierarchies and schedule plans written up in bar graphs in order to render the complexities of a large building site in

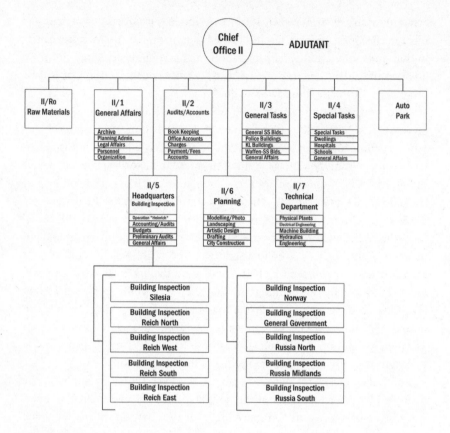

simple, visual form. Reading them at a glance, any civil engineer could become the master of time.

The quintessence of time and the firm grasp of the material world of labor and production were what made modern management so different from traditional forms of administration; Kammler's sensitivity to this made his style different from, say, management in the IKL or in Globocnik's staff. When Kammler warned, "All superfluous paperwork restrains the supervisory organs [of building firms] from the oversight of construction work," he was not bashing bureaucratic "pencil pushers" like Theodor Eicke. He was urging the statistical management of time and demanding concise information for command and control. Unlike Eicke, the truculent prison warden, he understood how to render daily observation in a standardized, statistical format "that permits easy overview" instead of the long-winded prose of traditional administration.[59] Such techniques allowed him and his managers to keep what was "out of sight" from becoming "out of mind."

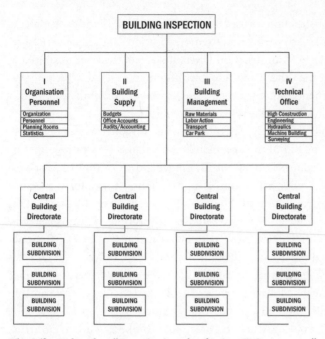

```
                          BUILDING INSPECTION


         I                    II                   III                  IV
   Organisation            Building             Building            Technical
    Personnel               Supply             Management             Office
  Organization            Budgets            Raw Materials        High Construction
  Personnel               Office Accounts    Labor Action         Engineering
  Planning Rooms          Audits/Accounting  Transport            Hydraulics
  Statistics                                 Car Park             Machine Building
                                                                  Surveying


     Central              Central              Central              Central
     Building             Building             Building             Building
    Directorate          Directorate          Directorate          Directorate

    BUILDING             BUILDING             BUILDING             BUILDING
   SUBDIVISION          SUBDIVISION          SUBDIVISION          SUBDIVISION

    BUILDING             BUILDING             BUILDING             BUILDING
   SUBDIVISION          SUBDIVISION          SUBDIVISION          SUBDIVISION

    BUILDING             BUILDING             BUILDING             BUILDING
   SUBDIVISION          SUBDIVISION          SUBDIVISION          SUBDIVISION
```

The Office II, based on "Organisationsplan für Amt II–Bauten 1941,"
RG-11.001M, 19(502-1-12).

Thus as timing and implementation of the New Order became a pressing concern to Himmler, Pohl, and the RKF, it was no accident that the SS turned to Kammler. One officer testified that almost no one could keep track of Kammler's flurry of activity in 1941 and 1942. He stormed in almost daily with new organizational charts and revisions of old ones. He also lured a cadre of civil engineers and architects away from the German air force to the SS and made sure that they were much like himself: they had elite education and careers managing large-scale construction projects. Before mid-1941, only a handful of high-ranking, technically trained officers had sat in the Office II. That number now doubled, the number of lower-level drafters and technicians quintupled, and Kammler made plans to increase the size of the Office II even further.[60]

In addition, Kammler did not simply lay out a national administrative structure and then expect it to run automatically, as had, for instance, Wilhelm Burböck (see chapter 3). Kammler enforced his managerial system through methodical surveillance. He distributed over forty standardized forms to all levels of the SS construction corps. Thereafter he could track the daily material conditions of on-site construction at remote locations and made sure to replace existing chiefs of Construction Directorates who had been exposed as incompetent. On the other hand, capable officers like Gerhard Weigel, who had led SS construction since the early 1930s, stayed on. Kammler was quickly convinced of their technical knowledge and dedication.

Kammler also dissolved the "New" Construction Directorates and effectively lifted building out of the Kommandanten's hands by creating Building Inspections (SS-Bauinspektionen, or BIs) independent of the IKL. Each inspection dispersed financial resources and supplies, maintained machine parks, and carried through steady audits of all projects, large and small. At the next administrative level under the Building Inspections, Kammler erected Central Construction Directorates (Zentralbauleitungen, or ZBLs). Typically, the ZBLs absorbed the existing personnel of the New Construction Directorates and acted as managerial clearinghouses with a flexible corps of trained engineers. Like Russian dolls that contain smaller and smaller concentric units within their shells, the ZBLs parsed out specific projects to Construction Directorates. In turn, Construction Directorates broke down into construction sites (*Baustellen*) for the supervision of individual buildings and, finally, into individual construction works (*Bauwerke*), to oversee discrete technical jobs (e.g., bricklaying, excavation, cement pouring, or carpentry). Previously the engineers at Flossenbürg had worked differently from those at Buchenwald, Sachsenhausen, or Auschwitz. From now on, they executed their tasks in the same way, using the same forms, subject to the same audits. Now any engineer could move from directorate to directorate and find colleagues who were used to working the same way within the same system. Their activities—and thus their experiences within the Office II—became normalized and subject to interchangeability.

The "Great Industrial Tasks" of the SS

To build the New Order, Kammler had to manage slave labor effectively. One of the chief impediments was the Inspectorate of Concentration Camps. Had Kammler been unable to reform the desultory organization of the IKL's Construction Directorates (which he inherited) his Building Inspections would have ended up little different from Wilhelm Burböck's ineffectual Labor Action Office—something that looked like a "modern" organization on charts but which had almost no effect on daily practices. Yet Kammler, unlike

Burböck, quickly made his organizational vision a reality. A short anecdote illustrates the difference in style between Kammler's hierarchical, tightly coordinated modern organization and the disarray of former operations. At the beginning of 1939, the chief of construction at Flossenbürg switched his electrical supplier and received this obsequious petition: "Last night I dreamed that you purchased all your electrical materials from [the rival firm] Weidner. Is this true? Have you really forgotten me, an old acquaintance and party member? I would rejoice if you would consider me once more in your contracts."[61] The former camp directorates had routinely concluded contracts by word of mouth in such personalized interactions, often without formal records. By contrast Kammler's modern methods were swift, blunt, and simple: he merely informed all directorates, "In no case is it valid that written records and approval is completely missing."[62] Whenever money changed hands, no matter how small the amount, the dealings became visible to Berlin on standardized forms. Should anyone fail to submit complete reports, Kammler returned them unread; should contracts be closed irregularly, he canceled them; should anyone make the same mistakes twice, Kammler appeared in person to set things straight. Before 1941 the SS had an accountability problem: sloppy reckoning kept waste and corruption alive but out of sight. Engineers could do business on the basis of their suppliers' "dreams." Under Kammler this would no longer do. He held his men accountable for supplies, moneys, and, above all, the quintessential element of time.[63]

Kammler not only imposed modern organization on his own engineers; he also made sure that they operated apart from concentration camp staff. As shown in previous chapters, the camp SS could wreak havoc upon meaningful work even when formally ordered not to do so by Pohl's managers. It was part of the "polycratic" nature of the SS in and of itself, which—not unlike any large, bureaucratic organization—evolved different professional identities in different, specialized institutional niches. Naturally conflict could erupt whenever functions or "organizational cultures" clashed, not least because heartfelt values were at stake.

As shown, the camp SS often used its control over prisoners to impose the primacy of policing above any concern for production, despite complaints to the contrary. But Kammler brought this to an end. He did so by exerting technical control, something the German Commercial Operations never succeeded in doing. "I ask you to support the administrative organs [of the Office II] entrusted to you," he told Auschwitz's Kommandant Rudolf Höss, "in accordance with the highest possible standards of technical competence." Then he made it clear that Höss was to stay out of the way:

Each and every intervention from a technically unqualified party hinders the planned completion of construction projects and therewith hinders the fulfillment of the Kommandant's own interests and construction demands. The Office II will approve the allocation of building supplies and the leader of SS building sites is accountable to me. The leader of the SS building site can succeed in meeting his objectives only when . . . planned building projects and their corresponding preparations are not interfered with.[64]

By March 1942 Kammler was also writing orders directing the lethargic and ineffectual Richard Glücks to meet the demands of SS engineers. His office requested the allocation of nearly 47,000 prisoners (political prisoners— mostly German—as well as POWs, Jews, and "otherwise incarcerated foreigners, etc.," as Kammler put it).[65] Kammler also requisitioned low-ranking SS men with technical skills (such as craftsmen and skilled laborers) from the Kommandanten's own staffs and reassigned them to Construction Directorates. Before 1941 most concentration camp staff could pull rank on SS civil engineers. Now Kammler's subordinates numbered among the highest officers next to the Kommandanten themselves. There was no question which branch of the SS now held sway over building sites; it was another instance of the ablation of the IKL's domain of authority that had begun years before.[66]

It is hard to overemphasize how much these changing policies and the excitement generated by the New Order reoriented SS slave labor toward the end of 1941. With the foundation of Majdanek, the consolidation of the DWB, and the recruitment of Kammler, Pohl was consciously striving to integrate vertically all aspects of settlement construction from supply to the management of building sites within one institution. Himmler was fully aware that the SS would be constrained by more than just money in these endeavors. One of the scarcest resources anticipated after the war was labor, especially in the construction sector, which is highly labor-intensive. Because the Wehrmacht had captured over 3 million Soviet POWs in 1941, Himmler first hoped to feed Kammler's Building Inspections with this "unlimited" supply of ready slaves as part of the spoils of the eastern campaign. The General Staff of the army agreed by September to send 350,000 into the SS's camps. In consequence, in November, Himmler ordered the expansion of POW camps at Auschwitz and Lublin to hold 150,000 each. By the end of the year, however, most of the expected Soviet prisoners perished due to mistreatment, starvation, and exposure before they had even arrived.

Therefore, on 26 January 1942 Himmler next issued orders to Richard Glücks, Inspector of Concentration Camps:

Now that Russian POWs cannot be expected, in the coming days I will send a large number of Jews and Jewesses into the camps that are to emigrate from Germany. In the next four weeks you must make appropriate arrangements in the concentration camps for 100,000 Jews and up to 50,000 Jewesses. In the next few weeks the concentration camps will be assigned great industrial tasks. SS Major General Pohl will inform you of the details.[67]

These Jews (to come predominantly from Germany) were caught in ongoing plans to secure a captive labor pool to build the New Order. The 150,000 "Jews and Jewesses" represented only a small portion of those rounded up at this time. There were already between 200,000 and 250,000 Jews in the district of Lublin alone (an estimate that is probably low). Himmler was not, therefore, planning to halt the genocide of the vast majority of Jews in the name of an expanded labor action. The major departure in policy lay in what kind of prisoners Himmler now intended to deploy to fulfill the RKF's dreams. While the decision to proceed with the "Final Solution to the Jewish Question" had been made over the past autumn, now Himmler ordered the preservation of some German Jews for the SS Labor Action because he could get no others. One thing is clear. The SS never intended to spare as many Jews for work details as it intended to kill outright, an intent reiterated in the protocol of the Wannsee Conference hosted by top officials of the Reich Security Main Office early in 1942.[68]

Six days after issuing his order to redirect "Jews and Jewesses" into the SS Labor Action, Himmler also released the "Guidelines for the Planning and Design of Cities in the Incorporated Regions of the German East," intended to raise support among outside authorities for the RKF's policies. He had also ordered Kammler and Pohl to organize SS work brigades in order to turn these "guidelines" into an accomplished construction program. Kammler had already produced the first blueprints for SS and Police Strongholds in August, which are still preserved in the City Archive of Lublin. By the turn of the new year, he and Pohl were working furiously on a proposal to go forward with construction by mid-1942. Notably, Kammler wished to deploy roughly the same number of prisoners in his Building Inspections that Himmler had ordered Glücks to collect at Lublin and Auschwitz (150,000 Jews and Jewesses). The HAHB estimated the SS's needs at 160,000 inmates without specifying what variety, and, in fact, Kammler was undiscriminating, suggesting "prisoners of war, Jews, foreign workers, and inmates." He asked for any and all prisoners available. These, then, were the SS's "great tasks," and

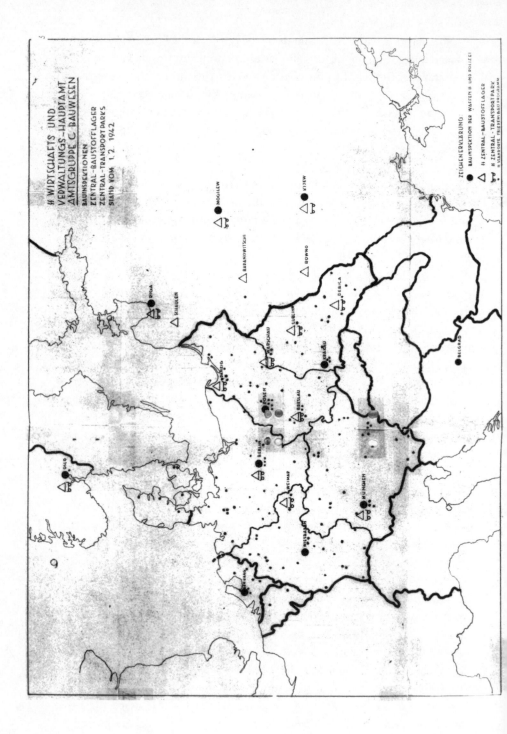

SS WIRTSCHAFTS UND
VEWALTUNGS-HAUPTAMT
AMTSGRUPPE C BAUWESEN

BAUINSPEKTIONEN
ZENTRAL-BAUSTOFFLAGER
ZENTRAL-TRANSPORTPARKS
STAND VOM 1.2. 1942

ZEICHENERKLÄRUNG:
BAUINSPEKTION DER WAFFEN SS UND POLIZEI
SS ZENTRAL-BAUSTOFFLAGER
SS ZENTRAL-TRANSPORTPARK
SS STANDORTE FRIEDENSBAU PROGRAMM

OSLO

RIGA
SCHAULEN

MOGILEW
BARANOWITSCHI
KIJEW
ROWNO

DANZIG
WARSCHAU
LUBLIN
DEBICA

POSEN
BRESLAU
KRAKAU

BELGRAD

BERLIN

WEIMAR
MÜNCHEN
NISSBAUM

PRAG

they involved fantasies of an imaginary peace despite the reality of ongoing total war.[69]

Meanwhile the scale of the RKF's General Plan–East burst all previous boundaries (even as the reality of war meant that the conditions for its realization were crumbling). In January 1942 Kammler submitted a budget of 13 billion Reichsmarks as a tentative future SS construction schedule, but Himmler found this figure much too low. Nothing better demonstrates the SS's spirit of unbridled optimism. The reality of the eastern campaign should have long taught them otherwise, but Himmler asked Pohl and Kammler to submit an even larger budget.[70] Taking Himmler's advice to think big, Kammler now projected a building schedule over double the initial volume. Still Himmler's handwritten notes chided the engineer's needless parsimony: he crossed out Kammler's estimate of 20 to 30 billion Reichsmarks and penciled in the margin that 80 to 120 billion would be more appropriate, over a trillion Deutschmarks in today's currency.[71] By the second week of February, Kammler and Pohl had finished a proposal for a vertically integrated construction organization. As a supply base, the WVHA Office Group W planned to produce and deliver its own raw materials, theoretically making SS construction independent of capitalist industry and state ministries alike. Kammler was to take charge of all SS civil engineering, and he laid out his organization along Europe's rail network so that construction machinery, workers, and raw materials could be transferred flexibly and speedily throughout the continent. Kammler's final report, which he and Pohl had reworked continually since the end of 1941, reached Himmler on 5 March 1942. Ten days later, Himmler announced in the General Government that the RKF would increase the number of SS and Police Strongholds from six to eighteen and expected to complete this work within the year.[72]

Kammler planned new mobile SS Building Brigades of more than 2,000 prisoners each as the core of his plan of action. He divided the brigades by technical tasks and further broke them down individually into three battalions, one for *Tiefbau* (construction below ground level), one for *Hochbau* (construction above ground level), and one for *Ausbau* (interior construction). Each battalion had four companies of 200 men and a complement of 10 security guards. In a sharp departure from standard IKL practice, Kammler's engineers had direct authority over the organization of work, not the camp

---

Map prepared by Hans Kammler showing the network of SS Building Inspections along the rail networks of Germany and eastern Europe. The dark circles are "Building Inspections." The triangles are "Central Building Supply Depots." The wheeled figures are "Central Transport Parks." From BAK NS19/2065.

SS. To this end, Kammler requested the call-up of all SS reserves with technical training and likewise the transfer to his command of all building engineers from other SS branches.

Kammler and Pohl had actually raised Himmler's ire with their report, for Kammler had criticized the German Commercial Operations. Perhaps the competent engineer sensed the holding company's dissolute management. Even if the DWB ran at full capacity, Kammler pointed out, the building schedule of the General Plan–East would far outrun the current output of the VuWHA. His report therefore urged Himmler to allow the SS Building Brigades to contract with private industry. He appealed to the general desire of all involved to "make use of appropriate technical achievements and rationalization measures" and "bring technological progress into action in its broadest dimensions," whether it originated within or outside the SS companies.[73] SS enterprises should indeed expand to eventually meet all the SS's demands, he allowed gingerly, but he and Pohl agreed that they should do so slowly, incrementally, and avoid the mistakes of rash growth. This still irritated Himmler, who marked "No, No, No" in the margins. Himmler had informed Pohl that he expected SS companies to produce 80 percent of all raw material needed by the Building Brigades and pedantically advised his industrial czar to abide by the "strictest principles of Prussian thrift" (after chiding that Pohl's budgets were too low). He wanted immediate expansion of the German Commercial Operations to meet all the SS's needs. "If we do not," he warned, "we will never get our barracks, SS schools, administrative buildings, nor will we get houses for our SS men in the Reich, nor will I, as Reichskommissar for the Reinforcement of Germandom, be able to erect the homes that we will need in order to make the East German."[74] If the SS possessed its own building brigades, its own raw materials plants, and its own slaves, total vertical integration would make it impossible for anything to stand in the path of the SS's visionary program.

With the changes culminating at the end of 1941, the labors that Pohl and his staff had undertaken since the fiascoes of 1938–39 seemed to be coming to fruition. After the formation of the DWB, and with the Office II–Construction under Hans Kammler making impressive strides to control SS building, in mid-January 1942 Pohl gathered both the Main Office for Budgets and Building and the SS companies within the Administration and Business Main Office into a new, centralized Business Administration Main Office (Wirtschaftsverwaltungshauptamt, or WVHA). This organizational change was to take effect on 1 February. Pohl had formerly sat at the head of three separate administrative offices: the original "V" (Administration) department within Himmler's Personal Staff, the VuWHA, and the HAHB. In reality, the boundaries had always overlapped. Now Pohl presented an organizational chart to Himmler that consolidated all branches under one umbrella, a significant simplification.[75]

Pohl split the former Office I into two office groups. Originally Office I had handled SS membership funds, budgets, payroll, and supply, including quartermaster and paymaster duties for the Waffen SS. Pohl placed budgets and personnel in Office Group A–Troop Administration, with August Frank at its head. Pohl established a separate Office Group B–Troop Supply, led by Georg Lörner. The importance of each of these two offices within the WVHA was declining in direct proportion to the rise of the Waffen SS's combat deployment. Pohl's Office Group A–Personnel and Payroll retained nominal authority to audit all SS budgets; likewise, Office Group B–Supply retained nominal oversight of all regional Waffen SS "Troop Supply Depots" (Truppenwirtschaftslager). But in each case, these duties—which had been the key motor in Pohl's initial rise within the SS—atrophied. It was long past the time when Pohl had first organized SS military administration. The Waffen SS had now grown out of his hands, in part because Pohl himself had directed his attention elsewhere—to the manufacture and engineering of the General Plan–East. Himmler had founded additional head offices independent of Pohl to coordinate SS personnel and its military units (these were, respectively, the Personnel Main Office and the Leadership Main Office [Führungshauptamt] that served as a general staff of the Waffen SS).

The Office Group B likewise suffered from the steady cannibalization of its top officers, not to mention midlevel personnel, most of whom went into combat service, where their skills were badly needed. In the field, however, they fell under the command of their Waffen SS battalions. By the end of 1942, Pohl even declared the Troop Supply Depots to be independent service institutions. The Office Groups A and B within the WVHA became skeleton crews whose affairs hardly reached beyond the personnel of the WVHA itself.[76] Even here activities dwindled as war made financial audit less and less important, for raw materials vouchers and quotas were replacing currency in dealings within Reich institutions. Even when other WVHA offices needed personnel, supplies, or funds, they often skipped over Lörner and Frank. Hans Kammler, for instance, went straight to industrial planners like Albert Speer.

The center of gravity within the WVHA now unquestionably shifted toward industry and civil engineering, that is, toward large-scale, technological enterprises. The key changes culminating in the WVHA had already begun in June of the previous year. Kammler merely became chief of the newly christened Office Group C–Construction, comprising the former Office II. Likewise, the entire VuWHA became the Office Group W–Business. Pohl retained

Office Group A
August Frank

Office A1 Budgetary Office
A1/1 Budgets of the Waffen SS
A1/2 Budgets of the General SS
A1/3 Saviage Society
Office A2 Payments and Accounts
A2/1 Paymasters
A2/2 Comptroller
A2/3 Fees and Debits
Office A3 Legal Affairs
A3/1 General Legal, Tax, and
    Contractual Affairs
A3/2 Property Rights
Office A4 Audits
A4/1 Audits of Accounts and
    Funds
A4/2 Audits of Quartermasters
    and Troop Accounts
Office A5 Personnel
A5/1 Replacement, Release,
    Records
A5/2 Promotions, Assignments,
    Transfers
A5/3 Training and Education

Office Group B
Georg Lörner

Office B1 Food Supply
B1/1 Planning and Acquisitions
    for Men and Horses
B1/2 Troop Supply Depots
B1/3 Cook Training, Alimentation
    and Nutrition Experimentation
Office B2 Clothing
B2/1 Clothing and Equipment
B2/2 SS Clothing Works
B2/3 SS Clothing Accounts
Office B3 Shelter
B3/1 Planning and Acquisition of
    Equipment and Appliances
B3/2 Shelter and Stores
B3/3 Automobile Park
Office B4 Raw Materials
B4/1 Raw Materials
    (Textiles and Leather)
B4/2 Clothing Purchases
B4/3 Price/Cost Accounting
B4/4 Contracts, Imports

Office Group C
Hans Kammler

Office C1 General Construction Tasks
C1/1 Buildings of the Waffen SS
C1/2 Buildings of the IKL
C1/3 Buildings of the German Police
C1/4 Buildings of the General SS
Office C2 Special Tasks
C2/1 Quartermaster Buildings
C2/2 Communication Buildings
    and Ammunition Depots
C2/3 Hospitals
C2/4 National-Political
    Educational Chapters
C2/5 Dwellings and Housing
C2/6 Industrial and Special
    Buildings
Office C3 Technical Office
C3/1 Engineering
C3/2 Hydraulics and Sanitation
C3/3 Machine Building
C3/4 Surveying
Office C4 Central Building Inspection
C4/1 Overnight of SS Building
    Authorities
C4/2 Budgetary Affairs and Audits
C4/3 Raw Materials and Stores
C4/4 Automobile Park of the
    Office Group C
**Office C5 Building Maintenance and
Economics**
C5/1 Property of the Waffen SS
C5/2 Property of the General SS
C5/3 Pre-Audit of Construction

Original organizational chart signed and approved by Himmler (submitted by Pohl) on January 15, 1942.

Office Group D
Richard Glücks

Office D1 Central Office
- D1/1 Prisoner Affairs
- D1/2 Communication, Security, Watchdogs
- D1/3 Automobile Park
- D1/4 Weapons and Equipment
- D1/5 Training of the Troops

Office D2 Labor Action of the Prisoners
- D2/1 Prisoner Labor Action
- D2/2 Prisoner Training
- D2/3 Statistics and Calculations

Office D3 Sanitation and Camp Hygiene
- D3/1 Medical and Dental Care of the SS (Totenkopfverbände, etc.)
- D3/2 Medical and Dental Care of Prisoners
- D3/3 Hygiene and Sanitary Measures in the Concentration Camps

Office D4 Administration
- D4/1 Budgets, Accounts, Payments
- D4/2 Supply
- D4/3 Clothing
- D4/4 Shelter
- D4/5 Legal, Tax, and Contractual Affairs

Signed and approved by Himmler on 3/3/1942.

Office Group W
Oswald Pohl

The following corporations were directly under the Office Group W Staff:
German Healthfoods GmbH, East Industries GmbH, Cooperative Dwellings and Homesteads GmbH, House and Realty GmbH

Office W1 Stone and Earth Industries (Reich)
- W1/1 German Earth and Stone Works GmbH
- W1/2 Allach Porcelain Manufacture GmbH
- W1/3 "Bohemia" Ceramic Works AG
- W1/4 Porag Porcelain Radiators GmbH

Office W2 Stone and Earth Industries (East)
- W2/1 East German Building Supply Works GmbH
- W2/2 Klinker-Cement GmbH
- W2/3 Golleschauer Portland Cement Factory GmbH
- W2/4 Unified Prague Building Materials Factories

Office W3 Food Processing Industries
- W3/1 Sudeten Springs GmbH
- W3/2 Heinrich Mattoni AG
- W3/3 Apollinaris Operations GmbH
- W3/4 Freudenthaler Beverages GmbH
- W3/5 Lesnoplod Orava Sojka a Spol OHG
- W3/6 German Foodstuffs GmbH
- W3/7 Smoked Foods and Preservatives Factory AG

Office W4 Wood Working Industries
- W4/1 German Equipment Works GmbH
- W4/2 German Masterworkshops GmbH
- W4/3 German Noble Furniture AG
- W4/4 German Home Design GmbH
- W4/5 German Plywood and Typefounding Works GmbH
- W4/6 Forestry and Sawmills GmbH

Office W5 Agriculture, Forestry, and Fisheries
- W5/1 Experimentation Station for Alimentation GmbH

Office W6 Textiles
- W6/1 Textile and Leather Utilization GmbH

Office W7 Publishing
- W7/1 Nordland Press GmbH
- W7/2 Volkish Art Press GmbH
- W7/3 SS Form Press GmbH
- W7/4 Lumbeck Corporation for the German Book Industry GmbH
- W7/5 Research Institute for the German Book Industry

Office W8 Special Tasks (Charitable Organizations)
- W8/1 Society for the Promotion and Care of German Monuments e.V.
- W8/2 Dolmens Foundation
- W8/3 King Heinrich Memorial Foundation
- W8/4 Resort for Natural Healing and Lifestyle e.V.

Consolidation of the WVHA on 15 January 1942 and the subsequent addition of the IKL as Office Group D by 3 March 1942. Source: BDK. Drawing by Steve Hsu.

its leadership in his own hands, an indication that this had become his darling. Contrary to much scholarship that postulates a mad dash at this time to control armaments industries on behalf of the SS, there can be little question that the WVHA carried forward the consistent motivating drive immanent since 1939–40. Pohl had oriented the work of the HAHB Office II and the VuWHA Office III–W toward the realization of the "peace building program" of the New Order and had no intention of diverting those aims now. Richard Breitman's excellent biography names Himmler the "Architect of Genocide," and that appellation should be taken quite literally. In Kammler's case, this was not even a metaphor. Engineering the genocide and constructing the New Order were one and the same thing within the WVHA of February 1942.[77]

Engineering Ideology

Kammler's branching Building Inspections with their nested Construction Directorates and mobile Building Brigades extended over the whole of central Europe, and the nature of their technological work on this scale necessarily demanded that Kammler place trust in technical men at building sites beyond his direct, personal control. Considering Kammler's insistence on rigid structure, it may seem that SS engineers were becoming mere cogs in the wheels of a giant bureaucracy. But Kammler underscored the fact that bureaucratic duties are double-edged: they demand subordination to large, impersonal organizations but also bestow the power to exploit the capacity of collective work. There can be no doubt that Kammler demanded subordination: "The Building Inspectors [the chiefs of the Building Inspections] are not to develop or to design but to audit, supervise, and compare." Nevertheless, he also stressed that within each engineer's domain, "he has to decide for himself and represents the Office II in all construction affairs with local Waffen SS and police authorities as well as the army and civilian authorities."[78] The memorandum of one Central Construction Directorate officer clearly shows his awareness of control and initiative: "In a recent meeting the chief of the office [Kammler] . . . has ordered that the head of each service division take responsibility in the future for all work in his division, likewise departments, and will be held responsible."[79] Far from being cogs in a machine, the SS civil engineers were the big fish in their ponds of regional authority, and they were conscious of their command.

In the SS companies, conflicts over the multiplicity of meanings within Nazi ideology had caused turmoil and failure, even when Pohl had recruited competent managers. The SS civil engineering corps was a marked contrast.

Kammler knew that his appeals to responsibility and accountability would ring hollow—as Wilhelm Burböck's had in the IKL, for instance—without the amalgamation of initiative *and* competence. In the absence of energetic participation from dedicated subordinates, his administrative hierarchy would remain a creature of SS filing cabinets. Kammler therefore sought to build a modern hierarchy of like-minded men in which consensus might act as a further stimulant to initiative and the fungibility of experience. All SS personnel sheets included a category for the estimation of "ideology" (*Weltanschauung*). Typically, this slot was filled with perfunctory statements like "longtime National Socialist" whose importance should not be exaggerated. Nevertheless, Kammler took the evaluation of dedication seriously. He even barred some from advancement on ideological grounds. Regarding one officer Kammler wrote, "Not fit for office chief." One of the reasons given was the lack of "ideological equipment."[80] In other cases, Kammler emphasized the dedication of those he wished to promote, often praising technical and ideological capacity in the same sentences. Kammler once wrote of an officer, "[He] is especially eager for action, has an outstanding comprehension of how to exploit developing possibilities, and possesses the old Staffel spirit."[81] By "Staffel spirit" Kammler meant the Schutzstaffel. He was referring to the collective identity which, as the chief of Office Group C, he wished to create and uphold. In this environment, the organization was the ideology and vice versa.

Did Kammler's corps heed that "spirit"? Careful study of collective biographies suggests his officers' identification with their work. (I have found data on thirty-nine, at least two-thirds of the elite.) Conscious ideological engagement can be proved for nearly half, eighteen in all. Engagement denotes those who made conscious statements or decisions of ideological sympathy. This number may plausibly be extended to twenty-six if one counts an additional eight officers who joined the Nazi Party before January 1933; for early membership implies identification with National Socialism before membership was fashionable. Naturally, this qualitative evidence must be carefully weighted. A large difference may be drawn between officers who promoted National Socialism and those who only sympathized or participated. Some SS construction officers gave mild but nonetheless unmistakable signs of sympathy but cannot, by any stretch of the imagination, count as overt fanatics; on the other hand, a full quarter were proven activists. Several officers participated in agrarian settlement clubs, others in overtly National Socialist organizations; others sat in jail for their political beliefs; and recall Gerhard Weigel, who suffered grave injuries in street fights with the Nazis' political enemies. They did more than just participate in the Nazi movement. Either by building organizations or by promulgating ideology, they put their time, creativity, and knowledge at the disposal of National Socialism and consciously sought to propel the movement forward. Because they are such obvious cases, like the activism of Kammler himself, let us start with weaker cases.

One engineer, whose family was forced out of Alsace, for instance, volunteered the information that "because of the French occupation authorities, my parents had to desert this German territory."[82] Perhaps he was not as fervid as men like Kammler or Weigel, but he clearly announced his resentment of the First World War and Germany's subsequent disgrace and territorial losses (dictated by the Treaty of Versailles). It is hardly likely that this SS officer had anything against Nazi rhetoric that sought to address his grievances. Nevertheless, for such men, evidence yields a differentiated conglomerate of issues in which, once again, National Socialist ideology displays its multiple currents.

Take for example the thirty-two-year-old former Organization Todt engineer, Heinrich Courte. Courte was like many new officers of the total war era: Kammler did not recruit him directly to the Office Group C–Construction; rather, the Office Group C snapped him up out of the mass of men who enrolled in the Waffen SS in 1942. Kammler bestowed the title of "expert leader" upon him, and he distinguished himself in the deployment of slave labor in the Building Brigades.[83] Courte's personnel file yields various indicators that he found himself at home among ideologically driven men like Kammler; on the other hand, *absolutely no evidence* suggests that he was alienated. At the Polytechnic University of Aachen, during what was probably the beginning of his political consciousness, he joined the Student Platoon of the SA and then switched after a few months to the SS. Here he took time out from studies to participate in special training for military construction. Second, after receiving his diploma in engineering, Courte went to work for the Organization Todt. Part of the "social structure" of the German engineering profession included a traditional association of the full mobilization of the nation's technical capacity with social progress (a sentiment by no means unique to engineers in Germany). It is therefore probably safe to assume that Courte believed Nazi construction projects such as those under the Organization Todt and the full employment of civil engineers they helped create were positive contributions to German society. Third and last, his curriculum vitae contains another hint of ideological sympathy. He took pains to explain his father's suicide as the result of the "general disintegration" of the Weimar Republic.[84]

Taken together Courte's biography remains ambiguous evidence. As a student, he may have succumbed to peer pressure and joined the SA. Ulrich

Hartung, Thomas Zeller, and Franz Seidler have pointed out that Organization Todt engineers discussed the very form of bridges, the vistas offered by parkways, the architecture of rest stops, and plantings on medians as symbols of a Nazi renaissance. Thus the "structure" of these organizations almost inevitably included exposure to such ideals, literally in the blueprints; nevertheless, even if Courte would have had to be extremely obtuse to miss the meanings attributed to his work in the Nazi media or among colleagues, the historian must still admit to such possibilities: perhaps Courte was oblivious. In addition, even his condemnation of the Weimar Republic may have been disingenuous, for SS racial theories linked the German "will" and racial supremacy to bloodlines, and suicide in the family could be viewed as a sign of poor genetic or spiritual mettle. Courte had an interest in explaining away his father's death in a way that would appeal to his superiors. Further, even if Courte held a wholehearted, genuine disgust for the Weimar democracy—a disgust widespread among technical men in the 1920s and 1930s—this would prove only a certain affinity for Nazism but not necessarily active belief.[85] In light of available evidence, one can only wager the conclusion that men like Courte interpreted aspects of their lives in ideological terms, and, judging from their service, they also willingly participated in Kammler's organization.

For twelve officers there is no hard evidence of ideological engagement. They may have been dedicated participants (although not activists) or they may have been mere fellow travelers. Nevertheless, the success of the Office Group C depended equally on cooperation and the *absence* of any counteractive dissent. It is therefore also significant that, in contrast to, say, the dissent of those within the SS companies, *no evidence* reveals any similar friction within the organizational structures initiated by Kammler, nor did anyone seek to undermine them. Because, as Kammler well knew, formal bureaucratic directives alone could provide only slender ties within his institution, at the very least it is safe to say that the vast majority of officers in the Office Group C had nothing against radical SS ideology.

For instance, as is hardly atypical in any large, impersonal bureaucracy, Kammler's subordinates broke and bent the rules. But they consistently did so to reinforce their organization. The Central Construction Director of Auschwitz, Karl Bischoff, defied the orders of his regional Building Inspector in order to deal directly with the Ministry of Armaments and War Production. He did so, however, to speed up the construction of Auschwitz and secure key raw materials.[86] It is important to remember that he could have chosen to act otherwise. Had he merely "followed orders" as his institutional niche formally prescribed, Bischoff could have easily chosen to let schedules fall by the wayside and might have even pointed out his fidelity to formal instructions by way of deflecting blame from his own office. In other bureaucratic niches of the WVHA occupied by individuals who did not readily identify with the SS, this is exactly what happened. For example, the chief engineer at an SS-owned factory near Auschwitz, a civilian and not an SS man, refused to deliver shipments to Bischoff's construction corps and based this obstruction on petty state and municipal regulations that prohibited such transactions— regulations over which the SS routinely ran roughshod.[87] By contrast, Bischoff ignored a steady stream of complaints from the municipality of Auschwitz in order to push ahead with the construction of the concentration camp.[88]

Bischoff himself belongs among the "participants" like Courte, not the "activists." While he had worked for the German air force, he had also served the German Workers Front as a Cell Leader. Ronald Smelser has noted the general ideological commitment of those who usually served as Cell Leaders.[89] Thus the structure of this organization would lead us to suspect Bischoff's commitment, but nevertheless no evidence speaks of his activism. At Auschwitz, on the other hand, he clearly refused to go out of his way to support the cause of Nazi fundamentalism. When queried by the Regional Party Leader (*Ortsgruppenleiter*) of the NSDAP and the Kommandant staff of Auschwitz whether his subcontractors were "free from foreign and Jewish capital" and if "the owners of the firms and their wives are of Aryan descent and Reich German," Bischoff blandly replied that he did not know and the Regional Party Leader should seek such information elsewhere.[90] The engineer was not exercised by such issues, yet when this same engineer requested the Gestapo to arrest some civilian workers for shirking, he added that they were "for the most part lazy Poles" (the German is somewhat stronger: *polnische Bummelanten*; an American construction site manager might use "jerk-offs"). Likewise others in Bischoff's command referred to foreign workers as worthless loafers who shirked "in order to live in their homeland of donothingness or alms and handouts."[91] Of note, the letters requesting Gestapo intervention to punish workers were very standardized. Not only the SS but also its civilian subcontractors routinely issued them. The most usual practice was merely to copy the same paragraph over and over while filling in new names. Those petitions to the Gestapo quoted earlier differed, however; Bischoff and his subordinates added unbidden slurs to existing stock phrases, a sort of racist flourish upon otherwise fungible information.

This unusual deployment of racial stereotypes cuts to the crux of issues surrounding human agency in bureaucratic structure. Earlier, Bischoff had lifted not a finger to "prove" the Aryan character of his subcontractors. Regarding workers, on the other hand, he and his subordinates were ready and willing to add extra slurs to routine requests to the Gestapo. Why? One answer has recently been advanced by Wolfgang Seibel, who stresses the ways in which bureaucratic context structures agency. Even when those structures represent established pathways of action, they are inert without the activation energy provided by initiative, motivation, and interest. Thus, at one level, even when ideological consensus exists that "something must be done" but institution-alized pathways are absent, little is likely to happen. At another level, however, when institutions and intentions reinforce each other—even competing in-stitutions or otherwise contradictory intentions—the consequences are usu-ally swift and decisive.[92] In other words, ideology mattered more, not less, because of the nature of National Socialist organizations.

When the Regional Party Leader appealed to Bischoff to uphold the racial hierarchy of National Socialism among his contractors, the engineer found himself called upon to take initiative that lay far beyond the bounds of his normal activity. In the Central Construction Directorate he already had plenty to do and no time to chase down what must have seemed extraneous gossip. Prying into the private lives of other engineers and businessmen was police work, work for which he and his staff had neither training nor re-sources. But this hardly meant that racial identity did not matter to him as soon as "business" made it relevant, and there seems to have been a consensus in his office that poor performance among prisoners or civilian workers was due to their nature as Poles, Jews, Czechs, or whatever "racial" category could be made to fit. Bischoff did not hesitate to mobilize bureaucratic organization by appealing to this consensus, and his office seems to have added racial slurs in its arrest requests to get prompter action or harder punishment; at the very least, the added sentences would have differentiated these requests in the eyes of local SS officials from otherwise standardized formulations.

Some may ask whether Bischoff really believed in such stereotypes or whether he merely used them cynically to "fit in." The point seems moot to me. He worked within a radicalized institution (the SS) within an already radical political movement (National Socialism) that associated efficiency with racial supremacy and inefficiency with "ballast existences" and "unneces-sary eaters." If such appeals could and did incite the Gestapo to a prompter response, this could only have served as "evidence" to confirm belief in racial tautologies. There is no necessary dividing line between pragmatism and fanaticism in such a context. These officers had clearly internalized Nazi racial supremacy to some degree (and likely to a much greater degree—they were, after all, engineering genocide). Bischoff was not a man to demonstrate in the streets on behalf of Nazi ideals; on the other hand, everything suggests that he had nothing against the most radical Nazi policies, and some evidence sug-gests he identified with them passively.

Perhaps nothing was more powerful than this passive acceptance and the lack of any contravening dissent in Kammler's organization, dissent that officers could have otherwise found ample opportunity to express in petty administrative obstruction. The actions of officers who did not express their ideological engagement still spoke volumes about their dynamic initiative in an organization that could construct gas chambers and manage slavery only through collective action. Ideology facilitated operations precisely because the maintenance of consensus never needed to be a heated topic of daily declarations and contention. It had become a matter of their collective identity as engineers of the New Order.

CHAPTER 5

MY NEWLY ERECTED HOUSE

SLAVERY IN THE MODERN

WAR ECONOMY

By early 1942 Pohl had centralized all operations under his command within the Business Administration Main Office in order to put the concentration camps to work both building and manufacturing the New Order. As total war quickly took precedence, however, the conflicts and inconsistencies within SS industrial management came a cropper. Only Hans Kammler and his SS Building Inspections were capable of providing essential services to the war economy by forging a mutual sense of purpose with competent industrial managers *and* by providing the knowledge and skills to bend the complex world of production to the Third Reich's needs. The SS's own slave-labor empire initially failed to render any similar service, even when industrialists and the Armaments Ministry alike solicited Pohl's cooperation.

Links to non-SS industrial concerns by no means formed quickly. Most industrialists were not particularly eager to use concentration camp prisoners. At the beginning of the Soviet campaign in the summer of 1941, many expected a victorious peace to arrive quickly enough to forestall economic stringency. The Economic and Armaments Office of the military giddily promised a huge booty from the conquered territories: a surplus of 7 million tons of grain and equally giant quantities of industrial raw materials. The fall and winter of 1941 should have brought caution and sobriety as the German army spread itself thin across the vast Soviet front. As early as January 1941, the Reich Minister of Armaments and Munitions had called for a reorganization of production, and by summer all authorities responsible for economic planning began to acknowledge shortfalls. In November, Reich Marshall Göring, head of the German air force, had tried to reorganize aircraft production and failed. At the same time the Red Army began its first counteroffensives. By December, the first T 34 tank appeared on the eastern front. When the new year dawned, Adolf Hitler released "Armaments 1942," a decree that officially announced the end of the Blitzkrieg and acknowledged the staggering attrition of the stalled eastern offensives. Only after this point did the straitened war economy slowly pull the SS and war industries closer and closer into a troubled partnership.

The crux of the war economy was not a material but a human resource: labor. Here too Germany had expected quick relief from "lightning victory" in the form of front soldiers returning to the factories. By the end of 1941, however, industry demanded 1.1 million additional workers just as the army announced new call-ups.[1] An increasingly desperate search for any and all sources of labor ensued, and here the SS held desperately needed reserves. The concentration camps became the object of a last grasp for some "inexhaustible" supply of workers. This was far from the first such desperate grasp for a "quick fix," however. In 1941 the German economy had turned to the millions of Soviet prisoners of war taken on the eastern front. Germany had captured 3,350,000 Soviet soldiers, what seemed like a limitless number (Himmler too had hoped to cull over 300,000 to build the SS and Police Strongholds), but almost none arrived at German factories. By the end of 1941, 1.4 million had already perished. The German army treated the Soviets so poorly that by March 1942 a mere 166,881 reached the Reich in any kind of condition to work.[2]

Fritz Todt, Hitler's Minister of Armaments and Munitions, quickly rose as the dominant industrial planning authority during this time. Unlike many of Hitler's other advisers, Todt acknowledged the labor shortage in its full technical and organizational dimensions and announced that only a thoroughgoing rationalization of production could increase yield. As a result, German industry began to achieve new economies of scale and scope.[3] Todt also secured Hitler's approval to conscript workers from the conquered territories, a policy that evolved into the Reich Labor Action in March 1942 under another engineer, Fritz Sauckel, the Gauleiter of Thüringen. The civilian populations of German-occupied eastern Europe seemed yet once more to offer limitless workers, and Sauckel began to conscript them forcibly. (This vast labor operation should be sharply differentiated from the SS Labor Action that had evolved since 1938.) Still the demand for workers—never the supply—proved inexhaustible. The German war economy turned to the SS's starving and beaten inmates only once all other resources had been depleted.

Total war after 1942 therefore marked the concentration camps' final phase of development. Before 1936, unemployment had prompted widespread opposition to prison labor; thereafter full employment removed these barriers, but the scope of SS industrial ventures remained a minute proportion of the German economy; only now did shortages lead to desperate calls from all sides to expand slave labor. Nevertheless, the link between the camps and the war economy grew slowly for one additional reason. The German Commercial Operations were occupied with totally different aims than armaments production. Many have falsely attributed to the SS a desire to expand into weapons production in order to secure a self-supply for the Waffen SS, but the SS was loath to abandon its fantasies of manufacturing a New Order. True enough, the SS already planned to enlarge its Labor Action in 1941, but Pohl did so in isolation and sometimes despite the war economy. The Waffen SS for the most part remained content with the traditional military-industrial relationship. Its generals came as customers, sometimes as consultants, but not as entrepreneurs. By and large, they also got what they wanted and had no need of "self-supply."

Industry and Ideology

As a rule, a peculiar breed of company first entered into slave-labor contracts with the SS. The pioneers were typically other state-owned or at least state-controlled corporations. During 1941 concentration camp involvement in non-SS factories was limited to the construction of a light-metals foundry at Fallersleben for Volkswagen (owned by the German Workers Front of Robert Ley), an aircraft engine works of Steyr-Daimler-Puch at Graz (a subsidiary of the Hermann Göring Works), and IG Farben's synthetic rubber plant at Auschwitz. The SS also assigned a small contingent of prisoners to the Heinkel Aircraft Works in Oranienburg near Sachsenhausen. Heinkel had been a private corporation but ended up under state ownership in the 1930s. Its plant at Oranienburg was brand-new, opening in 1937. Motives varied, but the overrepresentation of state corporations (a small, though important, minority of Germany companies) is perhaps no accident, for their management was much more likely to share some or all of the SS's fundamentalist goals— productivist anticapitalism, racial supremacy, modernization. Rainer Fröbe suggests that a generational turnover had taken place by 1942–43. New managers were now entering positions of power; some had known no other professional world than that created by Hitler's Germany. They had usually entered as young men into the top ranks of management after formal education in Germany's politicized universities or technical schools and had advanced in the forced pace of Hitler's war economy. Many owed their careers to tightly coupled networks of connections in which economics and politics had completely melted into one. The state conglomerate of the Hermann Göring Works, for example, was itself the creature of Nazi policies of autarky, and here the "coming generation" of Nazi managers was even more pronounced.[4]

The best-known case of slave labor in the German war economy involved the IG Farben synthetic rubber factory at Auschwitz, and this was owned and operated by a private corporation. But here the exception proves the rule. IG Farben's synthetic fuel and rubber development relied almost entirely on armaments contracts, and the plant at Auschwitz was brand-new just like the Heinkel Works of Oranienburg. IG Farben had agreed to invest in it partly to keep the state from founding a competing chemical firm of its own, something Nazi economic planners had indeed done in the case of the iron and coal industry, where private corporations demurred in the face of National Socialist autarky policies. The state had founded the Hermann Göring Works to accomplish its purposes. Thus, even in the case of IG Farben, the firms involved in the first attempts to use concentration camp prisoners outside the SS's DWB relied heavily or even depended wholly on state investments. They were the creatures of National Socialist economic policy.[5]

These projects also followed a distinct pattern: they were initiated by individuals outside both the SS *and* the regular channels of the Armaments Ministry. The SS reacted eagerly to the petitions of these outsiders. In so doing, it quickly set a precedent for general incompetence. IG Farben executives first contacted Himmler when the Reichskommissar for the Reinforcement of Germandom promised to settle "ethnic Germans" near the plant, and IG Farben hoped to use them as its work force. After this proved impossible, IG Farben tried to secure local Polish workers, then Soviet prisoners of war. Only when these attempts failed did the company's executives turn to Auschwitz. By mid-February they asked for prisoners, were promised 10,000, but received between 2,000 and 3,000.[6] This proved typical of the SS's efforts to oblige German industry: when approached Himmler would promise the moon and the stars. Pohl and the IKL could seldom deliver and usually disappointed all involved.

Ferdinand Porsche's Volkswagen followed a similarly twisted path, though his executives showed somewhat greater enthusiasm. Early in 1941, VW's management had toyed with the idea of putting to work Polish Jews from the annexed region of Warthegau (northwestern Poland). The Reich Labor Ministry, not the IKL, led these negotiations, yet the plans fell through when Hitler forbid any Polish Jews to enter the Reich. On 7 April the VW let the project drop, yet Porsche personally took up contacts to Himmler once again at the end of 1941. Of note, the dealings concerned a light-metals foundry not even directly intended for armaments production but for aluminum, air-cooled engines needed for Porsche's "peoples car," the famous VW Beetle. Porsche wanted the foundry in order to come closer to the goal of total vertical integration on the model of Henry Ford's River Rouge. On 11 January, Himmler and the automobile designer met with Hitler, who signed an order clearing the way for the foundry. Shortly thereafter, to secure the cooperation of the SS, Porsche offered the direct delivery of 4,000 amphibious transport vehicles (*Kübelwagen*) to the Waffen SS. In the previous month, he had already sent the Waffen SS chief of staff, Major General Hans Jüttner, a few prototypes. These negotiations have prompted some to insist that the SS "thus took a significant step toward the expansion of its industrial projects that had for so long been constrained to the marginal operations of SS business undertakings [i.e., the DWB]."[7] But such assumptions reduce these dealings to a pure contest for expansion and influence, evacuate the SS (or Porsche) of any ideological motivation, and ignore the fact that the DWB was of every importance to the SS as manufactory of its cultural fantasies. Himmler was clearly as enthusiastic as Porsche about the "people's car." The offer of high-tech gadgetry must have made the SS even more so. The fact that many of Germany's established private automakers opposed the Volkswagen concept likely also appealed to the Reichsführer's sense of anticapitalism. No doubt he hoped that the joint project in the name of a motorized Germany would bring the SS prestige. He quickly delegated it to his industrial expert, Oswald Pohl, who began a round of meetings with Porsche to hammer out details. Like the precedent of IG Farben, however, high-level industrial magnates had taken the initiative, not the SS.[8]

The metalworking firm Steyr-Daimler-Puch proceeded similarly. Its director, the colorful Georg Meindl, had established himself as a veritable wheeler-dealer among the "cream" of Nazi leadership in Austria. He incorporated in his person the kind of "polycratic" contacts that the Third Reich's fluid and changing institutions made possible. He had met Hermann Göring in the 1920s, and Göring had appointed him director of Steyr-Daimler-Puch shortly after the annexation of Austria. Meindl was himself a member of the SS and the Nazi Party. He had made the personal acquaintance of the chief of the SS Leadership Main Office, Jüttner (who acted as chief of staff for the Waffen SS), and the Higher SS and Police Führer of the Donau District, Ernst Kaltenbrunner. He had also made the acquaintance of Franz Ziereis, the Kommandant of Mauthausen, though how closely is unclear. Meindl made a point of getting to know the Nazi Party Gauleiter of the Gau Oberdonau August Eigruber and Siegfried Uiberreither of the Gau Steiermark. Wilhelm Burböck, a much smaller fish in this highly politicized and powerful web of Nazi businessmen and politicians, had come from the same milieu. Unlike Burböck, however, Meindl was a top manager with formal university training in national economy and political science. Beginning in early 1941, after labor shortages had become a problem at his airplane engine factory at Steyr (problems that generally grew more dire more quickly throughout Austria than in the so-called Old Reich), Meindl employed up to 300 prisoners as construction workers. He later contacted Kaltenbrunner about the possibility of using prisoners in production. Kaltenbrunner in turn intervened with Himmler directly as well as Ziereis. A satellite camp of Mauthausen at Steyr-Daimler-Puch resulted from these negotiations in the spring of 1942. This was a new precedent. Up till then, the IKL had insisted that workers walk to work sites—regardless of how far from the camps—and return each day. Meindl apparently used his personal contacts to change Mauthausen's practices and ease its paranoia for security.[9]

The only other case of note, involving the Heinkel Aircraft Works near Oranienburg, was perhaps even more portentous. Typical of the majority of the SS's future labor allocations, including Steyr-Daimler-Puch's engine plant, Heinkel assembled and produced parts for military aircraft. This sector was not only one of the first but also the most successful to deploy inmates. By the end of the war, 6,000 prisoners worked in munitions factories, 2,100 (mostly women) worked in tank production, yet 18,000 worked for aircraft-related firms (and this did not include the armies of prisoners set to work on the construction of factory spaces, the majority of whom also worked on projects for aircraft firms). So successful was Heinkel-Oranienburg that it, along with other aircraft manufacturers like Henschel, became a model for slave labor. In the fall of 1941, however, the firm used only a few unskilled workers. Like Steyr's engine works, the expansion of prisoners into production did not follow until the next year when Heinkel asked for and received permission to establish a satellite camp of its own.[10]

Thus rather than a war of "all against all" painted in so many histories of the Nazi regime, the truth was actually far more sinister. German managers often sought personal agreements with the SS, arranged as favors among a few individuals. Things like honorary membership in the SS cut across institutions and opened them up to what we might now call "networking." Increasingly informal favors became formalized in satellite camps and contracts. Polycracy thus led to business deals just as readily as to "internecine strife." In fact, the historian Mark Spoerer, who has surveyed thirty-three known case histories of industries that employed concentration camp prisoners, has found only one, the Akkumulatorenfabrik AG Stöcken, a battery factory for U-boats, in which any evidence suggests that the SS or *any state authority* forced a firm to deploy slavery against its will. On the other hand, twenty-two, including sixteen private corporations, eventually initiated such dealings of their own accord. It also bears mention that the evidence in the case of the Akkumulatorenfabrik AG is extraordinarily weak and stems from postwar interrogations by the British and attorneys in Hanover in which the firm's managers claimed that SS men "pressured" them into war crimes. Documents from 1943 show no such thing, though the chief of staff of the navy and Albert Speer's Armaments Ministry do seem to have pressured the firm to cooperate with the SS. Therefore, even if we indulge all possible gullibility, the evidence is still overwhelming that state ministries and industrialists alike came to the SS as viable partners.[11]

The Rise of Albert Speer

By year's end 1941, VW, IG Farben, Heinkel, and Steyr-Daimler-Puch were still exceptional cases. Why? we might ask. Now that some very prominent industrial managers were extending the opportunity, why did the SS not exert itself more to increase its profile in the armaments sector? If we take the SS's stated ideological intentions seriously, the answer is simple. Above all the SS's Administration and Business Main Office itself had other projects dearer to heart. Pohl and Kammler were just then, in December, drafting plans for the Building Brigades and laying the groundwork for a vertically integrated construction and manufacturing combine for RKF settlements. Again the statistics are informative. While the SS mustered no more than about 4,000 slaves for private armaments companies, it continued to employ upward of 10,000 in DESt alone.

Nevertheless, as Himmler gave free rein to his fantasies of Germandom and issued directives for an SS Labor Action over 150,000 strong, it was perhaps inevitable that the interests of the SS—as slave-labor lord—and those of state planners and German companies would eventually converge. Furthermore, the SS could hope to fulfill its dreams of a New Order only by integrating the DWB and the IKL into the total war economy. Otherwise the Armaments Ministry was beginning to shut down nonessential factories and transfer their resources to others. The increasing demands of war also threatened to cut off the easy credit that the German Commercial Operations had enjoyed since 1938. Starting in 1942, Pohl sought to provide labor to weapons producers, and, so doing, the SS sought to preserve a modern industrial infrastructure that it might reconvert to a "peace building program" once "final victory" was won.[12]

The rise of Albert Speer as Reich Minister of Armaments and Munitions served as the catalyst of this not-so-harmonic convergence. Ironically, he was the one man who always claimed after the war to have labored mightily to keep the SS out of the German war economy. Speer replaced the previous armaments minister Fritz Todt shortly after Todt's death in February 1942. Todt had paid a visit to Hitler on 8 February. Most secondhand accounts of this meeting agree that he wished to impress upon Hitler that Germany was hopelessly outmatched by the combined industrial might of the United States, the Soviet Union, and Great Britain. Supposedly this news was so unpalatable that the two ended up in a shouting contest, after which Todt's plane mysteriously crashed on takeoff. When Todt was killed instantly, many speculated that he was sabotaged as the bearer of bad tidings, but most evidence suggests nothing more than an accident.[13] Speer himself narrowly escaped death by declining Todt's invitation to fly with him. Whatever the future outcome of Todt's meeting with Hitler might have been, however, it is certain that big, structural changes in the German war economy were afoot, with or without Speer.

After 1945 Speer claimed to be an architect, an artist, but he hardly arrived in the Armaments Ministry as a freelancing aesthete. By 1942 he had long proved his skill in exactly those areas which suited him for the centralized command of industry. For years he had coordinated tight timetables for the production and delivery of raw materials on construction projects such as the monumental Führer buildings. He had deftly orchestrated contracting firms and even completed his projects ahead of schedule. The scale of activities under his control had grown steadily, and by 1941 his duties had expanded beyond construction to tank production. Like Todt, Speer understood the complexity and inseparability of machines, labor, and management. His initiatives managed all effectively. Todt had proposed the widespread rationalization of German industry, but Speer's organization first forged the statistics that allowed uniform assessment of productivity, supply, and demand. He also devised a new rationing system indexed to the real capacity of German industry.[14]

Human labor remained "the only completely unsolved administrative problem in my newly erected house," as Speer commented upon his ascension to the position of Minister of Armaments and Munitions.[15] Hitler recognized this as well, and, less than two weeks after Speer's entry into office, the Eastern Workers Decrees laid the foundation for the forced conscription of eastern European civilians (the Reich Labor Action). Thus the transition to total war intensified the pace of industrial rationalization and, at the same time, escalated the compulsion used by German management. Even before Speer entered into his vexed partnership with the SS, the Armaments Ministry and private German companies were already traveling down the "crooked road" to Auschwitz, a road they and the SS would litter with broken corpses. For that matter, Speer had followed the same path as early as 1938 and 1939 when he had helped the SS start up the largest brickworks in Germany to aid his monumental buildings in Nuremberg.

In the context of Speer's reorganization of the war economy, the Armaments Ministry approached the concentration camps in the consistent pattern set in 1941 by VW, IG Farben, and Steyr-Daimler-Puch: that is, those outside the SS, in industry or in the Armaments Ministry, took the initiative, *not* Himmler or Pohl. But once contacted the SS greeted the possibility of joint ventures with enthusiasm. A memorandum in March from Walther Schieber, who coordinated all subassembly for armaments production under Speer, mentioned that negotiations for concentration camp armaments factories had been under way "for a few months," almost certainly alluding to those with Porsche's Volkswagen, Heinkel, and Steyr-Daimler-Puch. (IG Farben had already received its prisoners.) To the chief of the SS Main Office, Gottlob Berger, Schieber expressly stated that he wanted to work with the SS "in all possible ways," and he had laid the groundwork for pilot factories at Neuengamme and Buchenwald by the first weeks of March.[16]

Speer sought after the war to characterize these dealings as an "infiltration." On the one hand, Himmler's minions' sole strategy was to "make the SS independent of the national budget on a long-term basis"; that is, the SS had no other motive, it would seem, than the love of sweet power.[17] In addition, Speer would have us believe that the SS forced slave labor upon him while his ministry harbored the defenders of the German economy, "technocrats in the widest sense of the word . . . a staff of industrial specialists who had already excelled before 1933 or who stood out after 1933 in a relatively free selection in individual factories."[18] Curiously, he also argued that a technocratic mentality led his ministry into criminal dealings through an amoral and blinkered pursuit of efficiency: "Today moral sensibilities are being suppressed everywhere and the human factor is being ubiquitously downgraded by technology. Obsessed with performance goals, devoured by personal ambition, people still tend to see human events in technocratic terms of efficiency."[19] He wished, in other words, to argue both "They made me do it" and "I knew not what I did." But such sentiments count on our gullibility and ignorance of the history of technology as well as modern organization, which never function outside of dynamic human interaction—including conscious moral deliberation—no matter how corrupt, perverse, or stupid. It lay in the "polycratic" nature of the Third Reich with its tumbling kaleidoscope of institutions that "technocrats" continually had to choose deliberately with whom and toward what ends they worked.

As mentioned, the first initiatives to expand concentration camp armaments factories can be traced to Walther Schieber, one of Speer's most trusted officials. Schieber had gained access to Himmler with a facility that shows how close contacts and ideological affinity lubricated the channels of formal communication between large and potentially agonistic state bureaucracies. He was himself a high-ranking honorary member of the SS. Thirty-five years after the war, Speer took pains to explain that he had therefore been one of Himmler's "spies," but during the war Pohl and Himmler both considered Schieber Speer's man, not their own. And Schieber was unquestioningly loyal to Speer. On the other hand, absolutely no evidence suggests that his various allegiances to the SS bothered Speer at all. In fact, like Meindl of Steyr-Daimler-Puch, Schieber was known for his tact. Many valued his services especially for the aplomb with which he soothed the egos of public officialdom and private industry. This had earned him the nickname "Angel of Peace," and in this capacity his ideological affiliations clearly did not damage his reputation.[20]

Schieber was also an experienced technological manager who had come from private industry. After finishing a degree in chemical engineering in 1922, he had started his career with IG Farben as a top manager (*Werkleiter*). He eventually founded an artificial fibers company and published articles about the importance of synthetic textiles to German autarky. Some might argue that he was only promoting the fortunes of his company, but the dichotomy between material gain and genuine conviction is a false dilemma in Schieber's case, for the man's ideological passion was unmistakable. In 1944, as the war closed in around him, he begged Speer in a handwritten note: "I ask you to tell the Führer that I have dedicated myself since 1931 and before with all my power to our idea—not from the point of view of an industrialist but from that of an SS man—and I will continue to do so."[21] Would an "infiltrator" write in such an open and pleading manner to his superior? Or is it more plausible that Speer knew of Schieber's values, his contacts, and his overlapping affiliations and valued what Schreiber's note referred to as "our idea"? Much earlier, before the war, as chief executive of the fibers company, Schieber had also conformed to racial dogma with a devotion that would be as laughable as it is hypocritical if the consequences were not so tragic:

A German firm in Chemnitz . . . has offered the Thüringische Zellwolle [Schieber's firm] a certain process [that] may be of decisive importance. . . . In order to decide on the matter, it is necessary to engage in three to four weeks of cooperation with Dr. Huppert. I have not researched the matter, but I am sure that Dr. Huppert is a Jew. It is not only a heavy burden for me personally, but also a burden in my professional capacity as a supervisor, to allow this man to work in our factory.

He then assured the SS, "Once our work is concluded I will also call a general assembly of all our personnel in order to clarify the presence of Dr. Huppert to all. Thus our simple workers will not be left in alarm [*kopfscheu*]."[22] Needless to say, these were hardly the words of a "technocrat" obsessed by efficiency alone.

It is likely that Schieber's ambitious plans for expanding concentration camp armaments ventures caught Himmler by surprise in early 1942. Himmler's ongoing discussions with Pohl throughout January gave no intimation of them; rather the SS had been preoccupied with the New Order. Nor did Glücks have any inkling of the comprehensive reorganization soon to follow as a result of Schieber's initiatives. Glücks had just disbanded the regional branches of the Office I/5–Labor Action under Wilhelm Burböck, and in early 1942 no concrete plan for slave-labor management yet existed. The Reichsführer SS, a canny political actor, no doubt saw Schieber's overture as a golden opportunity to increase the influence of the SS as is so often claimed, but this was hardly his only motive. Himmler genuinely wanted to serve the Third Reich in every capacity he could. "The greatest reserves of labor are hidden here," he insisted to the IKL, a conviction he held long after the working relationship with Schieber and the armaments industry had begun to sour.[23]

Nevertheless, to pull off any viable "labor action" the SS had to radically revamp its management precisely because it had never intended to venture into armaments production in any grand manner. In March Himmler quickly judged that Glücks would not be able to cope with the coming pace and scale that the Armaments Ministry would expect. The logical choice for an SS labor lord, one capable of meeting armaments moguls on equal terms, fell to Oswald Pohl. In contrast to Glücks, Pohl seemed to be a nimble organization man and a model of competence. Since 1938 WVHA officers had founded the German Commercial Operations and overseen the erection of the SS building authority. In each case, Pohl had responded with exactly the kind of "restless" dedication idealized in SS jargon, and on 3 March Himmler ordered him to take over the IKL, which Pohl subordinated to the Business Administration Main Office as a new Office Group D. The reorganization preceded Fritz Sauckel's appointment as General Plenipotentiary for the Labor Action by almost two weeks. Sauckel erected his own Labor Action in parallel, and it covered a different area of competence. No evidence suggests any conflict or competition between Sauckel and the SS over labor allocations at this time.[24]

The fact that Pohl had not included the addition of the IKL in the administrative restructuring of the WVHA of just one month before is further evidence that the need to convert the IKL into a labor exchange for war industries came as a surprise. But Pohl welcomed the new opportunity. After the war, he claimed that the Office Group D was the most important wartime duty that fell to him, and his statements during the war show that he felt it eclipsed all others. At the same time, he considered the armaments initiatives as a way to further secure the means of the peace building program: "The present mobilization of all prisoner labor . . . for the purposes of the war (increased armaments production), and for the purposes of construction in the forthcoming peace, now comes to the fore."[25] Of note, such statements acknowledged no dividing line between the "pragmatism" of war production and the ideological goals of the New Order. The New Order had always lain down a road that the Nazis intended to pave with blood.

Pohl's enthusiasm could not, on the other hand, conceal the fact that he had to embark on a haphazard organizational innovation. The WVHA gained control of the IKL only gradually and, even then, partially. The WVHA had immediately developed an organizational chart for the new Office Group D but did not officially announce changes until 13 March. The transfer took place without predictable scheduling or clear channels of communication. For the most part, Pohl was forced to take over existing administrative structures and personnel. The Office Group D–IKL retained its Oranienburg headquarters as well as the slothful and incompetent Richard Glücks. The leadership on location at concentration camps remained as well, where the WVHA asserted its authority only after a long delay, sometimes as late as May.[26] Although Himmler wished to present a facade of competence and willing cooperation to men like Walther Schieber, behind the scenes Pohl scrambled to establish command.

After an initial meeting at Himmler's headquarters, top armaments officials met with Glücks on 16 March. The SS demanded sole responsibility for security and insisted that all industries that wished to use prisoners had to relocate within the perimeters of concentration camps, a condition set directly by Himmler. On the other hand, no one in the SS showed interest in the ownership of armaments plants. All were agreed "that relocated armaments industries in the concentration camps will continue under the guidance of their individual firms, not only for production but under all economic considerations as well."[27] This should be no surprise since settlement construction remained the core motivation behind the SS's own industrial empire.

At this point, too, designs for concentration camp armaments factories remained almost the exclusive work of Schieber. Initiative on the part of the SS was notably absent. This actually annoyed Schieber, who quickly began to express irritation that the SS was dragging its feet. Besides paranoid demands for security, what had the SS brought to the table? Armaments Ministry officials must have wondered the same thing, for Glücks proffered only questionable estimates of the workers currently available:

Buchenwald: 5,000
Sachsenhausen: 6,000
Neuengamme: 2,000
Auschwitz: 6,000
Ravensbrück: 6,000
Lublin: to be filled.

"All the skilled workers and professionals will be sorted out from these prisoners and subsequently assigned to camps in which armaments production is to take place."[28] These figures, not to mention Glücks's liberal use of the passive voice, show two things. The first could not have gone unnoticed by shrewd managers like Schieber: the SS had no exact statistics on workers, only guesses rounded to the thousands. Second, unbeknownst to Schieber, Himmler had informed Glücks in January that "in a matter of weeks" the camps could expect 150,000 fresh prisoners. Where were they? What transports did arrive during this period were dying off as fast as they entered the camps. Glücks offered only a total of 25,000 workers but was, in fact, bluffing. He expressed what the SS hoped to achieve, not what the IKL could effectively muster. Speer's hard-nosed, experienced managers expected information based on technical and organizational reality. As plans for slave labor in armaments took shape, they quickly began to lose confidence in the SS.[29]

Putting the SS's House in Order for Total War

The problems ahead for the SS were not lost upon Pohl either, and he took immediate action. Over the summer of 1942, he embarked upon three main initiatives: a reform of the Kommandanten, the institutionalization of SS Economic Officers in all regional SS authorities, and a final reorganization of Burböck's now defunct Labor Action Office. Of note, he directed his efforts toward internal reforms of the SS itself, not toward the "takeover" of external industries. Even these changes came sluggishly and never quite satisfactorily.

Only in the latter part of April 1942 did Pohl first address the leadership of the concentration camps. He called all Kommandanten to a meeting in Berlin and informed them that the entire concentration camp system had to change. His tone was jubilant, for the future was at stake. Everyone had to rally for the war, he told them: "The collaboration of all authorities shall proceed without friction; the fetters of discoordinated administration must be shed in the concentration camps and will be hailed everywhere as progress."[30] Nevertheless, Pohl proposed few concrete changes and left the ultimate control over working prisoners under the boot of the Kommandanten, their subordinates, and their Kapos. In the main, this was wholly consistent with the WVHA's ethos of communal industry, which identified the secret of success in unified will and concerted action; Pohl exhorted all to work together.

He never clarified the limits of the Kommandant's mandate to police the labor force, however, and that which Pohl *did not* urge upon the Kommandanten speaks volumes. For the most part, he did not even address the direct supervision of production. He did not, for example, arrange a control mechanism for monitoring the output of the camps, freeing the Kommandanten from the very beginning of any accountability. He could have acted differently. He could have recognized the need to replace Kommandanten with experienced industrial managers, or he could have instructed Kommandanten to act in unison with industry in the same way that Speer's agents did in the parallel Reich Labor Action; but instead he utterly neglected the dimension of factory management.

Pohl did remove fully one-third of the Kommandanten, many due to incompetence. Hans Hüttig, Karl-Otto Koch, Karl Künstler, Hans Loritz, Arthur Rödl, Wilhelm Schitli, and Alex Piorkowski left the camps' service. This was an attempt to clean house, for only Loritz left for promotion. Pohl originally wished Hüttig to take over Natzweiler, but he moved instead into the service of Higher SS and Police Führer of Norway. The rest were simply dismissed. A sex scandal surrounded Koch, and this as well as other scandals also involved Rödl. Künstler and Piorkowski both had drinking problems, Künstler's apparently so bad that it offended Himmler personally. "If the RFSS once more hears about [his] orgies and drinking excesses . . . he will be locked up for years," Himmler's adjutant wrote to Pohl.[31] Like his boss, Pohl was motivated in these matters by a sense of indignity. The men he expelled had not acted like proper "SS men." They had transgressed against the ideals of community and good behavior that Pohl espoused.

As new Kommandanten, Pohl promoted Wilhelm Gideon, Anton Kaindl, Josef Kramer, Paul Werner Hoppe, and Fritz Suhren. Despite concern for propriety, these replacements demonstrated little care for production. As Karin Orth points out, this marked the first time that a large cadre entered the ranks of Kommandanten who had not been directly influenced by Theodor Eicke. Suhren, Kaindl, and Gideon came from the military administration of the Waffen SS, not IKL administration. Nevertheless, Kramer came directly from the "Dachau School," and some who had not served under Eicke were no less embedded within the tradition of terror that reigned in the camps. Hoppe had led guards at Lichtenberg and Hassebroek at Esterwegen. Suhren had been inducted into the detention camp of Sachsenhausen by Hans Loritz, one of Eicke's most favored men. Rather than recognizing that the new industrial tasks of the IKL demanded a new managerial style of leadership, Pohl chose these men for their assiduous loyalty to long-established SS practices. None had succeeded in industrial management, let alone technological management, and most had typically broken careers behind them. Kramer had trained as an electrician but had found no steady employment from 1925 to 1931. Hassebroek had trained as an office apprentice in a machine shop but was fired in 1931, after which he sold newspapers and then fish on commission. Suhren had "worked" in his family textile merchandising firm, likely a cover for his actual unemployment after he left a cotton mill (where he had managed inventory). Perhaps this past experience with textiles led Pohl to appoint him to Ravensbrück with its successful Textile and Leather Utilization GmbH, but since 1931 Suhren had done nothing but serve in various SS Sections as an administrator, where he proved his strength "managing correspondence."[32]

How little the WVHA actually changed IKL practice became all the more clear later in the war, when Pohl and his chief officer for the Labor Action, Gerhard Maurer, promulgated a new set of service regulations: "When the concentration camps were first erected after the seizure of power they had only one task: the custody of all persons who posed a danger to the security of the people and state. . . . In wartime it is necessary to convert them for operations in conjunction with the war effort in which they can have a decisive influence on the outcome of the war."[33] But how was this to be accomplished? The new regulations were silent. Such vagaries never really urged the Kommandanten to anything new. Pohl only asked Kommandanten to change their "attitudes," and his calls to reform would retain the air of ignored manifestos: preachy, enthusiastic, totally void of program.

At the end of July 1943, the WVHA issued a question-and-answer drill to be memorized and performed by the camp personnel who managed labor details. It was supposed to be a kind of Nazi morality play, but all the roles were typecast long before. To the question "Why are prisoners dangerous?" the response read, "Because they can destroy the unity of our nation, lame our power, threaten our victory. They threaten to make it possible for those at home to stab the soldiers at the front in the back, just as they did before [i.e., in 1918]."[34] The manual went on to define prisoners not only as enemies of the state, but as deviants, freeloaders, traitors who had maliciously withheld their contribution to the war effort. It also reinforced the general SS consensus that prisoners had to be driven harder and longer. Likewise it urged Kommandanten to adhere strictly to the codes of the "SS man."

The same doctrines, and their consequences, are evident in the second of Pohl's internal reforms. The WVHA also tried to reorganize forced labor under the Higher SS and Police Führer, the powerful officers who unified all functions of Himmler's organizations as regional representatives of the SS's authority. The Higher SS and Police Führer had become especially strong in the occupied East. Here few other agencies could challenge them, and they had initiated a hodgepodge of forced-labor projects using Jews, Poles, and the so-called work education camps (short-term detention centers for civilians supposed to be rehabilitated through hard labor). Like the Death's Head Kommandanten, Higher SS and Police Führer held their captives in squalor, victimized them through starvation, sickness, and arbitrary brutality, and murdered them outright. The same primacy of policing reigned here as in the concentration camps. In the summer of 1942, Pohl hoped to incorporate the Higher SS and Police camps, which he viewed as untapped labor reservoirs, into the IKL Labor Action. To do so, he proposed to create a new WVHA position, the SS Economic Officer, and, with Himmler's backing, he ordered the transfer of all regional labor operations under their jurisdiction as representatives of Office Group D.[35]

As he had when seeking new Kommandanten, Pohl handpicked and installed the Economic Officers, but he chose from cliques that usually identified closely with their Higher SS and Police Führer. These used their bureaucratic position to squelch the authority of the WVHA. In consequence Higher SS and Police Führer continued to act as they pleased in matters of industry. Once again, Pohl misread the problems faced by the SS Labor Action and followed his belief that cooperation and community spirit on location were the keystones of sound operations instead of cooperation *and* managerial competence. Just a few years before, when Pohl had sought out Kammler, Hohberg, or Schondorff, he had consciously looked for knowledgeable, experienced professionals whom he could reasonably expect to understand industrial enterprise. He had proved unable consistently to attract officers to the SS companies who shared the same sense of mission as veteran SS company men, but competence had mattered in the old VuWHA. Now, by contrast, Pohl proposed six, mostly older, longtime SS hacks as Economic Officers. Like the Kommandanten chosen at roughly the same time, they possessed scant industrial experience. Their dedication and length of service distinguished them more than any accomplishments. Most came from the military administration of the Waffen SS. Two were chosen from among existing HSSPF staffs, hardly a position from which to start administrative change, and two had already been inculcated into the administrative ways of the Death's Head Units. None had participated in the German Commercial Operations. Perhaps the general shortage of trained managers throughout the Reich and in the SS forced Pohl to make these choices, but his enthusiastic assurance that he had found six good men suggests he was simply trusting his instincts. He had evaluated their work earlier in their SS careers, praised them as "restless," "ruthless" (*rücksichtslos*), and "unwavering," and chose them because they fit his image of the good SS man.[36]

The WVHA's failure was nowhere more evident than in the General Government, the rump lands of Poland not directly annexed to the Reich. Here the responsible Higher SS and Police Führer was Friedrich-Wilhelm Krüger (HSSPF Southeast–Kraków), to whom Pohl assigned Erich Schellin. At first glance, Schellin appeared a likely candidate, for some experience set him apart from his peers. He had even worked before the First World War as a book-keeper in the brick industry, a sector in which the SS had substantial holdings. But Schellin had merely bounced from one petty firm to another and never held a steady job. Furthermore, he blamed his shattered career on the democracy of the Weimar Republic, especially the Social Democrats, and his fervid dedication to the SS seems to have been part and parcel of a resentment of sound business practice, much like that which reigned within the IKL.[37]

Schellin had already worked with Krüger long before the WVHA created the "SS Economic Officers." Their good relationship can be inferred from the many imbroglios that erupted around Schellin's intemperate character and Krüger's constant efforts to shield his subordinate from repercussions. Schellin found himself in hearings by 1941, when an anonymous report claimed that he had embezzled goods from Waffen SS stores, but his comportment offended more than any mismanagement. He was a mean and wild drunk known to pick brawls, which he often lost. Once he had sat himself down bellowing in the early morning streets of Breslau and woke the neighborhood by crying out for champagne. Objects had been known to fly out of his window and smash in the street below. Despite these difficulties—which included embezzlement—Krüger considered Schellin a capable organizer whom he could not do without, and there can be little doubt that Schellin identified with Krüger, his protector in these many indiscretions.[38]

Schellin's spectacular history of tactless behavior became an issue within the WVHA a short time after his appointment as SS Economic Officer, for this was the same behavior for which Pohl had cashiered two Kommandanten. A legal officer had reported (without a hint of irony), "SS Lieutenant Colonel Schellin suffers from circulatory system disturbances during the full moon, and at those times he is known to sweat heavily for several days. . . . At such times he is not entirely capable of work, and because this condition disturbs him, he vents his frustration in anger and often distasteful statements."[39] The indiscretions under the "full moon" regarded Schellin's womanizing, drinking, brawling, and general hell-raising. He called his superior officers "bureaucratic fatheads" and "lamebrains"; he used his authority to requisition automobiles and gasoline for personal recreation; and, after he once lost a woman (to his great consternation) in competition with an injured sailor in an SS cantina, he had shouted: "You get yourself a young girl like this one every night! Tell me, you've got to have a pecker like a bull, that all these women run after you. Isn't that so!?!"[40] For his part, Schellin believed that all his "work was correct." He wrote a lengthy, captious report laying all faults at the door of his accuser, whom he named "an outspoken materialist, who joined the SS only after the seizure of power out of self-interest." Schellin then had the audacity to conclude his defense by asking for a promotion. "It is completely beyond me why I am singled out and must still run around as a colonel."[41]

This behavior offended Pohl deeply, and he followed the disciplinary proceedings with a personal interest. But, oddly, he never showed any dissatisfaction with Schellin's ability to manage per se, and Schellin's behavior says as much about Pohl's judgment in expecting such a man to supervise the SS Labor Action as it says about the surreal corruption that was engulfing Nazi-occupied eastern Europe. Furthermore, it also demonstrates how little formal bureaucratic structures or economic rationality could bind such men to "good" behavior. In this case, the Higher SS and Police Führer Krüger took proceedings against Schellin under his exclusive jurisdiction in September 1943 and closed the lid on the case. Pohl could do nothing. The structural position of the SS economic Führer meant little without the conscious dedication of these officers to the WVHA or to skilled management. If Hohberg and Schondorff may serve as examples of managerial talents that failed the SS because they were at odds with the WVHA's ideological consensus (further explored in chapter 7), Schellin demonstrates that ardent Nazis did the WVHA no good without managerial knowledge. Both ideals and knowledge mattered because each was necessary to animate the other.[42]

Pohl's last initiative in the name of reforming slave-labor management actually proved the most effective. He had planned as early as March 1942 to resurrect the former Office I/5, this time under a seasoned business executive from the German Commercial Operations, Gerhard Maurer. Maurer led a new Office Group D2–Labor Action and represented a conscious attempt to infuse the IKL with modern business administration. He came to his new duties after overseeing the German Equipment Works, which employed over 7,000 prisoners at the time. By the beginning of 1942 he had also served as a liaison to the IKL, helped establish links to IG Farben at Auschwitz, and erected the entire industrial organization at the new Stutthof concentration camp near Gdansk. Maurer's dedication to the Nazi cause was unquestionable, and, like so many WVHA managers from the early days, he had come to work for the SS out of conviction. He quit a steady bank job in 1933 to work for a Nazi paper published in Halle and had already joined the NSDAP in 1930. In the heady summer after the Nazi seizure of power—the same summer that he went to work for the Nazi press—he also joined the SS. Thus before Pohl recruited him in 1939, Maurer had long merged his work experience with his political conviction.

Of course, Schellin and Burböck were also dedicated Nazis, but, unlike them, Maurer proved a capable, inspiring, and interventionist manager. He was not afraid to visit regional subordinates when they were neglecting assignments. After his wife died in a bombing raid on Halle in August 1944, he redoubled his concentration. He quickly established himself as the de facto chief of the entire Office Group D by continually absorbing tasks that were ignored and neglected by others. Hardly any important document emanated from the Inspectorate of Concentration Camps after 1942 without his initials in the letterhead. Even orders officially signed by Glücks, his nominal superior, had usually been formulated by Maurer, and likewise Pohl, who trusted Maurer almost unconditionally, also routinely signed off on his initiatives. Like all of Pohl's most active and creative officers, Maurer was a young man, only thirty-four when he took over the SS Labor Action. Somewhat unusual in the WVHA, he had no advanced education, but he did have extensive work experience in accounting and modern corporate administration. In the Weimar period, he went straight from secondary school to work as a factory bookkeeper and then switched to work in a bank.[43]

On the other hand, detained by other duties, Maurer could not enter his new office until the end of April, and not until the end of June did he finally manage to inform concentration camp Kommandanten of the new Office Group D2. Perhaps because of this late start, Maurer had little choice but to assume most of the basic structures planned by his predecessor Wilhelm Burböck. Maurer did implement one major change, however, a change that demonstrated the importance of interventionist management in modern bureaucracy. Whereas Burböck, the civil servant, did little to implement the measures he had proposed, Maurer, the modern manager, went about making them a reality by imposing systematic, statistical surveillance as a means of centralized control. Maurer began to track and coordinate the camps' working populations, and here his efforts quickly met with the eager cooperation of industry. This, of course, had already been suggested by Burböck, but only Maurer took effective action to implement it. Where files had been kept before, they had varied from camp to camp. Maurer homogenized his system on preprinted cards that Labor Action managers could transfer in an interchangeable format. He also systematically extended control over population statistics in a way that Burböck had never considered: he introduced standardized forms. Burböck had relied on prose reports that were impossible to compile, and in consequence Office I/5 never knew how many prisoners were working or where. Maurer's forms allowed his Office D2 to enforce the uniformity and interchangeability of information between disparate camps needed for any modern management. Statistics from myriad sites soon became manageable in polyglot.[44]

Maurer's forms are easy to overlook, for what could be more banal? Yet his methodical definition of the information he wanted to manage says much about what he considered to be the key problems with the SS Labor Action. When distributed on 1 July, they had nine entries:

1. Sick
   a. Ambulatory
   b. Bedridden
2. Invalids
3. Reported Sick to Doctor
4. Interrogations
5. Releases
6. Conditionally Able
7. Arrests
8. Quarantine
9. Entries[45]

Curiously, these statistics covered almost exclusively the physical condition of prisoners, not the material world of production. Of course, this was not without its pragmatic reasons, for mortality rates in some camps reached over 10 percent a month, and this endangered production. Once collated in graphs, statistics on population and death could yield comparisons and trends at a glance and enabled the Office D2 to spot camps where unusual numbers of workers were sick or detained. A low-level officer under Maurer reported how this worked in practice:

Death announcements were ordered alphabetically in a well-equipped cellar room. . . . I performed this activity daily. Later I collected the incoming population reports that came by mail from the individual concentration camps. I collated these together by the following criteria: age of prisoners, entry date, population, history of transfers, causes of death, special rubric

SB [*Sonderbehandlung*—that is, systematic extermination]. I further had to manage the files: reports from the concentration camps about individual escape attempts and privileges allowed individual prisoners. . . . My activities collating the populations of camps were collected into a general statistical report every month.[46]

In other words, these forms functioned just as any "functional" tools of administration, but Maurer's forms also incorporated the eugenic doctrines popularized in the Third Reich. From top to bottom in the WVHA hierarchy, SS officers blamed the poor performance of working prisoners on "poor human material," which they construed in medical terms and tied to an assessment of race. The late Detlev Peukert's last essay has suggested the "genesis of the final solution from the spirit of science," in which specious scientific claims call for "identifying, segregating, and disposing of those individuals who are abnormal or sick."[47] One does not have to take sides in what are popularly known as the "Science Wars" to recognize that in the Office D2 a perverse spirit of science found one of its most radical expressions in daily management. At an SS cement factory near Auschwitz, for example, the regional Labor Action officer routinely complained that incoming prisoners were "unfit to work" (*arbeitsunfähig*). When a contingent of 882 Jews from Auschwitz arrived, emaciated and worn down by harsh labor and exposure, he pleaded, "Exchange (*Austausch*) is of the utmost urgency, for these prisoners are completely unproductive and present an unbearable burden on the *Lager*." The explanation he gave for their worsening condition was typically racial: they were "mostly of Hungarian origin—not in the slightest measure meeting the demands [of the work camp]."[48]

Reactions were no different higher up the ladder of SS labor administration. At Gross-Rosen, which consistently posted one of the highest death rates among the camps, the chief executive officer of the SS's stone quarries complained that "the prisoners sent by Oranienburg [the IKL] represent for the most part exceedingly poor human material."[49] Atrocious conditions of shelter, provisions, and grueling work warranted no mention. Pohl used similar phrases to conduct negotiations with the Reich Ministry of Justice. In 1943 he complained that between 25 and 30 percent of prisoners transferred from the ministry to SS custody died in transport. "It is certain to assume," Pohl confided to Himmler, "that Mauthausen [the destination camp] received the worst human material."[50] This was the milieu in which Maurer had pursued his professional career for almost a decade. There can be little doubt that he embraced its medicalized, racial explanations for why prisoners weakened and died. His own initiatives confirm this, for he defined mortality rates as a medical problem *of the prisoners* instead of confronting the obvious fact of catastrophic mismanagement governing supply, shelter, or sanitation in the IKL, not to mention violence. Maurer's new forms originally carried the heading "*Prisoners Unfit to Work or Prisoners Unavailable for the Labor Action*" (Maurer's italics), and so Burböck's original term "unfit to work" now reappeared in refined form.

With these forms Maurer went further than Burböck in one crucial regard. He began to shift responsibility for the camps' working populations onto camp doctors. The IKL doctors were in a unique position to aid the Office D2. Nominally members of the camp medical staff belonged to the Kommandant's administration, but they were simultaneously subordinated to the "Reichs Doctor," the highest-ranking SS medical officer on Himmler's Personal Staff. They could therefore appeal to the authority of the latter to get around the former or vice versa. In other words, they wore two hats. Specialized training also set them apart from the Kommandanten, not to mention their generally higher social standing. In practice, medical staffs also worked in separate buildings (the prisoner sick bays, or *Häftlingskrankenbauten*), away from regular administrative officers.[51] All of this helped the Labor Action because it allowed doctors to operate in relative independence from the Kommandanten's primacy of policing.

The doctor's professional identity seems to have included no ingrained resistance to production or modern administration like the Death's Head Kommandanten. Well before the Nazi period, German medical professionals had processed statistical forms on the "fitness to work" of World War I veterans, for example, or standardized insurance claims. Maurer's forms now bid them turn such experience upon the physical conditions of prisoners' fragile bodies, and it is clear that most earnestly attempted to preserve prisoners who were "fit to work." On the other hand, the SS doctor's new job as labor manager went hand in hand with the *destruction* of the living. Along with medical orderlies, doctors took responsibility for culling the "unfit to work" and often for killing them as well (e.g., with injections of phenol). They disposed of sickly inmates without compunction and often with great alacrity. In fact, purging the sick was justified as a measure that protected working inmates from epidemics.

In some cases, the SS also managed to make the camps' sick bays places that prisoners wanted to come to die. Yves Béon, a survivor of the Dora concentration camp which manufactured the V-2 rockets, writes of his comrades who pleaded with one eighteen-year-old prisoner named Louis not to go to the infirmary, but the boy retorted, "You tell me that if I go to the infirmary, I'll not come out alive. Well, I already know that; I'm not stupid, and I couldn't care less. At least, perhaps there they'll let me die in peace—and horizontally."[52] Of course, not all prisoners committed themselves willingly to the camp infirmaries and sick bays. On the other hand, camp doctors nominally remained there "to heal" and held discretion over extra rations and special permission slips entitling prisoners to days of rest. Perversely, Maurer's new system therefore enlisted a measure of complicity from the WVHA's ultimate victims through psychological trickery. Either as a bastion of last hope or as a domain of easy death in a hell where prisoners were routinely beaten, starved, and tortured, the sick bays became effective stations for culling the "unfit." Statistics of the "fit" became steadily more accurate in equal measure.

And there can also be little doubt that the collated graphs of the Office Group D2 began to serve the purpose Maurer intended. The IKL now began to get an overview of mortality and morbidity for the first time and with shocking results, even to those accustomed to the grisly concentration camps. Incoming prisoners were dying, like the Soviet prisoners of war before them, almost as fast as they arrived. In December 1942, after Glücks (still officially Inspector of Concentration Camps) noted that 70,000 incoming prisoners had died out of 136,000, he sent a directive to all camp doctors:

> The best doctor in a concentration camp is that doctor who holds the work capacity among inmates at its highest possible level. He does this through surveillance and through replacing [the sick or injured] at individual work stations. Statistics for those able to work should not merely be a figure on a piece of paper; rather they must be regularly controlled by camp doctors. . . . Toward this end it is necessary that the camp doctors take a personal interest and appear on location at work sites.[53]

Thus Office Group D ordered doctors to intervene directly at factory work sites, an unusual move since production supervision has never been part of medical training. Yet by relying on doctors, Maurer, whose code in the letterhead of this order clearly marked it as his initiative, was merely acting consistently with his conception of what constituted the problem with production: it was a medicalized problem of managing "human material." The order also demonstrates that the lives of prisoners meant nothing to the Office Group D2, for Maurer specifically ordered the camp doctors to remove and replace weak, sick, or injured prisoners. The fate that awaited the "unfit to work" was common knowledge. As a foreman once snorted to a Jewish laborer near Auschwitz, "Whether you stinking Jews work or not, you'll go one way or another into the crematorium and go through the oven."[54]

Nevertheless, the SS doctors and their statistics now allowed Maurer to demand specific results. Consider, for example, a letter written to Buchenwald at the end of 1942 about a transfer of 150 unskilled inmates to Auschwitz. The Kommandant had merely used the opportunity to dispose of unwanted prisoners and loaded up the rail cars with sickly inmates—in keeping with long-established IKL practice:

> By 12/4/42, out of the total transported from Buchenwald, 18 prisoners have died. Another 3 are currently lying in the prisoners' sick bay suffering from various afflictions. Of the remaining 129, 22 are bodily weak; 3 have foot injuries, inflammations, and swelling; 1 has no left arm; 1 has a deformed wrist; 3 have frostbite on their fingers. From the entire transport, only 100 are fit for work, two-thirds. Of these 2 percent were skilled workers [while the request was for unskilled].[55]

Maurer invited the Kommandant's explanation. From his desk in Berlin, Maurer could now monitor the transactions between camps hundreds of miles apart. Just two years before, Phillipp Grimm had called attention to similar problems, but the Kommandant of Buchenwald and Dachau had easily dismissed his objections. By contrast Maurer now used the statistical information gathered by the doctors at Auschwitz to hold the Kommandant of Buchenwald responsible. If nothing changed after his inquiries, he, like Kammler, was wont to appear personally to set things straight.

As a direct result, camp conditions stabilized during a brief period from the end of 1942 to the beginning of 1944 (when the disruption of German transport and industry due to bombing as well as the swelling population of the camps eroded what fragile order the Office D2 could impose). Mortality rates among working prisoners dropped from around 10 percent a month throughout 1942 (17.2 percent in December at Mauthausen) to around 2 to 3 percent in 1943. Yet Maurer's modern managerial initiatives can hardly be judged humane on these grounds: the very fact that Office Group D judged a 2 to 3 percent mortality rate tolerable speaks for itself, and Maurer had, from the start, coupled the preservation of working prisoners with the liquidation of the sick and weak. The statistics are further deceptive because the camps' population was increasing geometrically throughout this period. The absolute number of deaths never fell. At its height, Maurer's Office Group D2 would handle over 700,000 prisoners. Most who arrived at Auschwitz, for instance, went immediately to their extermination and never even entered Maurer's balance sheets and graphs. At an absolute minimum, the camp complex of Auschwitz-Birkenau alone slaughtered over a million prisoners. Neither the modern methods of management nor the WVHA's productivism contributed to a more humane treatment of prisoners. Maurer had merely introduced the elements of modern management into the institutions of genocide to balance systematic extermination with industrial labor.[56]

Maurer could have acted differently had his intentions been different. For example, denying the most basic sustenance was one of the chief tools in the hands of Kommandanten who wished to destroy prisoners. Maurer could have monitored the supply of foodstuffs throughout the camp network just as he tracked disease and death, yet after mid-1943 Pohl and Maurer placed responsibility for a secure food supply in the hands of local Kommandanten.[57] What better exemplifies an option that the WVHA *did not* take or things that officers like Maurer could have done differently—if, that is, life had mattered to them? By contrast, the WVHA did appoint a special officer in 1942 to evaluate benefits programs and food rations for German civilians in its employ: this was none other than Kurt Wisselinck (introduced at the beginning of this book), who went out of his way to keep leftover potatoes out of the hands of working prisoners and demanded statistical records in order to enforce his initiatives. Thus the WVHA was perfectly capable of managing supply, shelter, and other basic necessities for a healthy work force, which underscores a fundamental point. When given the chance, WVHA officers generally took initiative to reinforce the SS's barbarous system.

If the SS was indeed "polycratic" in its structure and intentions, camp doctors proved so effective because they could bridge the competing interests that surrounded slave labor. By culling prisoners, they meshed seamlessly with the primacy of policing among the camp SS who aimed at destruction. But because the sick bays and infirmaries operated in isolation from the Kommandant's staff, doctors could also engage in the modern administration of statistics without interference. This, in turn, meshed with the WVHA's emphasis on modern administration and productivism. Last, because doctors' training covered only the bodies of prisoners, they did not interfere with the technical management of industry. They helped prepare the way for a division of labor between the WVHA and Speer's Armaments Ministry that grew in importance after 1943: the ministry relied upon the SS to manage the prisoners' bodies—getting rid of those who had been worked to death and supplying fresh replacements. Meanwhile the Office Group D left technical management to industry. The WVHA could allocate prisoners to Speer's ministry, IKL doctors could gather statistics on the "unfit to work," Kommandanten could senselessly murder valuable workers, and all could believe they were making "decisive" contributions to the war economy. Here the "organization" was the ideology and vice versa precisely because so many of the contradictions of Nazism were effectively subsumed within Maurer's organizational innovation.

Should we believe, as those like Wolfgang Sofsky would urge us to do, that core institutions like the Office Group D, which spread the wreck of human life in the camps, truly articulated an insane logic of absolute power "grounded upon itself"?[58] Or does not Maurer's example suggest that it is far more credible to believe that the otherwise ridiculous act of putting doctors in charge of production sites, of assigning incompetents to oversee prisoners' sustenance while assigning competent managers to keep food out of their mouths—that all this followed from conscious commitments grounded not in the pursuit of power alone but in the SS's racial management and Nazi fundamentalism? SS ideology was multifaceted and even sometimes contradictory, but it hardly inspired arbitrary actions or a mere mad dash for control dictated by some demonic logic of "power."

The Armaments Ministry's First Pilot Projects

Maurer's and Pohl's reforms dragged on through early 1943, but nevertheless a sham confidence arose during the WVHA's frenzy of activity throughout 1942 and into 1943. Pohl could point to ever new initiatives as proof that the SS was getting things done, meeting demands, and smashing through old roadblocks. His emphasis on "restless" activity obscured the fact that industrial managers and DWB officers alike were colliding head-on with the entrenched mismanagement of the IKL and, further, that he had consistently mistaken ideological "unity of will" for the confluence of managerial consensus *and* technological competence needed to manage factories.

The reality of managing modern industry persisted, however, and Pohl's rhetoric of community and "progress" could not whisk it away. While Pohl was involved in his many internal reforms, over the course of 1942 the Armaments Ministry had already embarked on pilot projects within the concentration camps. Walther Schieber was struggling to start up armaments factories at Buchenwald, Neuengamme, and Auschwitz, with uniformly disappointing results. An investigation into one venture, the Gustloff Works at Buchenwald, demonstrates how quickly conflict arose despite an initial spirit of mutual optimism.

We should first of all avoid the erroneous characterization of the resulting donnybrook as a clash between "technocratic" management (founded in "pure," pragmatic reason) and dogmatic "ideologues" (as so often construed by the postwar memoirs of the Third Reich's "crisis managers"). For one thing, the Gustloff Works was not just any firm but boasted a proud Nazi heritage and advertised its status as "the first National Socialist industrial foundation and its most modern model factory" in the *Journal of the Four Year Plan* as well as in the SS journal, the *Black Corps*.[59] Its eponym, Wilhelm Gustloff, was an ardent Swiss Nazi shot dead in Bern by a Jewish student named David Frankfurter on 4 February 1936. To honor Gustloff, the Nazis had given his name to one of the first "Aryanized" companies of the Reich, in this case a firm formerly owned by the Jews Arthur and Julius Simson. The Simsons' company, the Suhler Weapons and Vehicle Works, had received the dubious privilege of being the only Jewish firm to receive contracts from the German army after the Treaty of Versailles. That Jews should be entrusted with defense contracts, of course, enraged the Nazis. The national press had pilloried the Simsons since the 1920s, accusing them of embezzlement and demonizing them as the spearhead of a world Jewish conspiracy to emasculate the German armed forces. By 1935, the Gauleiter of Thüringen, Fritz Sauckel, finally succeeded in throwing the Simson brothers in jail and appropriating their company. Sauckel formed a "communal foundation" (*Gemeinnützige Stiftung*) to manage the brothers' factories. Shortly thereafter, the Swiss Gustloff was shot down in the streets of Bern, and Sauckel decided to name the foundation after this Nazi hero.

Sauckel, an engineer, declared that this quasi-nationalized armaments company would "promote the well-being of the employees and fulfill the first tasks of efficiency and quality . . . and thereby, beyond this, participate in the solution of general tasks in the area of economics, technology, society, and culture."[60] One of the foundation's corporate lawyers also declared, "The Gustloff Works comprise nothing like a capitalistic concern. . . . rather its members [*Gefolgschaften*] comprise a unitary, like-minded, and ideologically bound group of comrades."[61] Thus the Gustloff Works had tread a path illuminated by racial supremacy, enthusiasm for Nazi modernization, communal industry—in short, a path paved by Nazi fundamentalism very similar to that of the WVHA. By the 1940s both Walther Schieber and Fritz Sauckel sat on Gustloff's board of directors. Thus, when conflict arose between the SS and Gustloff, it did so among equally committed Nazis, not between the SS and "pragmatists."

Given the broad basis for mutual consensus between the two organizations, it should be no surprise that things first proceeded quite smoothly. Schieber's meetings with the SS resulted in a progressive schedule to expand carbine manufacture at Buchenwald. In full-blushing optimism Schieber hoped that work could begin in no less than three to four months, that is, by the end of July 1942. The WVHA was even more optimistic and expected to swing into production almost instantly. On 6 April, Glücks ordered the Kommandant of Buchenwald to ready 300 prisoners for production in twenty days.

In July, the factory's director and the WVHA closed a contract, negotiated at the highest levels of the SS. Himmler personally took part in the meeting along with Pohl.[62]

The WVHA had gone out of its way to follow Schieber's lead throughout the summer. When the Gustloff's director demanded new work halls, the SS agreed to build them at its own expense. Schieber told Pohl that he wished Buchenwald's production to expand "as soon as possible" to a total 75,000 pieces a month and was apparently "very satisfied with the overall development in our concentration camp," as Pohl relayed proudly in June.[63] In keeping with the usual division of labor between the WVHA and armaments firms, technical planning was left to the Gustloff Works, including the design of factory space. The ownership of all machinery remained with the Gustloff Works, as did the oversight of installation. Hans Kammler's SS Building Brigades scheduled construction to begin on 13 July. No conflicts had appeared at this point, but plans were already running behind Schieber's original schedule.

Shortly thereafter the first problems arose, initially over technical and managerial coordination. Gustloff's director Fritz Walther confronted all the headaches typical of plant start-up, not to mention the entirely new problems of managing slave labor. Compounding this, subcontractors began to miss deliveries of needed machine tools. After beginning late, work stalled. In mid-June, a full two months after Glücks had naively ordered start-up, the Kommandant of Buchenwald had to telegram Himmler that original plans for two work shifts were "technically impossible."[64] Simultaneously, it became harder for Pohl to reach Schieber directly, most likely because Schieber had more important projects to occupy his time.

Difficulties now began to awaken the suspicion of the SS: while Buchenwald was a minor project within the entire economy, the SS was convinced that it was nothing less than decisive for the war effort. In March 1943 Himmler received a report that compounded these suspicions. The original target of 75,000 would be cut back to 20,000 pieces. In order to meet the army's wishes, the ministry had sidelined carbine production and wished to convert Buchenwald to the assembly of a new weapon, a machine gun. This would naturally take time and entail new target outputs. Like his Führer, however, Himmler nevertheless remained fixated on numbers, and insisted on the original figure of 75,000 set for carbines. He remained oblivious to the fact that Buchenwald had shifted to an entirely different weapons system. Himmler, Pohl, and the Kommandant of Buchenwald arrogantly began to suspect the flagging energy, will, and initiative on the part of Gustloff's management as well as Schieber himself. Ignorant of the technical organization of production, the SS quickly overlooked the real requirements of factory start-up and did not even acknowledge its own mismanagement of prisoner allocations. Himmler genuinely believed that the SS had fulfilled its half of the bargain, and on 5 March he wrote to Speer directly to urge more action and less talk. Why had Speer's officials allowed such delays? Why did Buchenwald lie idle under the supposedly capable hands of Schieber and Gustloff's management? "I am asking for the speedy delivery of the necessary machines. Work space and work force are already available."[65]

Himmler continued to object to the failure to meet the original quota of 75,000 guns a month, and at the end of a long letter to Speer he requested: "In the entire matter I ask you if it might not be better when the firm transfers its technical personnel to us so that we may work at the tempo to which we are accustomed, such that we may work as self-responsible people."[66] And, if nothing else, Himmler suggested that Speer should at least install some other useful production in the factory halls instead of allowing the Gustloff Works to dawdle. Dismissive of all reports of managerial difficulty, confirmed in his belief in the "restless energy" of the SS, incapable of believing that any other organization knew the Third Reich's interests better, Himmler had jumped to the conclusion that all delays and excuses were humbug. "At this time I believe that we should appear often and unannounced at the factory in order to force the pace of work with the bludgeon of our word and to help at the work site with our own energies," he declared to Pohl.[67] This was nothing other than the primacy of policing, and Himmler naively hoped to apply it to Gustloff's management.

Himmler's letter was, then, a real suggestion to break the agreed division of managerial tasks concluded between the WVHA and the Armaments Ministry one year before. Then the SS had pledged to leave its hands off technical supervision. Now Himmler wanted to extend the "bludgeon of his word" to Gustloff's managers, whom the SS suspected of subverting war production. The brutality of Himmler's language has understandably misled even careful historians to believe that the SS's motive here was control and autonomy in armaments production. Nevertheless, it is interesting that the SS did not, for instance, request ownership of Gustloff's machine tools, nor is there any record that the Waffen SS demanded a proportion of the guns produced.[68] Himmler took pains to convince Speer that the larger good of German war production was his only concern:

I cannot help the impression that the Wilhelm Gustloff Works, which has taken over the representation and technical support of this factory in the camp Buchenwald, somehow feels that this factory is a thorn in its side.

The firm has the silliest fears that these assembly halls could later become competitors to the Gustloff Works. I have clarified once again to these gentlemen that in the coming peace I have completely different ambitions than to become a competitor in this sector. I see the tasks of the SS in peacetime, as you well know, in the area of settlements and every other sector where I might promote families well endowed with children and whose healthy lives I might encourage.[69]

Thus despite Himmler's bullying, his letter expressed his continuing long-term commitment to the grand cultural crusade of the New Order. Yet Speer reacted quickly to assuage what he thought was the SS's true desire: "I consider it right that the SS should maintain its own carbine assembly in peacetime. . . . I consider it more tactful, however, when the current relationship between all parties remains as it is."[70] But Himmler had never asked for gun factories in peacetime. He was far more preoccupied with settlement dreams, as he *had* expressed.

The next months only confirmed and aggravated Himmler's impatient desire for action. To the Armaments Ministry, however, the problems with slave labor at Buchenwald proved typical of the other pilot projects as well. Schieber had arranged for Krupp to install a fuse plant at Auschwitz; Siemens, an assembly plant at Ravensbrück; and the firm Franke, a Howitzer plant at Neuengamme. Similar difficulties arose in each case.[71] At Buchenwald, the factory often sat idle. Long-ordered machinery did not arrive. Worse yet, the Gustloff Works began to contract out to other firms.[72] The Kommandant of Buchenwald sent Himmler a steady stream of telegrams insinuating that the main culprits were Schieber and Gustloff's director. "Work halls stand at their disposal as do machines; we have the best will and also the workers," he wrote in August. "In spite of this we cannot produce either carbines or the G43 [machine gun]."[73] "The G.W. [Gustloff Works] has also avoided the assembly of weapons, although the largest part of the necessary machines have already been assembled in the Buchenwald halls."[74] "Should SS Brigadier General Schieber question whether Buchenwald has reserved factory halls for the war-decisive assembly of the G43, let it be mentioned that there is still enough space in the Buchenwald works."[75] The SS's productivism led Himmler and Pohl to see organization itself as a manifestation of united Nazi community; it was therefore wholly consistent, when faced with delays and crumbling plans, to diagnose the problem as a "lack of will" and perhaps even treasonous opposition to the *Volk*. In addition, because the SS lacked competent produc-tion engineers, the reality of knotted managerial impasses, juggled priority ratings, and the difficulties of adjusting assembly to forced labor were all literally beyond the comprehension of Office Group D, Office Group W, and Himmler himself.

Despite misunderstandings and mutual antagonism, the SS nevertheless remained more or less committed to its settlement goals and did not begin a marauding crusade to take over the German economy or even individual armaments factories. And no matter how heated mutual misunderstandings became, there is no reason to construe Himmler's statements to Speer about settlement goals as a rhetorical camouflage. In internal correspondence on the eve of the German invasion of the Soviet Union, Himmler had informed Pohl—to whom he had no need to conceal "hidden agendas"—"You know well my opinion that our economic interests have quite narrowly bounded goals. They are essentially confined to tasks which make possible the increase of good and worthy blood [i.e., the "German" race] through nutrition and housing [i.e., SS settlements]."[76]

Much evidence speaks for the continuity of these intentions. When Himmler actually did camouflage his intentions, he did so to hide settlement projects long after authorities had forbidden "peace building." In the spring of 1942, for example, at roughly the same time the Buchenwald project began, Himmler had run into problems with the Eastern Rail Road, the state-controlled rail lines of the General Government and, notably, those of Lublin. Here the SS's gargantuan plans called for a new concentration camp (Majdanek) with 150,000 prisoners as well as settlements for 60,000 SS men. This would have more than tripled the population of Lublin. The city's water supply and other infrastructure threatened to break under the added burden. Unsurprisingly the demands also strained the Eastern Rail Road to the limit. The Lublin district counted as the breadbasket of occupied Poland. It was important not only for military traffic toward the front but also for desperately needed agricultural transports. Therefore the Transport Ministry informed Himmler that it would postpone all future shipments for his grand settlement projects until the conclusion of hostilities. Himmler tried disingenuously to insist that the SS's construction served to supply combat troops with needed equipment:

> This prisoner of war camp [Majdanek] contains the necessary emergency buildings for the occupation of prisoners of war and likewise other prisoners in which, under the orders of the Führer, necessary equipment shall be produced for the units of the Waffen SS and Police in the East. Among other things, this shall serve the purpose of reducing, even eliminating, the burden of supply industries in the Old Reich.

In closing, Himmler informed the ministry that "for the peace, Lublin will become the primary location of the Waffen SS and Police," but then claimed that this construction had already been postponed until final victory. He concluded tersely, "Thus your demands . . . have already been met."[77]

Although Himmler assured the Transport Ministry that everything afoot in Lublin was in the name of urgent military supply, SS men were right then planning to expand the "strongholds" from six to eighteen. Himmler had just met with civil administrators in Kraków and declared "that he lay much worth upon . . . the quickest possible renovation of the historical city center of Lublin and . . . the same with the marketplace in Zamosc. . . . Just as quickly the families of German farmers from Transistrien should be further settled in the region of Zamosc." All was "to be carried out within this year."[78] The WVHA had conceived the industries founded in Lublin as support for these projects, and one of its largest operations, a textile processing plant in an abandoned airplane hangar, took in the clothing from the murdered Jews of Belzec and other killing camps, cleaned and searched garments for valuables, and then redistributed them to "ethnic Germans." Thus rather than an independent basis in armaments, Himmler and his SS men were pursuing genuine ideological commitments to the regime's New Order, even as total war encroached.

To return to Himmler's tiff with Speer, we should recall that the armaments minister had taken up Schieber's initiatives in March 1942 and personally requested Hitler's approval for concentration camp production. Due to the bungled pilot projects and, no less, Himmler's irritating intervention, by 20 September Speer turned around and asked Hitler to limit the SS's influence. Speer's notes are worth quoting at length, because they reflect the mutual, often willful misunderstandings that had quickly arisen over concentration camp factories like those of Gustloff at Buchenwald:

Called to the Führer's attention that—beyond a small amount of work—it will not be possible to extract meaningful production from the CC [concentration camp]. Here there are neither (1) necessary machine tools nor (2) necessary buildings, as opposed to industry, where both are on hand for two production shifts. The Führer agrees with my suggestion, according to which various factories will be moved outside of population centers because of the danger of air attacks. They shall transfer their existing labor force to factories remaining within the cities in order to fill second shifts, and then the factories moved outside population centers will receive replacement workers from the CC, enough for two shifts.

I have called to the Führer's attention that I see in the demands of RFSS Himmler a desire to exercise a measurable influence over these factories. Likewise, the Führer considers such an influence undesirable. The Führer agrees that the RFSS expects to derive advantages for the equipment of his military divisions by putting CC prisoners at the disposal of armaments industries. *I suggested to him that Himmler be offered a percentage of the weapons produced in the CC.* It would be a share of something like 3 to 5 percent, with which the Führer agreed. The percentage will be reckoned according to the total work hours contributed by prisoners. The Führer is prepared to order the supplementary disbursement of such devices and weapons after being presented a list.[79] (emphasis added)

This audience with Hitler is often cited as evidence of Himmler's persistent encroachment, no less, of Speer's diligent attempts to brake SS expansion as a "state within a state."

Yet, to the contrary, little had changed from the year's beginning in the sense that a relentless labor shortage continued to draw the SS ever closer to the center of the war economy. We should not forget that these were days of hurried reorganization. September had thrown the entire German war effort into tumult. Summer offensives had stalled, and the Wehrmacht had failed to secure the Soviet oil fields of the Caucasus. Franz Halder, chief of the army's general staff, resigned. Hitler cashiered the responsible general of Army Group A, Field Marshal Wilhelm List, in order to take personal command of operations. By 15 September, the Red Army launched its first counteroffensives (on the upper Don) that would lead, eventually, to the destruction of over 280,000 Axis troops and their equipment in the cauldron of Stalingrad. All of this created an air of urgency in September. Everyone—not only Speer and Himmler—sought new solutions to long-standing problems.

At exactly this time, the Plenipotentiary for the Reich Labor Action, Sauckel, first began to miss promised deliveries of conscripted workers from the East. As Schieber soon reported, Sauckel even falsified and exaggerated estimates of those deliveries he was making. Thus, what some had formerly considered an "inexhaustible" supply of labor was now dwindling to a trickle. Speer was just then planning to introduce a two-shift workday in order to exploit the only untapped resource left at his disposal, time. But two shifts, of course, also demanded double the number of workers. Therefore he proposed that factories consistently targeted for bombing raids should be moved away from population centers; at the same time, their laborers were to remain to fill out second shifts in secure factories. Speer hoped to rely on Himmler's camps to fill factories relocated outside of population centers.

If Speer would have us believe that he was heading off the SS's "infiltration," it is curious that five days *before* his late September meetings with Hitler he also received Oswald Pohl and the chief of SS engineers, Hans Kammler, with whom it "was agreed upon that inmates of concentration camps will be made available for armaments industries."[80] An armaments inspector (no lackey of the SS) made this clear to the general governor of occupied Poland, Hans Frank: "The Reichsführer SS *as well as Reich Minister Speer* and the director of the [Reich] Labor Action, Gauleiter Sauckel, are placing increased worth on the action of Jews fit for work" (emphasis added).[81] Speer's September discussions with Hitler and the SS therefore represented plans to increase, not curb, the SS's Labor Action. This proceeded within the context of intensified, if vexed, collaboration.

Speer's offer of 3 to 5 percent of all production was a device of the minister's own making, as his own notes from his audience with Hitler record; no evidence suggests that it was a "concession" to SS demands. With Hitler, Speer openly discussed fears that Himmler wished to leverage the Waffen SS into a position of "independence in armaments." To be generous to Speer, this was not without some small foundation, for the SS made no secret of its quest for economic independence, albeit only in the building sector. Local construction officials at Kattowitz, near Auschwitz, for example, accused the SS of trying to become a "state within a state" because it would not subordinate itself to regional development plans.[82] Fear of the SS's infiltration also seems to have been widespread amid the general paranoia of the Third Reich's top leaders, and Speer apparently genuinely believed the SS's petitions for autonomy in construction betrayed a secret agenda to become self-sufficient in every conceivable field. One could believe this, however, only by discounting the SS's genuine goals, namely the New Order.

There can be no doubt, as Jan-Erik Schulte has claimed, that the SS experienced a kind of "armaments euphoria" in late 1942. Who would expect the self-appointed bodyguards of the National Socialist People's Community to do anything else than seek to contribute their utmost—especially when so many other authorities were bidding them to do so? Himmler also held fast to his belief that the "bludgeon" of his word could force the pace of war production to new heights. Nevertheless, the SS hardly threw itself into a mad dash to control the war economy or make the Waffen SS self-sufficient in weapons production.[83] Pohl's own report to Himmler of his meetings with Speer expressed enthusiasm for cooperation:

All are agreed that the reserves of labor in the concentration camps must be deployed for large-scale [*Grossformat*] armaments tasks only. . . . In order to take over large, self-contained armaments tasks, we must drop one of our basic demands. We must no longer stubbornly require that all production be enclosed within the bounds of our concentration camps. As long as we are content with small crumbs [*Kleckerkram*]—as you yourself, Reichsführer, have correctly characterized the extent of our work up to now—we may hold to this demand. If we wish to take over an enclosed armaments factory with 10,000 or 15,000 prisoners in the days to come, it is not possible to erect such works *intra muro*. The work must, as Reich Minister Speer correctly maintains, lie on the open fields. Then we may put up an electric fence around it; then we can provide the necessary number of prisoners; and then the factory can run as an SS armaments works.

The entirety of SS armaments factories will then represent a contribution to the armaments program, and may, as Reich Minister Speer himself states, be placed directly under you. In which case you will answer to the Führer directly. Currently operating factories will be vacated and their [civilian] workers transferred to bolster the capacity of other works. Then these remaining, empty factories will be filled 100 percent with our prisoners. . . . Accordingly, *Reich Minister Speer expects the immediate mobilization of 50,000 Jews who are fit to work* for enclosed factories that are currently being emptied.[84] (emphasis added)

Here, Pohl first appealed to Himmler to ease security requirements for forced labor. In this, he actually opposed his superior's expressed intentions in honor of Speer's requests. Second, Pohl emphasized Speer's statement that some few armaments factories might be transferred to the SS's control. Pohl then proceeded to the most pressing action item. Namely Speer was now asking for twice the number of prisoners promised in the spring. Regarding these 50,000 Jews, Pohl added cynically, "The necessary laborers, primarily from Auschwitz, will be taken out of the eastern immigration. . . . The Jews who are fit to work must therefore interrupt their immigration to the East and undertake some armaments work."[85] It is therefore no accident that Speer approved the release of scarce building materials for the expansion of Auschwitz at this very meeting.[86]

The Armaments Ministry had also agreed to place "SS armaments works" directly under Himmler, and Himmler indeed seems to have eagerly desired this new responsibility. The past pilot projects had disappointed him no less than Speer and Schieber. He thought the SS was capable of more. For their part, Speer's officials followed up quickly. On the day after Pohl and Speer's meeting, for example, Krupp received a telegram from the Reich Ministry for Armaments and Munitions offering concentration camp labor, and a few days later Krupp requested 1,100 to 1,500 prisoners.[87] Pohl had dutifully reported some minor conflicts to Himmler: "Our attempts up to now have run aground on an obstacle which, to my great surprise, I have discovered in the immediate proximity of Reich Minister Speer. Here the name [Karl Otto] Saur plays a key role."[88] The exact nature of Saur's objections are unknown, but he got over them three days later when he selected specific factories as "SS armaments works." These, however, never came to fruition, likely because it was easier to pursue the path taken by Krupp and transfer prisoners to existing factories under competent technical management. In all these dealings discussion of any 3 to 5 percent cut of production yield was conspicuously absent. Likewise, Pohl made no mention to Himmler of any intent to make SS companies the direct suppliers of the Waffen SS.

On the other hand, Pohl's report hardly avoided talk of genuine motivations. Only half of his communication to Himmler dealt with armaments, while exclamation points sprinkled the second half, far overreaching any excitement over proposed "SS armaments works":

> We are ready!
> I have discussed the organization of Building Brigades. . . . At the discussion meeting today that took place in the WVHA, State Council Dr. Schieber and Ministerial Council Dr. Briese representing the Reich Minister Prof. Speer also took part. In our Building Brigades I see the beginnings of our later peace Building Brigades that will develop and build.
> It will work![89]

Pohl made no secret here of specific long-term intentions, but these had never wavered from the program of the New Order. The Armaments Ministry even agreed to place orders for building materials with the German Equipment Works, and, as Pohl told Himmler, by fulfilling these duties the SS would be assured a seasoned organization ready to embark upon the peace building program after the war. These were the initiatives that the SS brought to the table, and, once again, Speer approved them.

Taken in by Speer's self-glorifying claim to have staved off an SS grab to control armaments production that the SS never really made, some historians have credited Speer with high political genius and a "skilled negotiation strategy."[90] Likewise, even Gregor Janssen, whose tremendous research of Speer's ministry is unparalleled, comments that "[Speer] had made an infiltration in his sector as hard as possible for as long as possible for Himmler's SS."[91] This was, of course, Speer's version of the tale, and it suggests that only pure power was at stake, a drive to increase institutional independence and prestige for its own sake. Above all, it obscures the potent influence of ideological commitment to the Third Reich across institutions. Constant attempts at cooperation *as well as* power politics drove these negotiations. Rather than a war economy in which the SS was a feared "infiltrator," taking the SS's motives seriously yields a different picture: one in which others sought out the SS and saw

Himmler as a legitimate ally, one in which men like Pohl or Schieber operated not as "spies" but as networkers.

"Polycratic" governance always gave rise to as many partnerships as conflicts. Speer cooperated with the SS's security service, for example, in order to enforce the measures put forward by his ministry, especially when these proved unpopular with other Nazi potentates. He and his wife entertained high-ranking SS celebrities at home, and Himmler came to visit him on his birthday. Speer also trusted deputies like Walther Schieber, who were esteemed SS men, as his liaisons, and they were no doubt valuable precisely because they had both professional ties and ideological sympathies that Speer could use to his advantage. During the Armaments Ministry's struggles to preserve the industrial might of Hitler's Germany, their commitments rarely gave Speer pause for thought; indeed, many of these men continually produced the results he wanted. By constructing a false dilemma, a supposed opposition between pragmatists and ideologues or between technocrats and fanatics, the fabric of the everyday management of Nazi barbarism has been whitewashed of the complicity of "ordinary" German managers. During the Third Reich, it was the SS's network of slavery that came to seem "ordinary," a barbarism that relied as much on the services of the civilian engineer as it did on the concentration camp guard.[92]

By the end of 1942, moreover, the reforms afoot within the Office Group D2 were beginning to render the beaten and starving bodies of concentration camp prisoners in the form of modern, statistical information. Armaments Ministry officials could use this information and they did. The WVHA would not and could not overcome all the destructive influences entrenched in the IKL, which still ran roughshod over many labor sites. But with Gerhard Maurer at the helm of the SS Labor Action, the Office Group D could now oversee allocations of slaves to private and state industries. If the camp SS continued to kill and beat up prisoners, Maurer's statistics proved an effective tool for circumscribing them with numbers. Camp personnel still zealously eliminated the "unfit," and so his system appealed to their own conception of their work as punishers and purifiers of the Third Reich. Meanwhile Maurer's officers, with the help of doctors and orderlies, kept rigorous count of the "fit" in order to make fresh deliveries to the production lines of German firms. The end result was a relatively stable number of working prisoners in the camps beginning in late 1942. The WVHA began to accommodate *both* the brutality of the Death's Head Units and private German industrialists' demands for fresh workers.

CHAPTER 6

THE HOUR OF
THE ENGINEER

As Pohl struggled with the SS Labor Action, Hans Kammler had consolidated a supple, competent corps of civil engineers for the "peace building program" of the New Order, but those foundations were laid in the midst of a massive transition to total war. Heinrich Himmler's utopia was constantly postponed as the Red Army proved more tenacious than the German leadership had ever imagined. The war had to be accommodated and Kammler accommodated it well. In the minds of many SS men, adjusting to war meant a more extended fulfillment of their principles: the glorification of war, the Führer principle imposed in the trenches. And, especially in the East, war provided a kind of moral theater in which the Aryan was pitted against the "Asiatic" races. It was an acknowledged precondition for the New Order that Germany must make a clean slate of the conquered lands of Poland and the Soviet Union. The engineers of the Office Group C threw themselves into this effort with passionate intensity. They built the crematoria and gas chambers of the first death camps in the Lublin district and then those at Auschwitz in the midst of total war.

Wartime also seemed to open up opportunities for the SS's brand of productivism and anticapitalism in a way that appealed to the engineers. Whereas Kammler had ridiculed the peacetime economy for what he called the "salesman's point of view," the frenzy of total war now suspended all limitations of traditional markets. Normally, construction calls for sober, long-term, fixed investments, but increasingly cost no longer mattered. The German state now sought to mobilize every last drop of capacity in the construction sector, and buildings became akin to short-term investments like disposable goods: thrown up rapidly, abandoned along moving fronts, reduced to rubble, and built anew. Quotas and raw yield began to replace prices as the quantifiable measures in accounts. This was the kind of economy that Kammler and his cadre had wished for all their professional lives. They could now work to the limit of their creativity and engagement in an economy dictated solely by the Nazi struggle for national glory and survival.[1]

Amid total war, the engineers under Kammler's office came into their own in what might be called the Third Reich's "Hour of the Engineer." The Office Group C's corps of engineers had skills that the war economy demanded, and they constantly drew projects into their ken. Their strong sense of collective identity *along with* their sound technical abilities meant that they could offer what other branches of Pohl's Business Administration Main Office could not: a devastatingly effective modern bureaucratic hierarchy that could manage ever more complex tasks and get them done on time. Kammler quickly developed a specialty in the management of slave labor on far-flung construction sites and also achieved a level of cooperation with like-minded men outside of the SS. Only Gerhard Maurer came close to achieving any similar success in the WVHA's other office groups. (In fact, a close relationship with Maurer was also important to the SS corps of engineers.)

Himmler eventually tried to use Kammler's expertise, especially his vital role in the V-2 rocket program, to leverage the takeover of "miracle weapons." Yet at the level of daily industrial management, complicity and cooperation were far more common than histories of intrigue have conditioned historians to believe. But before turning to the well-known story of the SS's grab for the rockets, the Messerschmitt 262 fighter-bomber, and other celebrated high-tech weapons, we must first turn to Kammler's rehearsals for total war. Reich ministries, municipal authorities, and industrialists alike began to petition the WVHA to build their buildings. The SS never had to "infiltrate" the armaments sector; the armaments sector came looking for the SS engineers. If we make the credible assumption that the multiplicity of institutions in the constantly changing war economy created as many opportunities for cooperation as for struggle, the reality is somewhat less suited to spy novels and accounts of an SS bid to "control the economy," but it is also more sinister. Himmler, Kammler, and Pohl entered German industry as invited guests, mutual players, and willing allies. In the era of total war, the SS rapidly became a normal business partner in the Nazi economy, and it was slavery that brought the SS, private industry, and the supposed "technocrats" of the Armaments Ministry together. If this chapter is about the hour of the engineer, it is also about the growing routinization of horror that filled every minute of that hour.

Rehearsals

The first war projects undertaken by the Office Group C were modest, but each project met the demands of its customers. As already mentioned Ferdinand Porsche first requested Kammler's services to build a light-metals foundry at his Volkswagen Works in Fallersleben. Of note, this was not yet a war project but had more to do with Porsche's dream of a vertically integrated factory for the "people's car," the famous VW Beetle. Porsche's primary objective had been to secure prisoners to construct the plant. He also wished to get building supplies through the SS that Volkswagen could not obtain alone. Industrialists like Porsche saw construction as a necessary but temporary start-up phase. It demanded a different array of mobile machinery, raw materials, and workers, which had little to do with normal factory operations. Cooperation with the SS would thus free his hands to deal with matters of design and factory production, his specialty and primary concern. Kammler's Building Brigades had begun to establish expertise and a captive labor force.

Himmler and Porsche both lobbied Hitler for the permission to go ahead with a light-metals foundry (for casting aluminum engine blocks and other items). After Hitler granted approval, Himmler left Kammler no time to dally: construction on the foundry had to be completed in order to start operations in the fall.[2] Initially these affairs followed the same pattern set by the WVHA's relationship to industries like Gustloff or IG Farben, but Kammler established a different precedent. Here again outsiders to the SS had asked for its services. Porsche had gone directly to the Reichsführer SS, bypassing the regular channels of the Armaments Ministry and the army. As already shown, he proposed to equip Himmler's Waffen SS with amphibious vehicles as a special sweetener to the deal. In return, the SS had pledged eager cooperation and made exaggerated promises. Kammler's role differed in this case, however, for whereas the Inspectorate of Concentration Camps and the WVHA failed to live up to the standards of industry, Kammler delivered. He completed the Fallersleben project, with only slight delays, by the late fall of 1942. Both Albert Speer and Porsche declared themselves "satisfied with the progress of construction."[3] Some historians insist that the entire project came to life "in the framework of plans to erect an SS industrial empire."[4] But this makes sense only if we insist that the SS was only out for raw power. It ignores the fact that the SS's industrial empire had already long been in the making and conformed to a cherished program of settlement development and other cultural crusades. RKF planners and WVHA business executives alike shared an enthusiasm for automobiles, but nowhere did car factories figure in their utopian plans.

Despite mutual cooperation the foundry never came to fruition. Even before completion, the Armaments Ministry slated the Volkswagen Works for heavy tank production and canceled the production of the vehicles for which Porsche originally intended the foundry. Hitler had also withdrawn his direct support after losing interest. Such changing priorities are hardly unusual in any war economy, but the project remained significant as a kind of "dress rehearsal" for the Office Group C. The SS had mobilized its Construction Directorates and Building Brigades for heavy industry for the first time. Furthermore, Kammler had begun to form an effective partnership with Office Group D2 (under Gerhard Maurer) for the supply of prisoners to his building sites. He even seems to have intervened in an extraordinary move, with Pohl, to place a technically competent Kommandant in charge of "Arbeitsdorf," the satellite camp of Neuengamme set up for the foundry project. The Kommandant, Martin Weiss, was a trained electrical engineer whom Pohl had found among the ranks of the camp SS. Kammler saw to it that productive work mattered at Arbeitsdorf and violence was curtailed, at least to the extent that it disrupted labor sites.[5] Kammler (and Weiss) could also speak the same language of modern managerial statistics that was common parlance in the Armaments Ministry and industry. No altercations arose between the Office Group C and VW as they did between Gustloff, Glücks, and Pohl.

Perhaps Buchenwald's Gustloff Works best underscores the difference between Kammler's organization, which brought a consensus of purpose *and* technical competence to its tasks, and the IKL, which also nurtured consensus, to be sure, but one that worked to destroy any meaningful production. As Walther Schieber struggled with the SS to bring the first pilot factories into operation at the Gustloff Works, Gustloff had not only asked for new production halls; the firm had also requested a new rail spur to connect its original factory in Weimar to the camp perched in the hills above the city. While Gustloff engineers began to marvel at the inability of Hermann Pister, the Kommandant of Buchenwald, and began to fear Himmler's inability to understand the technical difficulties of factory work, and while Glücks failed to deliver skilled prisoners in the numbers specified by the firm (as shown in chapter 5), Kammler quietly completed the necessary construction. He began in July 1942 and predicted completion in June 1943.[6] According to the customary division of responsibilities—the same agreed upon previously with Porsche—Kammler left architectural planning to Gustloff's engineers and erected the factory halls in Buchenwald according to their specifications.

Here, too, the Office Group C rehearsed the skills that would propel it to the center of Hitler's total war economy. Its engineers continually forced the pace of construction and even pushed through political obstacles over the rail spur from Weimar. When the Gauleiter of Thüringen, Fritz Sauckel, objected to the ugly swath it cut through the picturesque wooded hills of his capital city, Kammler petitioned for the support of Speer and Himmler to override Sauckel's influence as well as objections from the German Railroad.[7] Thus while Gustloff and the IKL bickered, cooperation with the firm and Hitler's armaments minister was more usual at construction sites. Kammler proved adept at mobilizing both building materials and political clout in unison.

He also began, apparently for the first time, to substitute labor-intensive means of construction when he could not get labor-saving machinery. Here, too, the contrast is instructive. Engineers of the German Commercial Operations, as shown in chapter 2, repeatedly opted for "sweet machines" even when these proved totally inappropriate. Kammler applied modern management with no such scruples (even when it meant abandoning modern construction methods temporarily—in the service of high-tech projects). Time was his demon, and he aimed to beat it by any means necessary. "A steam shovel has been clogged with mud and broken down," he reported. "Nevertheless, manual laborers and increased numbers of prisoners have been sent into action. We will try to make up the loss of work through multiple shifts of work."[8] The civil engineers were more flexible than the factory engineers under Erduin Schondorff. When the Office Group C had labor-saving machines and the fuel to run them, it used them to accelerate construction; when machines could not be had or broke down, it drove prisoners as brute manual laborers without hesitation and again accelerated production. Kammler finished the rail spur on time, to the satisfaction of both Gustloff and Speer.[9]

By later standards, projects like Fallersleben or the rail spur at Buchenwald were small. Yet Kammler had successfully mobilized concentration camp crews and simultaneously focused his resources on multiple projects in a nationwide managerial hierarchy. Whereas Pohl's industries increasingly began to falter in the face of raw materials rationing and internal disputes, demand for SS construction management grew. Men like Schieber and Speer came to see Pohl as a fool, but Kammler began to earn a solid reputation within the larger universe of armaments engineers, planning experts, and private industrialists. SS civil engineers developed an expertise that few other Reich institutions could match and many needed.[10]

Within the SS, Kammler also made it clear that he would brook no interference from incompetent management, and here his own initiatives on behalf of the war effort could not have differed more from Pohl's. During these "rehearsals," Kammler's firm control over all SS construction was challenged only once. As shown in chapter 5, over the course of 1942 Pohl had created new SS Economic Officers (SS-Wirtschafter) within the staffs of the regional Higher SS and Police Führer (the HSSPF). They were to serve the WVHA in all matters of SS industry, forced-labor allocations, and concentration camp administration, and, initially, in construction as well. As exemplified by the brawling, drinking, foul-mouthed SS Economic Officer Erich Schellin, in practice those appointed by Pohl often failed the WVHA. They were usually technical incompetents, neglected production, and identified themselves as cronies of their Higher SS and Police Führer. The same might have happened to Kammler's Office Group C once these economic officers took charge of building. Pohl's directives of July 1942 gave them sweeping power over Kammler's Building Inspections, making "the SS Economic Officer . . . the highest construction authority for projects within the jurisdiction of the Higher SS and Police Führer." Pohl's directive went on to remove the regional Building Inspections of the Office Group C from the direct supervision of Berlin and to place them instead under each HSSPF's Economic Officer. Potentially, this stripped Kammler of the control he had sought to establish since 1941, yet Pohl did add, "Technical directives are issued by the SS WVHA."[11]

By reinforcing Kammler's authority over technical matters, Pohl provided the chief of SS engineers with a vice grip that he quickly used to hold on to construction within the fiefdoms of the Higher SS and Police Führer. As already intimated, none of the SS Economic Officers possessed technical training. Pohl had chosen them for their long-standing service to the SS, but for the same reason none could challenge either the technical competence of Building Inspection officers or their well-oiled organizational machine. Furthermore, Kammler had succeeded in creating a hierarchy of subordinates that identified with his institution. As a rule, they did not switch their allegiance to serve incompetent henchmen of the HSSPF. On several occasions SS civil engineers received assignments to serve SS Economic Officers but did not integrate themselves into the administrative structures of the HSSPF. Instead they immediately began to fill out Kammler's standardized forms and reported directly to Berlin. Kammler also intervened in the affairs of the SS Economic Officers in order to assure that regional SS and Police Führer did not meddle with construction.[12] The IKL or the German Commercial Operations might have done the same thing—at least hypothetically—but, without an interventionist technical staff that identified with the WVHA's overarching goals, Pohl proved incapable of imposing control.

Scarcely two months after Pohl issued his orders establishing the SS Economic Officers, the WVHA issued new instructions reinstating the authority of the Building Inspections and releasing them from the Higher SS and Police Führer. The HSSPF continued to maintain building groups, but after 1942 Kammler integrated these too within his branching hierarchy (as the equivalent of Central Construction Directorates). Himmler repeatedly reinforced the supreme authority of the Office Group C in any and all matters of construction and also removed the SSPF Globocnik from settlement development in order to put Kammler in his stead.[13] From this point on, the authority of Kammler's organization within the SS only increased. Eventually it became unassailable.

The SS and the Rocket Team

Kammler's assistance to the Volkswagen Works foreshadowed the Office Group C's future trajectory. As SS business executives testified after the war, "The Reichsführer SS Himmler and, following his example, Pohl supported all inventors on principle."[14] Porsche was one of the most renowned automobile designers of the century, and he had invited the SS's participation in one of the most futuristic industries of his day. The German Commercial Operations had long indulged that impulse in the form of the needlessly complex "dry press" of the German Earth and Stone Works or the mass-production equipment of German Noble Furniture. It should thus come as no surprise that, besides Volkswagen, Kammler's engineering corps came to the timely aid of the one research and development project that would receive the most attention from Germany's military establishment, from Adolf Hitler, and indeed from the postwar world, namely the V-2 ballistic missile. The aura of futurism clung to these rockets, and who could deny, even today, that they represented the most advanced and portentous weapons developed in Hitler's Germany? In the entire world of the 1940s, only the atom bomb rivaled the V-2 in this regard. Himmler, so easily blinkered by technological wonder, could not help but be attracted to this new weapon, and Hitler, who compulsively filled his head with endless statistics on the German army's latest hardware, shared in the hype.[15]

Yet as Dieter Hölsken has pointed out, only when the systematic deployment of conventional German military might failed and failed spectacularly did Hitler turn his hopes toward fabulous "miracle weapons" in order to keep waging the war he had begun in 1939. Hitler chose to fast-track the V-2 project only in late 1942, that is, only when Germany began to lose and lose decisively, defeated at Stalingrad and retreating in North Africa. In fact, until 1944 the ballistic missile, when it had been known at all, was known as the A-4. Werner von Braun's research and development group had simply made a straightforward abbreviation of "Aggregate 4," as opposed to "Aggregate 1" through "Aggregate 12"—each referring to specific missile designs. But the "V" in the popularized propaganda name stood for *Vergeltung*, that is, "retribution." (The "V-1," or "Buzz Bomb," was the world's first cruise missile developed by the German air force.) This wordplay with "miracle" or "vengeance" weapons bespoke the atmosphere of high-tech desperation that drew the SS into the project. Outmatched and outproduced by the combined strength of the United States, Great Britain, and the Soviet Union, at sea, in the air, and on land, the German armed forces began the full collapse that would end with the fall of Berlin on 8 May 1945.[16] Hitler and his closest devotees like Himmler clung to the belief that fabulous inventions could somehow wreak revenge and reverse the tide of ruin that now began crashing down upon the Third Reich.

When Wernher von Braun's secret experiment station at Peenemünde on the island of Usedom achieved its first successful launch on 3 October 1942, General Walter Dornberger, the rocket project's champion in the German army, lost no time in capitalizing on the effort. He sought Speer's and Hitler's approval to make the transition to "mass production" just days after the first success. Getting Hitler's full support came slowly, however, but once it did come it provided one more charm that brought Speer and the SS closer together in an ever more fatal attraction. Speer had assigned a production expert to Peenemünde in mid-1942 (Dietmar Stahlknecht) and had directed other resources to von Braun's "rocket team" long before, even against the wishes of his predecessor (Fritz Todt), who had expressed measured indifference to rocket development. Ominously the first launch, late in 1942, fell shortly after Speer and Himmler's negotiations to increase the SS Labor Action in order to feed "self-enclosed" armaments plants. It was not long before the production experts at Peenemünde began to petition the SS for prisoners as well in order to make the modern manufacture of this high-tech weapon possible.[17]

Von Braun and Dornberger had talked about "mass production" since the project's conception, but it was not without reason that they came to be known as "rocket scientists" after the war, even though they were really "rocket engineers." Many of them held degrees as Diploma Engineers or Doctor Engineers, and von Braun actually resented the American press's habit of referring to him as a "scientist" after the war. Nevertheless, these engineers worked primarily on research and development. Industrial engineering was foreign to them. They possessed superb technical training and, in the case of von Braun, even genius, but they did not understand the key components of the production floor, its scheduling, or the problems of scale that a factory, unlike a laboratory, requires. Arthur Rudolph, chief engineer of the Peenemünde factory, joked about this after the war:

And so one day von Braun and Dornberger and I were sitting together. . . . out of the blue sky [Dornberger said], "I want to build a . . . production plant for the V-2 and the coming big rocket, and you will do that." I [replied], "Dornberger, for heaven's sake, I'm a development man, and you leave this up to industry; don't bother us fellows in development with your new ideas." And von Braun was of course saying the same thing, even harsher than I did.[18]

While the rocket team engineers often expressed enthusiasm for modern manufacturing, they did not understand it all that much better than Oswald Pohl. This can be seen in early efforts to prepare for "mass production." In February 1942 von Braun's team set a grand target of 5,000 rockets within the year, a figure it did not come close to meeting. Some bandied about figures of 150,000 rockets a year (including the Führer). Their naiveté was evident in some of their pronouncements: "It is to be presumed that these numbers [5,000 a year] represent a minimum. In the case of a higher demand it would then be necessary to increase the essential quantities of needed materials by the corresponding multiplier."[19] At least initially, the research and development team believed it could simply ramp up manufacture by linearly multiplying orders for supplies by whatever factor increased output demanded. The idea that larger scale also brings increasing organizational complexity, the need for hierarchical managerial bureaucracies, and close attention to scheduling, supply, and distribution; or the idea that manufacturing would demand—in addition to design for technical performance—choices to facilitate quick, reliable assembly by a large work force; all these considerations were absent until Speer's ministry undertook such steps at the end of 1942.

The Armaments Ministry brought to the rocket project the kind of industrial management Speer had been imposing upon the entire German war economy. Peenemünde was refitted for maximum production in a minimum of time. By the start of the new year, Speer had secured Hitler's personal approval for the serial production of 500 rockets a month, and the project now went forward under the aura of the Führer's watchful enthusiasm. Speer had also appointed Gerhard Degenkolb, a key future ally of Kammler and an engineer from the machine-building firm DEMAG. Degenkolb formed a Special Committee A-4 (after the prototype rocket Aggregate 4 developed at Peenemünde, the name for the V-2 used in internal correspondence). Degenkolb came fresh from revamping the production of German locomotives, in which he had forced the pace of output to 1,000 per month. Making railroad engines may seem an odd start for a new "rocket team" member, but Degenkolb had succeeded by reorganizing suppliers, shutting down inefficient plants, and streamlining subassembly. Most important for his later success in disciplining Peenemünde's research and development team, he had cut down on ongoing innovations in order to rationalize production. Most of his efforts, like those of any production engineer, were directed at an entire complex of suppliers, producers, and designers instead of the act of invention, research, or development.[20]

The urgent need to bring production engineering to bear on the project marked a crucial transition. SS engineers under Kammler formed close relationships with the new specialists chosen by Speer and Degenkolb, tighter bonds than would have been possible if the rocket team had remained in the firm control of von Braun and Dornberger. Both of these men developed an increasing antipathy toward Himmler's organization, and many veterans from Peenemünde also quickly developed a resentment toward the railroad man, Degenkolb, as well. They considered him ignorant of the rocket's elegant detail, unappreciative of their work, and even crude. General Dornberger, while granting him a certain intelligence, described Degenkolb in the most unflattering terms, as a man with "a completely bald and spherical head," whose "soft, loose cheeks, bull neck, and fleshy lips revealed a tendency toward good living and sensual pleasure."[21] But Degenkolb made his organizational talent felt immediately. Engineers in charge of subassemblies broke down manufacture of the entire rocket into 20,000 components, making each manageable as a unit; they then brought them all together in confluence on one main assembly line. The Special Committee A-4 introduced numerous branching subcommittees, each dedicated either to specific technical problems or to issues of supply and labor organization. From now on von Braun had to comply with the conditions of steady-flow, large-volume production.

Degenkolb also imposed disciplined scheduling for ongoing innovations. From now on those in control of manufacturing would allow alterations in progressive production batches only in the interests of serial production. Whatever improvements von Braun's research and development team failed to get ready by the set deadlines would be left out altogether until the next "batch" run, regardless of whatever enhanced performance laboratory tests might promise. Von Braun and his colleagues had been accustomed to tinkering with the assembly of each rocket right up to its preparation for launch. In other words, they had talked about "modern production," but they had really adopted a traditional, craft style of continual personal intervention as they fashioned each "Aggregate 4." It should be no surprise that Degenkolb took "mass production" out of their hands. He added the RAX Works in Wiener Neustadt to the expanding facilities at Peenemünde. The Zeppelin Works in Friedrichshafen had been chosen as an alternate production site in the winter of 1941–42, and he kept it but canceled General Dornberger's plans to build a new plant from scratch in eastern Poland. He recruited a production engineer from the Henschel Works, Albin Sawatzki, and shifted subassembly work to

Germany's most experienced modern corporations like Borsig and German General Electric (AEG).[22]

Labor remained an intransigent problem. Peenemünde had actually relied on compulsory labor as early as 1939 and 1940 for construction, first a few hundred Czechs and, later, Polish workers. These had worked for a special construction bureau subordinate, ultimately, to Speer's authority. Preceding the advent of Degenkolb's Special Committee A-4 and the acute labor shortages of 1943, the rocket project never availed itself of concentration camp inmates. Hitler had originally forbidden the use of any foreign workers, let alone inmates, in rocket assembly, because he feared sabotage and a breach in the program's secrecy. By 1943, however, there were no other sources of laborers to turn to, and once again the SS found itself invited into armaments production. At first the SS followed the pattern established by all previous SS ventures with armaments firms. Only this time, Himmler's sense of wonder over high-tech gadgetry and modern production leapt all bounds. But there was also one other crucial difference. Because of Kammler's slave-labor brigades, Himmler could now offer services that the rocket specialists could use. The SS men of the Office Group C could meet people like Degenkolb or armaments officials eye to eye on matters of technical competence. In addition, Maurer had, by then, gone a long way toward solving some of the problems caused by Himmler's propensity to promise slave laborers that the SS could not deliver. Now the SS was poised to deploy real managerial talent. Kammler's construction bureaus, linked as they were to the manpower reserves of the concentration camps, also stood out as reserves of untapped capacity. Degenkolb was preparing blueprints for factory halls that Kammler's corps of engineers would build out of concrete, steel, and the wrecked lives of countless prisoners.[23]

Although he later came to loathe Himmler, General Dornberger had already sent out feelers to the SS at the end of 1942. Nothing about the SS seemed unusual or unpalatable to him at the time. He hoped that Himmler would help him and von Braun gain an audience with Hitler. Like Porsche or Georg Meindl before him, Dornberger approached the SS informally through networks in which ideological sympathies and mutual interests overlapped. He first did so through a friend of Gottlob Berger's (the very contact used by Walther Schieber and one of Himmler's most trusted Main Office chiefs). On a second occasion, Dornberger also sent one of his subordinate officers to see Himmler personally. This was Gerhard Stegmaier of army ordnance, the military commander of Peenemünde's Development Works. Neither Dornberger nor Stegmaier held any formal connection to the SS, not even honorary membership; and neither was in the Nazi Party, as all Wehrmacht soldiers were officially forbidden from joining. Nevertheless, Stegmaier in particular was known to be a committed National Socialist. According to Dornberger's recollection, Himmler responded positively to the overture: "Should the Führer decide to give the project support, then your work is no longer an affair of the Army Weapons Office or of the army in general. You then belong to the German people [*Volk*]. I will undertake your protection against sabotage and treason."[24] Himmler's motives at this point were likely no different than they had been at the Gustloff Works, namely to force the pace of what he considered "war essential" projects, and it was characteristic that he mentioned "the *Volk*" in this context. The unusual technological glamour undoubtedly also spiced his excitement. Himmler was not alone in believing that the key to ultimate victory lay in the unified will of the people and the Führer, and he further believed that technological prowess manifested the superiority of the German race. Here was a miracle weapon that seemed to confirm these beliefs: nothing would be impossible, no matter how outmatched Hitler's armies might become in Europe, once the German "spirit" had been unleashed in this project.

In 1943 the Special Committee A-4 picked up on Himmler's enthusiasm where Dornberger had left off. To gear up for mass production, Degenkolb needed factory space. Both to build and to run the factories, he needed workers. Thus in April several members of the rocket team as well as Degenkolb's subcommittee for labor management took initiative and requested 2,200 concentration camp inmates for the RAX Works. The Zeppelin Works in Friedrichshafen, having already applied for prisoners independently of the Special Committee A-4, got them from Dachau by February. In the first days of June, Peenemünde also requested an additional 1,400. The Special Committee A-4 intended to use these first prisoners for construction but soon considered using forced labor in assembly itself. In the middle of April, Arthur Rudolph, the rocket project's assembly expert, had visited the Heinkel aircraft works. Heinkel had already enjoyed great success working with Gerhard Maurer and the IKL, and Rudolph's report showed his own enthusiasm for future cooperation with the SS:

The Heinkel Works have taken up contact with SS First Lieutenant Maurer and requested the assignment of prisoners from the concentration camps. The petition was agreed to and the action began as an experiment in one hall in August 1942. At first 300 men were put into action. The best experience has been produced. . . . After a time, remaining halls were filled with prisoners so that now, in the current month, the last hall is being converted to prison-labor operations.[25]

Rudolph went on to submit a provisional plan for Degenkolb's labor committee that detailed how to put prisoners to work in rocket production. Thus since the end of 1942, first Dornberger, then Degenkolb and Rudolph had sought out Himmler as a likely partner and promoter. The SS readily promised all the support that the Inspectorate of Concentration Camps could provide and agreed to the usual division of responsibilities. The SS would supply prisoners and guards and manage construction; the rocket team would direct factory supervision and retain the means of production. These were, in fact, the same working relations that Rudolph had observed at the Heinkel Works. Of note, there was no mention whatsoever of any percentage cut of overall output for the Waffen SS.

Mittelwerk and Dora-Mittelbau

If Kammler's first construction projects on behalf of armaments producers had been rehearsals, the rockets were the main event. At first, in July 1943, the sum total of prisoners allocated by the Office Group D to the V-2 project involved around 3,000 workers, a considerable number but a minority of the total labor force at Peenemünde, Zeppelin, and RAX. This was roughly on par with the SS's allocations to other industries outside the DWB. But just one month later the Special Committee A-4 asked the WVHA to increase its allocations of prisoners by a power of ten and, so doing, made the entire rocket program dependent upon the SS. The crucial change came after a rash of allied bombing raids hit each of Degenkolb's production centers. A first attack had already occurred at Peenemünde in July 1940, but it had been comical. A lost Royal Air Force bomber had dropped its payload on a nearby farm, killed an enemy cow and incinerated a haystack, then turned tail and flew home. Nothing would provide comic relief over the course of 1943, however, as air raids became both more systematic and more numerous. In the second week of August, RAF squadrons appeared over the RAX Works in Wiener Neustadt and bombed Peenemünde in a night raid, destroying test facilities and killing one of von Braun's closest assistants (Walter Thiel) along with nearly a thousand workers whose barracks had been mistakenly marked by reconnaissance. A raid on the Zeppelin Works in Friedrichshafen had already occurred on 21 June.[26]

These raids far from crippled the project but threw top Reich officials into a fit of alarm. The RAF had launched them along with many others aimed at easily identifiable major factories, and British intelligence had no knowledge at this time of the secret weapons under development in these specific plants.

Nevertheless, the Germans could scarcely help drawing the conclusion that the British knew about the rocket and had targeted it directly. Himmler, who saw his main task as the guarantor of Reich security and who had, after all, acceded amiably to Dornberger's personal request for support and protection, visited Hitler the morning following the Peenemünde attack. He expressed his suspicion that somehow, some way, security must have been breached. Albert Speer also visited the Führer's headquarters in Pommern the next day and suggested transferring the entire project to underground tunnels, where it would be impervious to bombing. Hitler agreed. Hitler himself was apparently in a state of excitement, and Speer recorded, "Every measure must be undertaken to carry forward A-4 production and the construction of assembly works. The Führer orders the mass deployment of resources from the concentration camps in cooperation with the Reichsführer SS for the corresponding construction."[27] By "resources" Hitler meant, of course, the manpower reserves of the concentration camps. Himmler had fueled the Führer's expectations with typically exaggerated promises, offering "skilled workers and even scientific specialists."[28]

At Hitler's order Speer and Karl Otto Saur, the leader of Speer's Technical Office, met with Himmler on 20 August in Hochwald to discuss the transfer of the project to underground caverns. Degenkolb's Special Committee A-4 chose the anthracite deposits of the Harz Mountains near Niedersachswerfen and Nordhausen. Here an "Economic Research Society," a state company originally founded by the Reich Finance Ministry, had excavated an extensive grid of tunnels to stockpile fuel supplies for the German army. Now the tunnels would house a rocket factory. Degenkolb wanted the move completed by the beginning of the New Year, and, with Speer's help, he requisitioned the space against the Economic Research Society's initial resistance. The Special Committee A-4 began to issue tight timetables for the transfer of machine tools from Peenemünde and Friedrichshafen. On 1 and 21 October Allied bombing raids hit Wiener Neustadt again, after which Degenkolb also folded the RAX Works into these schedules.[29] The rocket program's chief production engineer, Albin Sawatzki, drafted blueprints for the rational layout of transport lines, parts storage, and assembly stations in the tunnels. In some of the larger underground galleries, he planned double and triple decks.

Renovating an underground fuel dump as a modern factory in the time foreseen by Degenkolb, still intent on producing thousands of rockets in the new year, required large numbers of construction workers. There were almost none to be had. By late 1943 the specially appointed General Plenipotentiary for the Reich Labor Action was continually falling short of his set quotas for conscripted eastern civilians, and everyone involved recognized that Himmler's concentration camps held the only large reservoir of labor left in the Reich. In this context, the rocket team turned to the SS, and Himmler lost no time rising to the occasion. The number of prisoners transferred to the V-2 projects quickly rose from the thousands to the tens of thousands.[30]

All of this came in the context of Hitler's demands to make German factories invulnerable to air raids. A search for solutions had begun as early as April 1943 and marked the increasingly surreal desperation that began to characterize the entire German war effort. Hermann Göring's Air Ministry formed a "Special Staff" under a civil engineer named Bilfinger and at first argued (along with Hitler himself) that vulnerable industries had to be encased in concrete, bomb-proof bunkers. The plan was ridiculous because the sheer quantities of cement, steel, and labor required would have placed a tourniquet on any other essential construction. Various authorities and individual companies, on their own initiative, also scrambled to protect their production halls from Allied attack. One historian has aptly labeled this phase a "wild industrial transfer" (*wilde Verlagerung*) because it proceeded pell-mell. Managers broke up production lines and distributed them underground in the existing cellars of breweries, in subway tubes, train tunnels, old mine shafts, or even scattered tenement basements. Others sought refuge by decentralizing their works under camouflage. Still others slyly continued production in unprotected, partially destroyed factories where damage looked extensive enough to convince Allied reconnaissance that the buildings had been shut down.[31]

Although it was by no means obvious at the time, German industry would soon look to the V-2's relocation in the huge subterranean halls of the Harz Mountains as a model. The eventual successes of the underground V-2 factory also promised to counter tendencies toward decentralization (an alternative backed for a time by Speer). Centralized, vertically integrated war industries appealed powerfully to Hitler, fascinated as he was by the modern factory, gigantism in general, and the belief that a unified nation required unified production plants. V-2 rocket production, newly centralized in an enormous assembly line deep in the Harz, satisfied that vision.

Industry also came to look to the rocket factory's dependence on slave labor as a model, bringing the SS and armaments producers into an ever closer embrace. Even though Speer requested only construction work, in this case Himmler also began to act as if the SS was going to take charge of manufacturing: "I communicate in this letter," he wrote to Speer on 21 August 1943, "that I as the Reichsführer SS am taking over the production of the A-4 according to our agreement of yesterday. I have discussed the task in its entirety with my people and am absolutely convinced that we can fulfill the given assignment. I have transferred the task to Lieutenant General Pohl and, as the responsible project leader under him, Brigadier General Dr. Kammler."[32] This unmistakable grab for the entire project was recognized as such immediately by Speer. Agitated, impatient, agog with the high-tech project, Himmler was determined to step in and run things the way he thought they should be run. He genuinely believed that the armaments minister had been coddling private industry and operating sluggishly; no less he believed that "miracle weapons" could win the war.[33] He had also been recently emboldened by his discussions with Hitler, who demanded that rocket production proceed in utmost haste. As ignorant of technological management as he was firmly convinced of the miraculous creativity of the German "will" supposedly embodied in the SS, Himmler had also likely convinced himself that Pohl and Kammler could push aside dithering research and development engineers, impose a needed attitude change upon self-interested industrialists, and bring the crash V-2 program to fruition.

Yet it was also characteristic of Himmler's arrogant presumption that he never really considered the complex tight-wire act of technological orchestration that Speer, Degenkolb, Dornberger, and von Braun had carried on thus far. This was not the first time that Himmler had made wild claims nor would it be the last, but the fact of the matter was that the RFSS was, of necessity, detached from the details of daily industrial operations and had little direct power over them. In the polycratic environment of Reich industrial planning, modern managers like Kammler and Maurer mediated the SS's involvement in the rocket project at the midlevel. Himmler, essentially a policeman, lacked the expertise to do so. Producing missiles meant managing a web of relationships between the army, private industry, and state ministries. Everyone directly involved, including Pohl and Kammler, knew that the SS could never muster the technical competence to run serial production of the V-2 and acted accordingly. The SS's involvement, swiftly enacted by Pohl after Himmler's order, did not reach beyond the allocation of concentration camp prisoners through Maurer and the supervision of construction through the Office Group C. Pohl and Kammler left rocket production untouched, quite in keeping with all WVHA involvement in any armaments project up to that point. No one ever demanded ownership of the factory halls.

On behalf of the Armaments Ministry, Speer flatly ignored Himmler's letter. With the deftness of high-class bureaucrats versed in the delivery of deliberate snubs, he did not even bother to respond to the pompous "communication" of "immediate" action until a studied delay of exactly three months:

Recently I have entrusted the promotion of certain urgent construction projects to Brigadier General Dr. Ing. Kammler, who will carry out special construction tasks. This occurred because I have been promised additional reserves of labor through the allocation of concentration camp prisoners. I emphasize that the practical enactment of these construction projects is to be carried out under the closest scrutiny of the Head Committee for Construction [Hauptausschuß-Bau, i.e., the Armaments Ministry's established control bureau], which is responsible to me for planned and rational fulfillment of all building assignments in the Reich. For the same reason, the assignment of leading personnel [i.e., like Kammler] also lies under the control of the Head Committee for Construction and occurs only according to its closest scrutiny. The right to dispose and direct the services of private firms and all resources within the German building sector must remain with the Head Committee.[34]

Speer intentionally left any discussion whatsoever of assembly out of this short letter. The number of times he mentioned "construction" alone suggests that he was trying to drum the division of tasks into Himmler's skull. Assembly, subassembly, and supply continued, as it had since late 1942, under the direction of Degenkolb, Sawatzki, and Rudolph—who, as we shall see, all struck up perfectly collegial working relationships with Kammler unaffected by the rhetoric exchanged between Nazi heavyweights at the top.

Whatever ruffled feathers between Speer and Himmler, Kammler's regional Building Inspections had already gone to work. The Inspectorate of Concentration Camps erected a new satellite camp of Buchenwald, named Dora, and by October of the next year this camp became its own, self-contained concentration camp. It expanded continually to serve the Construction Directorates in the secret Harz Mountain location and came to be known by the vague code name "Mittelbau." Maurer had sent the first prisoners on 28 August (mostly political prisoners and some prisoners of war). Dora's three biggest satellite camps (Heinrich-Rottleberode, Hans-Harzungen, and Erich-Ellrich-Juliushütte) alone contained 10,000 prisoners transferred by late 1944. A total of 60,000 worked in the entire Mittelbau complex.[35]

Work proceeded without Himmler's interference on the strength of cooperation built on understanding among his competent subordinates. Himmler's power plays were far from over, but they remained ever after confined to machinations for political recognition and the plaudits of Hitler. Regardless of Himmler's pretensions, his power to disrupt the cooperation of engineers within the military-industrial complex essential to the V-2 was limited precisely because the multiplicity of contacts at this middle and lower level was

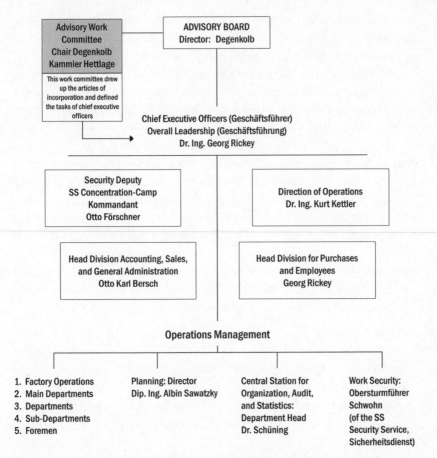

Corporate organization of the Mittelwerk GmbH. Drawing by Steve Hsu.

more durable than petty politics played out in the vestibules of the Führer's headquarters. For example, no one welcomed the SS into the rocket team with more enthusiasm than Gerhard Degenkolb.

Back in 1942 Degenkolb had proposed the foundation of a company, the Adolf Hitler GmbH, to hold the necessary patents and coordinate contracts for the V-2 project. At that time this plan had been tabled, but at the end of September the Armaments Ministry moved to convert the rocket program into a modern corporation. The Special Committee A-4 gave it the nondescript name Mittelwerk GmbH—literally the Middle Work Ltd.—yet it remained under the ownership of the Reich Ministry of Armaments and War Production. Like other vanguard industries that sought partnerships with the SS, Mittelwerk GmbH was a state corporation with private legal form. It came into official existence in October with headquarters located in Berlin's Bismarckstraße 112. Degenkolb chaired the board of directors and one of his first acts was to invite Kammler to join him on a special council of three (Dr. Karl-Maria Hettlage was the one additional member).[36]

Degenkolb and many of those within the Special Committee A-4 wanted the Mittelwerk GmbH to express the National Socialist spirit through its organizational form, its technological achievements, and its managerial philosophy. They acted much like Hans Hohberg, Walter Salpeter, or Pohl had done in forming an "organic" holding company, and they shared much of Kammler's own painstaking care over matters of communal spirit within his Office Group C. At Mittelwerk, however, the ideological impulse clearly came from outside the SS and was managed by seasoned men from private industry, not the dilettantes of Pohl's German Commercial Operations. Degenkolb and Kammler believed Nazism to be a key unifying element in creative work, one that in turn enhanced the power and scope of modern institutions. At the end of December, they submitted their finished articles of incorporation for the Mittelwerk GmbH in passionate if turgid prose:

> Work Comrades! Factory operations will be conducted in the spirit of a factory community [*Betriebsgemeinschaft*]. The factory community is a part of the national community [*Volksgemeinschaft*]. The fulfillment of duty is a law of honor for each comrade of the people and demands each man's highest personal dedication of energy and ability. Only he who does his duty fulfills the intended task of the Führer on behalf of the German people.
>
> The factory's Führer [i.e., the managers, as micromanifestations of the will of Hitler] and the followership [the Nazi word for employees: *Gefolgschaft*] together embody the factory community. They [the manager Führer] are all comrades in work. With their resources they constitute the most valuable part of the operation. A spiritual commitment and a mutual feeling of responsibility are the foundations of the factory community. He who possesses German blood can become a work comrade. To the chief executive officers [*Geschäftsführer*] falls the responsibility of leadership [*Führung*] for the organization of business and the further development of the entire operation. The destiny of the entire operation is the destiny of the entire factory community. This destiny is therefore the communal task of all work comrades, who must uphold operations with their last reserves of strength and productivity. It is the unconditional duty of all work comrades to follow the instructions of the chief executive officers.
>
> The highest foundational principle of the NSDAP is "Communal inter-

est comes before personal interest," and it is the highest goal of the factory community. The life and productivity of the factory community allow the realization of this high command in action. . . . From acknowledgment of the fact that the factory community is a part of the national community and that the family is the germ cell of the national community follows the correct bearing in life and work. . . . For the greatest portion of his life each productive man stands at his work post; his bearing at work is built upon the recognition that his life's work is not a commodity that can be freely disposed of but rather that his life must uphold the duty to his people and to his family that is constituted by his blood.[37]

This was a recipe for the productivism endemic to the SS: an inclination to see production as a salutary act that yielded up the German soul (rather than profit). Degenkolb and Kammler intended their managers and civilian workers not only to achieve maximum productivity on assembly lines or construction crews; Führer managers had to feel themselves to be the embodiment of the Nazi "will" based in German "blood," while subordinate employees, in their turn, also had to devote all their obedience to the "will" of their Führer. The factory community was thus to become a microcosm of the nation under Hitler. It was not enough for workers to make rockets or managers to supervise; they had to embody the spirit of National Socialism as well.

Not every top manager at Mittelwerk felt himself to be the "germ cell" of Hitler's "national community." One finance expert returned his draft of Degenkolb and Kammler's document to the Special Committee A-4 in January with the austere understatement, "The section concerning the factory community is somewhat unclear."[38] In fact, tension began to mount between an axis of fanatic Nazi engineers around Degenkolb and Kammler and others including Wernher von Braun and General Dornberger who were not Nazi fundamentalists. Yet if Degenkolb and Kammler began to alienate some, among themselves ideological commitment formed one of the key totems of mutual trust as they came to see each other as like-minded men.

Their alliance was most firm between the construction engineers of the SS and the production engineers of the Special Committee A-4, who had begun to alienate the researchers around Dornberger and von Braun long before the SS had any presence in the project. Kammler and Degenkolb could count Diploma Engineer Albin Sawatzki as one of their camp, the Mittelwerk manager most responsible for organizing assembly lines, parts supply, and subcontracting work, including slave labor in production. Sawatzki had come from the Henschel Works, which had already rationalized production using forced labor and worked closely with Gerhard Maurer's Office Group D2. In fact, Kammler originally asked that Sawatzki (not an SS man) be appointed to the board of directors. This was vetoed by Degenkolb, who gave the Henschel engineer even greater responsibilities, "with all the necessary power . . . and full responsibility for the initiation of series assembly in Mittelwerk, and in addition the full supervision of production at all outside firms that are participating in the A-4 program."[39] Sawatzki came from industry, as did Degenkolb, though their new powers derived from the Reich Ministry for Armaments and War Production. Kammler's authority derived from the WVHA. Potentially conflicting loyalties abounded, yet ideological commitments and common competence allowed them to bridge partisan affiliations and concentrate on the work before them.

In 1944, the very year that V-2 rockets began to land on their targets in England and Belgium, Franz Neumann published his monumental work, *Behemoth*, which compares the Nazi state to Thomas Hobbes's mythical beast of that name. The Behemoth exists in a state of uncontrolled malice, constantly expanding in all directions, making ceaseless, chaotic war with itself and all in its path.[40] We should not forget, however, that sinews of common cause still existed in the heart of this beast, which lurched forward on the support of modern factories manned by midlevel managers who proved capable of at least enough mutual effort to arm it with the most futuristic weapons of the day.

Less than Slaves: Labor at Dora-Mittelbau

Up to this point, I have used the term "slave labor" somewhat carelessly, for as Benjamin Ferencz pointed out long ago, concentration camp prisoners were "less than slaves."[41] Most true slave systems treat their workers as some kind of capital investment, whereas the SS murdered its workers with near reckless abandon even when the Reich faced crippling labor shortages. Of course, there have been historical cases of slave systems whose murderous intent approached that of the concentration camps, such as the bullion mines and sugar plantations of the Angolan slave trade. For example, in 1792 Governor Almeida e Vasconcelos of Brazil consciously treated African slaves as depreciating capital because they constituted a "commodity that died with such ease."[42] He systematically traded slaves for stable commodities like sugar, cotton, bullion, or lumber, which were then invested back in greater numbers of slaves whose value had quickly depreciated over time. Thus Vasconcelos's example made clear in the eighteenth century that nothing "rational" or what some might call "modern" in economic calculation could possibly serve as a guarantor of the sanctity of life. Nevertheless, Frederick Ordway, a journalist who wrote several celebratory histories on von Braun and his "rocket team," maintained that civilian scientists and engineers had acted rationally and therefore had helped save prisoners. Typically he blamed all brutality on Hans Kammler and the SS. In one of Ordway's interviews Arthur Rudolph "recalled" that "as time went on, the number of forced laborers was reduced" due to his and others' efforts to resist the SS.[43] The number of prisoners increased over time, and neither Rudolph nor Speer "recalled" their own initiative on repeated occasions to bring the concentration camps into the project.

As von Braun's research and development team entered into its pact with the SS, destruction, not preservation, prevailed in the tunnels of Mittelwerk. Of course, the SS did not employ a "cost-benefit" analysis like Vasconcelos had as it began to manage extermination through work, not least because Nazi ideology in general and the WVHA esprit de corps in particular discounted the "salesman's point of view." But as Kammler's Construction Directorates turned their formidable talents upon the V-2 rockets, they quickly brought modern management to bear on both killing and slavery, and Kammler completed his initial assignment in record time. Throughout the project he was in constant contact with Speer's steering committee for Reich construction, and in mid-December, Speer wrote to praise his effort:

> The leader of the Special Committee A-4, Degenkolb, reports to me that you have brought the underground installations in Nie. [Niedersachswerfen] to completion out of their raw condition in the almost impossibly short period of time of two months. You have transformed them into a factory which has no European comparison and remains unsurpassed even in American conceptions.
>
> I allow myself to express my highest recognition for this truly unique act, with continued request for the support of Herrn Degenkolb in this beautiful form. I will also, upon occasion, communicate this well-earned recognition to the Reichsführer SS Himmler.[44]

The "beautiful form" of Kammler's work came at a record cost in human lives. The SS engineering corps ground down over 10,000 concentration camp inmates in the tunnels of Mittelwerk. This number does not count those who died on death marches when the various satellite camps of Dora-Mittelbau were evacuated in the last months of the war. In fact, as Michael Neufeld has pointed out, the V-2 rockets are probably the only modern weapons system to have claimed more victims in its building phase than in combat (about 2,500). According to a survivor, Yves Béon, the camp Dora employed a special prisoner named Jacky to roam the tunnels in search of discarded corpses.[45] This is just one sign that, to accomplish the construction, the Office Group C not only had to supervise the technicalities of its building sites; it also had to supervise murder.

Kammler did so by putting modern managerial techniques into practice to maximize his most vanishing resource: time. The SS construction corps separated its projects into two categories. In the first, *A Projects*, Kammler gathered all existing underground tunnels that needed electrical installation, concrete reinforcement, ventilation, water mains, or connecting shafts to bring them up to the specifications of Mittelwerk's production engineers. Kammler designated as *B Projects* tunnels that had to be bored and outfitted from scratch. Rigorous prioritizing allowed the Office Group C to put the maximum amount of space at the disposal of industry in the minimum amount of time, and once Kammler's Construction Directorates brought individual tunnels to completion, he transferred their workers and engineers to the next most urgent projects. Early in 1944 he had finished 980,000 square meters to augment existing production space of 120,000. "Our methods have proved themselves," he summed up to an admiring audience in Speer's ministry.

> We run 72-hour shifts [a week] according to our characteristic methods without difficulties. . . . For all these projects I had to pump in an additional 50,000 political prisoners. Now [German] miners are being continually replaced and indeed by the prisoners transferred from A Projects to the B Projects. Now our projects have less and less to do with [German] tunneling specialists and miners. We can get a lot done by supervising tunneling firms and construction firms if we get the necessary machines.[46]

Kammler's boasting pointed to three, interrelated aspects of his success. First, Office Group C engineers had learned to manage forced labor effectively. The Construction Directorates replaced skilled civilians by "pumping in" prisoners, most of whom were untrained, weakened, unmotivated, and dying. Second, the SS was acting as an "arranger," orchestrating both forced labor and machines in a cooperative effort with multiple tunneling and excavation firms. Last, the SS had to manage machine operations in conjunction with this labor force. That is, as already shown in chapter 2, nothing proved inherently incompatible in the mixture of slavery and certain kinds of machinery. The difference between the shambles made of production at DESt and the speedy construction of Mittelwerk lay in the nature of the managers. Whereas the DESt was riven with conflicting visions of the "modern" factory and the "modern" concentration camp, not to mention no small degree of technical incompetence, Kammler's engineers could ply their expertise and, no less, maintain good relationships with other Reich organizations and private firms.

They did not, as did Erduin Schondorff, retreat from on-site management; rather they intervened to adjust work processes.

Furthermore, the "characteristic methods" Kammler referred to in his communication to the Armaments Ministry involved the conscious but methodical deployment of brutality. "Stand up, you fat Jewish pigs," screamed one SS man at his laborers at one site. As a survivor testified, the guard then went up and down the rows of prisoners just before roll call was counted and "slugged each one. Those able to remain standing were still usable; those who fell over from the blow were as good as dead."[47] Such methods combined the customary brutality of the Death's Head Units with the need to secure accurate statistics on those "fit to work." It would be an exaggeration to claim that Kammler's influence suddenly transformed labor management throughout the concentration camps or that "rational" techniques merged seamlessly with wanton brutality; nevertheless, the labor supervision on the SS's prestigious building projects did become systematic—though by no means less deadly—as Kammler's engineers drove their prisoners worse than slave drivers.

Kammler and his subordinates were well aware that their "methods" were built on absolute terror, and he recommended that private industry emulate them. When a member of Speer's ministry complained that prisoners had sought to escape a labor detail, Kammler retorted:

> It's always the case with these people [i.e., prisoners] when they notice that they are not being driven hard enough. I let 30 hang in special treatment [*Sonderbehandlung*]. Since the hanging things proceed in a little better order. It's the old joke: if people notice that they are not being held in a firm grip then they try to get away with all possible things.[48]

Although not all private firms emulated the SS, many did. Prisoners consigned to Building Brigades not only had their spirits broken; they usually collapsed bodily, often dying on the job. After work shifts, especially during the first months of construction, the prisoners had to sleep on the tunnels' concrete floors or in the dirt of unfinished shafts. These were alternatively cold and dank or hot and humid depending on where they were located. The noise of continual blasting and the din of machines made rest nearly impossible. Seventy-two-hour shifts a week, at grueling, physical labor, wore down even the hardiest. "Dora is filled with physically devastated men," recounts Yves Béon. "Those who, in civilian life had a comfortable layer of fat, soon notice an apron of skin across their stomachs. Others with flat stomachs and nothing to lose become skeletons within weeks."[49] The danger of disease

posed the biggest threat. Prisoners received slim rations, making them suscep-
tible to infection, and they originally had no access to sanitary facilities. Some
washed their faces with urine. Later, when baths were installed, the prisoners
could use them only once a week. Many refused because of the danger of
catching typhus. As soon as any individual could go on no longer, he was
declared "unfit to work" and liquidated according to Maurer's system.[50]

The Office Group C continually sought to speed the progress of construction by taking advantage of the SS's direct access to fresh prisoners
to replace the dying. "Beat them, and when you beat a thousand to death,
that doesn't matter, you'll get a thousand more," Kammler's officer Wilhelm
Lübeck told a foreman.[51] Kammler's relationship to Maurer's Office D2–
Labor Action was therefore crucial. Immediately upon Maurer's entry into the
IKL in the summer of 1942, Kammler's organization established a close relationship with him. In fact, Kammler could often get prisoners when other
branches of industry went begging. First of all, he had the direct backing of
Himmler, and, at the next highest level, the Mittelwerk-Mittelbau stood under
the orders of Hitler himself. This gave the Office Group C the political leverage
to override any foot-dragging that emanated from IKL old-timers or from
rival war projects desperate for labor. Most important, however, Kammler
made sure that Death's Head Kommandanten did not interfere with technically sound construction. As during his first "rehearsals" at Fallersleben
with Martin Weiss, Kammler also played a part in appointing Dora's Kommandant, Otto Förschner, and made it clear that he would brook no obstruction. In addition, the Office Group C erected a special operations director
(*Betriebsdienstleiter*) within concentration camp Construction Directorates
to test and select skilled construction workers among inmates. These "directors" also combed the camps' SS personnel for those with previous experience
in the building trades.[52]

It is also little known that the Reich Security Main Office (RSHA) in each
concentration camp often obstructed production out of paranoid concern for
security. Each camp had an RSHA Political Office which usually showed a zeal
for murder that overmatched the camp SS. At Dora Kammler put an end to
this by insisting that security at Mittelwerk be managed by an engineer named
Helmut Bischoff, who worked out of an enlarged architectural office in the
factory tunnels. Kammler recognized immediately that any surveillance aimed
at preventing "sabotage" would be useless unless carried out by the watchful
eyes of the technically competent.[53] Taken together, the systematic controls
Kammler imposed upon his organization, the competent personnel he appointed, and the spirit of consensus that focused their concerted action

meant that Kammler's engineers could reliably deliver what civilian managers wanted, when they wanted it.

Eventually a twofold division of labor evolved at Mittelwerk that accommodated efficient production *and* extermination through work. Construction Directorates divided unskilled manual laborers from semiskilled prisoners who worked with labor-saving machines or carried out intricate installation work. A contemporary sketch drawn by one of Mittelwerk's civilian managers gives a visual image of this division. It shows a mass of workers laying the concrete floor of an underground factory hall. Pouring large slabs of concrete demanded, then as now, the concerted effort of many individuals. Some tasks could not be accomplished without skilled workers, whereas others could be completed with brute manual labor. For example, construction firms and SS engineers drove unskilled prisoners mercilessly at transport work. Prisoners had to lug cement sacks or push wheelbarrows between storage areas and work sites, but at automatic mixer stations care was taken to appoint either a German civilian worker or a skilled prisoner who could monitor the quality of concrete during preparation. After mixing, unskilled prisoners again pushed the concrete in tip carts, wheelbarrows, or formed bucket brigades, but they delivered it to skilled overseers who directed the pour.

The illustration by Werner Brähne captures the key features. Some few, skilled workers do the finishing work in the foreground with cement "floats" on long poles. In the background the masses of the unskilled prisoners toil at transport work in an anonymous horde where so many fell to exhaustion or the bludgeon and expired. Meanwhile, SS construction engineers and private firms afforded skilled prisoners some small modicum of protection. As long as inmates proved useful at their tasks they were treated a notch above, as valuable slaves rather than expendable raw material. They received extra rations; and civilian managers and SS officers alike often, although never always, shielded them from beatings. Thus a desperate stratification among prisoners themselves often unconsciously reinforced the interests of management. These were the prisoners of Primo Levy's "gray zone," from which many survivors came: they existed simultaneously in the service of and victimized by the whole brutal system of management that enmeshed them. Always desperate to survive, they were conscious of their "privileged" positions which enabled them to make it from day to day by getting extra rations.[54]

As already mentioned, industrialists from outside the SS drove prisoners on the rocket's assembly lines, where they too incorporated the SS's distinction between skilled and unskilled prisoners. They also helped establish the terms upon which extermination through work in construction could pro-

Drawing from the fall of 1943 by Werner Brähne, civilian drafter at Mittelwerk. Printed with permission from private collection of Manfred Bornemann, Hamburg.

ceed in tandem with delicate assembly tasks. Sometime in late 1943 or early 1944 Dora began to differentiate sharply between construction prisoners under Kammler and assembly line workers under Sawatzki and Rudolph, a functional distinction that quickly became a matter of life and death. Of the tens of thousands of prisoners who lost their lives at Mittelwerk, relatively few died on the assembly lines of the V-2 rocket. The sheer technical complexity and sensitivity of the weapons system could tolerate neither sick and dying prisoners on the job nor the high rate of turnover caused by the steady removal of prisoners declared "unfit to work." Human skill mattered, and managers took initiative to preserve it. Therefore factory prisoners received higher rations than construction prisoners, even though they carried out relatively lighter tasks. On the assembly lines, attrition fell and mortality rates stabilized. The rocket, as a technological system, was not compatible with arbitrary murder.[55]

The changes probably took place around 10 December, when Speer visited the works. Later Speer claimed that he wished to act in a humane way: "The conditions of these prisoners were in fact barbarous, and a sense of profound involvement and personal guilt seizes me whenever I think of them."[56] He does seem to have commanded extra rations and building materials for passable accommodations. What he did not "recall" was that this slight amelioration of life at Mittelwerk extended only to production prisoners. Especially when they worked at unskilled transport work—heavy labor, which demanded more calories than assembly—prisoners died by the thousands. After Buchenwald transferred the Italian prisoner Alberto Berti to one underground construction site, he found that the rations were only one-half those of the main camp. Few lasted long under such conditions.[57] In January 1945 Dora-Mittelbau added a new subcamp, the Boelcke-Kaserne. Here other work camps and commandos sent the "unfit" to get them out of the way. The Boelcke-Kaserne was essentially a death camp, though it used no gassing. It simply left prisoners to expire.[58]

The division of labor between skilled and unskilled, construction and production prisoners also served the ends of manufacture in another way. Violence and murder offered a powerful incentive to workers in the delicate production process. Assembly managers like Sawatzki or Rudolph implicitly relied on the threat of transferring prisoners to the deadly underground construction brigades. The supervision of all work in the underground tunnels was of a piece, and much evidence suggests that the systematic parsing of construction and assembly prisoners spread systematically from the V-2 project to other SS underground building sites. As a rule, Dora's construction prisoners were not used as production prisoners once they finished a hall but were transferred to new tunneling projects. This kind of system was later applied at Leitmeritz, an underground installation for the production of tank engines in what is now the Czech Republic.[59] And high-level correspondence within the WVHA from 1944 also mentions that "the installation of passageways and land-clearing work and similar tasks demands the mass action of such prisoners that cannot be gainfully deployed on any other tasks but by a little earth moving can accomplish something en masse."[60] Differentiation of unskilled, weakened construction prisoners as a special category—a category essentially marked for death—had become common WVHA policy by this time. This process deployed extermination through work in the fullest sense of the term. It gave precedence neither to murder nor to production, but accommodated each in equal measure.

The management of extermination through work also led the SS to choose suitable machinery. Forced labor on SS construction crews followed its own logic just as the complexity of the rockets logically demanded the (relative) preservation of prisoners on the assembly line. Kammler personally preferred to use labor-saving machinery like power shovels when he could get them, but he was perfectly willing to use "low-tech" alternatives to force the pace of construction. SS engineers were well aware that tunneling, like concrete pours, could go forward with low-skill, labor-intensive means, and the German war economy enjoyed an abundance of exactly those machines which made labor-intensive tunneling the most practical choice: hand-held drills, hammers, and pneumatic shovels. Unlike other tunneling machines, these devices were not labor-saving; rather they enhanced the output of each individual worker and could still be deployed by labor-intensive crews. Furthermore, their operation did not demand highly specialized skills. They played the equivalent role in construction commandos that fast sewing machines did at TexLed (see chapter 2). As a result, workers were easily replaced; they quickly became interchangeable and expendable. Kammler and his construction officers forced prisoners to work at the tunnel face with these machines and with primitive tools like shovels, chisels, and picks, or even with their bare hands—loading stone and earth into tip carts, for example.

Of course, a single pneumatic drill or jackhammer hardly seems to embody Nazi impulses toward slavery, but when the system is seen as a whole, it is clear that they were used in order to accommodate the barbarity of production under the SS. For example, in one case, by no means unique, managers developed an effective method that capitalized on labor-intensive modes of tunneling. Normally, excavation would begin with a single pilot tunnel driven into the rock face to clear enough space for a light-rail spur to convey refuse to the pit head. Mining firms might have employed labor-saving devices like miniature power shovels to speed progress at the face. In contrast, the upper diagram in the accompanying figure shows a system used at Kahla near Jena for sandstone excavation adapted for the special deployment of the one resource the SS had in abundance, slaves. It involved four, simultaneous pilot tunnels driven into the rock (in this case sandstone). What exactly made this an appropriate system for slave labor? First of all, it was technologically sound. It reduced the risk of cave-ins because work gangs could set up steel-reinforcement girders in the side tunnels as excavation continued simultaneously in the center (illustrated in the cutaway in the lower half of the figure). Second, it betrayed the labor-intensive nature of the project, for it maximized the exposed portion of the face that manual workers could dig out, deploying four times the number of workers. Workers were equipped with relatively light, inexpensive, and durable hand-held pneumatic chisels or jackhammers in order to pound forward through the rock. On the other hand, labor-saving machinery like the power shovel would have proved impractical in this case because the capital costs of four such machines would have outstripped the benefit derived from any extra speed. Prisoners were driven mercilessly to work as hard and as fast as possible until they could work no more. While skilled operators received limited protection and marginally better rations, the division of labor ground down unskilled prisoners: they became a kind of raw material expended like

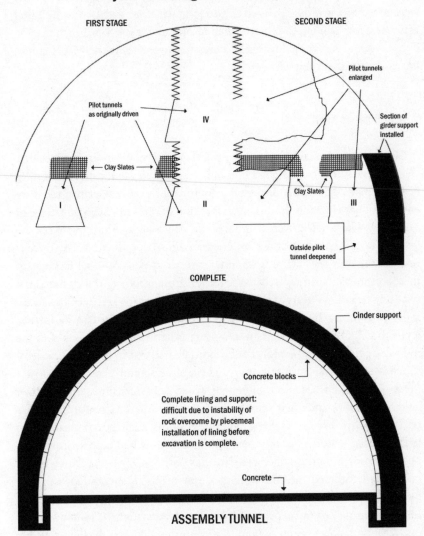

Project Reimahgbau   Jena/Kahla

FIRST STAGE

SECOND STAGE

Pilot tunnels
enlarged

Pilot tunnels
as originally driven

Section of
girder support
installed

IV

← Clay Slates →

Clay Slates

I

II

III

Outside pilot
tunnel deepened

COMPLETE

Cinder support

Concrete blocks

Complete lining and support:
difficult due to instability of
rock overcome by piecemeal
installation of lining before
excavation is complete.

Concrete

ASSEMBLY TUNNEL

Mining engineer's diagram of tunneling technique that maximized the pace of construction using labor-intensive means. Drawing by Steve Hsu based on Colonel W. R. J. Cook et al., Underground Factories in Central Germany, CIOS File No. XXXII-17, Deutsches Museum, Munich.

fuel. But the system undeniably worked: Kahla proceeded so quickly that it came to the attention of Hitler himself. In addition, this system was not the work of the SS alone but emerged amid a joint endeavor with private building firms before spreading to other tunneling sites under the SS's direction. The Kahla system thus unified technical necessity (overcoming the friable stone) *and* concentration camp labor.[61]

The Fighter Staff

No amount of crash armaments projects could make up for the fact that Germany was now outmatched on land, sea, and air, and, just as Kammler began to make his office indispensable, bombing runs over German territory entered a phase of white-hot intensity. A flood of raids in 1944 made those of 1943 seem like a trickle. In February, the Allies began the "Big Week," a systematic and relentless attack on German aircraft production. "Big Week" really lasted only six days, but when it was over 75 percent of all facilities in these sectors had been severely damaged. Historians have correctly noted the relatively transient effects of these raids, for the Germans managed to salvage most of their machine tools, and production soon returned to its former levels. But in February the trauma among Nazi war planners was genuine indeed: by the end of the month all production sagged in unison because of the chaotic upheavals in distribution and supply. A scramble to relocate armaments factories ensued, a scramble that would only intensify after the Allied landings at Normandy that soon precipitated the Nazis' retreat from industrial regions in France, Holland, and Belgium.[62] As already discussed, the highest-ranking leaders of the German war economy had been seeking various measures to protect factories since as early as April 1943, and Hitler had actually ordered the encasement of war plants in gargantuan concrete bunkers. But only a pell-mell scramble for safe havens had ensued as numerous committees and expert staffs produced few systematic initiatives. The one exception, now held up as a model for future action, was the high-prestige rocket program hastily ensconced beneath the Harz Mountains.

On 1 March 1944 Hermann Göring ordered the centralization of all existing committees working on schemes for bomb-proof factories into one "Fighter Staff," with the immediate objective to increase fighter airplane production. The most pressing concern was to reestablish some defense against Allied air raids, but more generally the Fighter Staff intended to set the pace for all transfers of industry into bomb-proof facilities. It was in a unique position to do so, because aircraft production brought together a flexible union of experts from almost every sector of the total war economy as well as the Air Ministry, the Armaments Ministry, the SS, and the Reich Labor Ministry. If the Third Reich was polycratic in structure, the Fighter Staff contained that multiplicity within its very composition. In fact, this is what made it so effective, and it was not unlike the supergovernmental, superindustrial organizations recently written about by Thomas Parke Hughes.[63] Although the Fighter Staff owned no factories, it scrupulously monitored design, quality control, and production yield. To tackle any given problem, it could bring all the varied expertise of German industry and state ministries to bear, and its members could also mobilize contacts that cut across any individual institution.

On 5 March, Speer met with Hitler to discuss its organization and emerged with the Führer's full support, which empowered the Fighter Staff to roll over any and all authorities that stood in its path. Karl Otto Saur, who headed Speer's Technical Office, was vested with the power to shut down inefficient plants and commandeer their materials, men, and machines for other firms. During the next six months he muscled through the obstruction of regional party officials, municipal civil servants, and private industrialists alike, and more than once he relied on the SS to arrest those who failed to act on his directives.[64]

Hitler had now abandoned his earlier conception of "invulnerable" concrete bunkers and instead ordered impregnable superfactories. A touch of the old megalomania remained from the Führer buildings of the late 1930s, as these factories were christened the "great building projects" (Großbauvorhaben). He wanted all aspects of aircraft production integrated into modern production lines within cavernous underground factories. Each was supposed to provide 600,000 square meters, and Hitler decreed that an output of 40,000 planes must proceed by the end of the year. Thus, as the Third Reich entered its last desperate months, it is nevertheless interesting to note a certain continuity. Hitler's architectural policies had progressed from one gargantuan, unrealistic program to the next. In the 1930s it had been the Führer buildings; in 1939 he had made Himmler the Reichskommissar for the Reinforcement of Germandom whose New Order projected settlement investments from 80 to 120 billion, something like 50 to 80 percent of Germany's gross domestic product in 1942; now the great buildings took on apocalyptic proportions as the Führer's preferred fantasy. At every step the SS had sought to pose as the Führer's dutiful master builder. Speer had always posed as his chief architect. As in the 1930s, so too in 1944, Speer assured Hitler that these megalomaniac projects were doable, while Himmler assured him that the SS would help construct them with its slaves.[65]

The great building projects received the highest priority within the Fighter Staff, despite the objections of some who believed them to be unrealistic. Saur immediately acknowledged that construction was a key and quickly demanded the aid of the SS's civil engineers with their concentration camp prisoners. The first great building was to be Mittelwerk-Mittelbau, where Speer relayed orders to Kammler to prepare an additional 80,000 square meters to accommodate the transfer of a Junkers Aircraft Works squeezed in alongside rocket assembly. Göring petitioned Himmler directly for 36,000, then 90,000, and then 100,000 concentration camp prisoners:

> [I] ask you to put at my disposal for air force armaments as great a number of concentration camp prisoners as possible, for these laborers have shown themselves to be very useful up to now. The situation of the air war makes necessary the transfer of industry under the earth, where circumstances lend themselves especially well to the combination of labor and incarceration with concentration camp inmates. These measures are necessary in order to secure the development of self-contained aircraft fabrication of the most modern kind.[66]

Himmler, Hitler, and Göring imagined the vast, subterranean great building projects as both ideal penal colonies and modern factories. Relatively few watchmen could seal the tunnel entrances and exits. As long as inmates remained within the caverns at all times, guards could allow them to move freely between work sites according to the needs of production.

The SS's high-profile role in the Fighter Staff filled Himmler with pride. In the 1930s he had called for the fusion of the concentration camps with industry to create "a completely new, modern concentration camp for new times, capable of expansion at any moment"; early in 1942 he had expressed incredulity to Pohl and Kammler that one could not get 200 percent of the work out of a prisoner that one could get out of a skilled civilian laborer: "It is simply normal and plain obvious that one can get double the amount."[67] The underground caverns of the great building projects seemed to fulfill these hopes after so many false starts and failures. He undoubtedly felt that the SS was finally being called upon to contribute to the war economy in the way he had always believed possible. In response to Göring's requests he boldly wrote:

> The tasks of my Business Administration Main Office are not fulfilled only with the allocation of prisoners to the Air Ministry, for SS General Pohl and his colleagues shall take care to create the necessary work tempo and thereby exert some influence over production results. I should add here

that through an expansion of our responsibility one certainly might expect a greater yield through a higher tempo.[68]

This was as ridiculous as Himmler's earlier letter to Speer regarding the SS's role in V-2 rocket production, and likewise all involved scrupulously ignored him. By 1944 Himmler sought to intervene in the affairs of various technical-development offices in the army, air force, and Armaments Ministry. His motivations were the same as they had always been. He believed that other authorities were either neglecting or actively suppressing vital contributions to the war effort and that the SS could make the war economy go faster and produce more and more dazzling weapons. The RFSS also still believed that "unified will," "the bludgeon of his word," and scrupulous policing were the key to disciplined labor organization. But he had absolutely no idea of what daily management actually meant on the factory floor. If this policeman continually sought to raise his voice about production, it was the voice of a dandy or rather the voice of a fascinated child who cannot wait to drive a car but is too short to see over the steering wheel or reach the pedals with his feet. Those who were forced to listen to Himmler's fulmination about technological or managerial operations either sought to brush him off respectfully or ignored him completely—including competent SS men like Kammler.[69]

It was thus typical of polycratic organization in the Third Reich that the same leading men who found Himmler annoying and importunate quickly formed close bonds to the technical men in management whom the SS put at their disposal in the Fighter Staff. To represent the SS within Saur's organization, Kammler organized a Special Staff–Kammler, patterned upon the Office Group C–Construction, which broke down into Special Inspections and local Construction Directorates (called Führungsstäbe, literally Leadership Staffs). Kammler quickly came to recognize colleagues like Saur as like-minded men and vice versa—small wonder considering that the staff around Saur was not composed of "apolitical" technocrats too concerned with technical detail to understand the politics swirling around him. These were seasoned Nazi managers enthusiastically committed to Hitler's Germany, and Saur introduced them with the following words:

Today we have brought together our entire Fighter Staff. These are men of practical experience, activists; above all these are men with a healthy human understanding. . . . The hardness [härte, i.e., hard-mindedness, a Nazi vocabulary word for toughness] grows out of the necessity of the moment. If you possess a Führer personality; and if, as men of responsibility, you possess the necessary courage, the necessary decisiveness and are at all times ready to stand accountable; and when you are ready to act on your own initiative without expressed orders or directions; then I have no doubt that we will come away from the current danger point.[70]

Saur referred to the spirit of "activism" and the "Führer personality," which he believed the members of this staff embodied in their daily work; at the same time, he wanted their dedication to stem from their mutual commitment to common National Socialist ideals.

In almost ritualistic pronouncements the members of the Fighter Staff asserted their personal energy, abilities, and proven accomplishments. They announced themselves as men who did not dither, who attacked problems and enacted immediate solutions: "I have done several emergency actions at the request of Saur," boasted one member.

These have all been projects that first bordered on idiocy. The men involved spoke to me like idiots. I nevertheless persevered. Planning wasn't even necessary. We laid our plans inside of three days. I never asked what the special committee had planned; I just began to give orders and things began to happen. . . . "You must do this and you have to name all obstacles that stand in your way." Then within 24 hours they showed up and named their difficulties, and I took care of them. Wherever it seemed as though it was impossible, it was doable nevertheless.[71]

With such words, facing each other across the table chaired by Saur, these men came to know each other as members of a heroic community of Führer managers. In such an atmosphere there would be no need for rigidity and no room for those who "just followed orders." The Fighter Staff was a self-consciously hierarchical and modern organization, but that did not mean that it conformed to Max Weber's "iron cage": it relied on mutual trust, participation, and the initiative of its members, whom Saur expected to solve their tasks quickly, without argument, and to solve them decisively.

Part of the collective identity formed within the Fighter Staff included a commitment to the modernization of German industry, which the members assumed to be synonymous with the mission of National Socialism. "One has essentially too seldom addressed the core question of methods regarding this problem," Saur told them. "One hears much too much about the old master-craft principles. We have to come to grips here with technology . . . and not with the methods of the past."[72] Or as he put it in a kind of philosophical lecture:

I come to the chapter, Capitalism. Here it appears as follows: In this matter I am an old and true man of dedication, and I have grown out of the struggle for National Socialism. I know well what damage results from

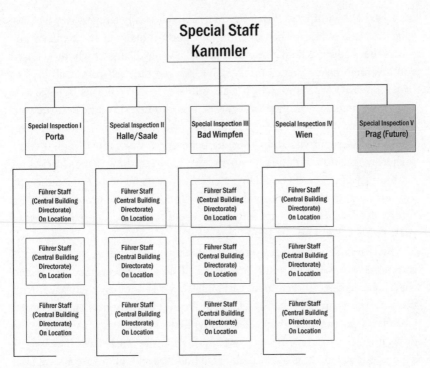

```
                          Special Staff
                           Kammler
        ┌──────────────┬──────────────┬──────────────┬──────────────┬──────────────┐
┌───────────────┐┌───────────────┐┌───────────────┐┌───────────────┐┌───────────────┐
│Special Inspection I││Special Inspection II││Special Inspection III││Special Inspection IV││Special Inspection V│
│    Porta      ││  Halle/Saale  ││  Bad Wimpfen  ││     Wien      ││ Prag (Future) │
└───────────────┘└───────────────┘└───────────────┘└───────────────┘└───────────────┘
```

Führer Staff (Central Building Directorate) On Location	Führer Staff (Central Building Directorate) On Location	Führer Staff (Central Building Directorate) On Location	Führer Staff (Central Building Directorate) On Location	
Führer Staff (Central Building Directorate) On Location	Führer Staff (Central Building Directorate) On Location	Führer Staff (Central Building Directorate) On Location	Führer Staff (Central Building Directorate) On Location	
Führer Staff (Central Building Directorate) On Location	Führer Staff (Central Building Directorate) On Location	Führer Staff (Central Building Directorate) On Location	Führer Staff (Central Building Directorate) On Location	

The Special Staff–Kammler. Drawing by Steve Hsu.

capitalism. I want to tell you one thing though. One should not always follow phrases; rather one should examine the thing in its detail. I can tell you all: Private capitalism is a darling little fellow compared to state capitalism and state capitalism is a grace in comparison to the form of Gau capitalism. For in the Gaus there are no supervisory organs and no rules with which one can point the way. The Gaus operate according to jealousy and self-interest in a way that is grotesque. Therefore, if we must, then we should save our energies and work with the educated men [i.e., of private industry] with whom one can complete a decent job. I say nevertheless: It's clear to me, if we can't do it in this war, then certainly afterward the rallying cry "self-responsibility of private industry" is going to be replaced by the self-responsibility of the German engineer.[73]

Saur aimed his derision at the Gauleiters, the Nazi Party representatives established in every region of the Reich who formed the main pillar of support within the Nazi movement for traditional, craft-style production. Not all of them opposed modern industry (e.g., Fritz Sauckel), but they were known as an ossified, corrupt group of hacks.[74] It should be no surprise that such a group existed within the Nazi movement, which was heterogeneous like any modern mass-political phenomenon, but during total war the modern impulses that had existed all along in National Socialism almost wholly eclipsed contravening ideals. Speer's ministry had backed the "self-responsibility" of large, modern corporations in private industry to carry out its policies as a conscious snub to Gau interests, and Saur continued this attack. Moreover, Saur not only intended the Fighter Staff to be a model for the underground transfer of German war industries; he believed its work must fulfill a comprehensive mission of national reform. He wanted to show that a closely knit community of engineers could bring Germany to its highest performance, its greatest heights of technological progress, and break new ground toward a future, ideal society. His Nazi engineers would transcend the self-interest of party cronyism and private capitalism alike.

The tenets of Nazi racial supremacy also propped up these self-conscious assertions, providing an additional token of these engineers' self-confirmed superiority as like-minded men. Although they did not busy themselves with tracts on the nature of genetic heritage or Aryan virtues typical of, say, Joseph Goebbels's propaganda, they often denigrated incompetent distributors for doing business "in a Jewish way."[75] At another meeting the head of the Organization Todt stated, "It is our task to bring building sites up to a technical level above that of the Congo Negro."[76] The Fighter Staff repeatedly associated poor organization or primitive technological means with "inferior" races, while, as a body of "Führer personalities" of the "master" race, it proclaimed the modernization of Germany as its goal; no less, it claimed the right of Nazi engineers to lead Germany to the culmination of its history in a Thousand Year Reich.

This community built upon the same kind of ideological dedication and engineering competence that Kammler had instilled in his own offices, and thus it should be no surprise that Kammler quickly found members of the Fighter Staff whom he could trust and who trusted him. Recognition was also quickly reciprocated. Erhard Milch, the General Master of Aircraft Production who at times became genuinely irritated with Himmler, began to call Kammler "a new pillar of the German building economy" with his "army of concentration camp prisoners."[77] Saur, remember, had originally opposed the SS Labor Action in the war economy in September 1942 (see chapter 5), but now he praised the SS's chief engineer: "It is self-understood that the assignments of Kammler must continue with all the energy of this man."[78] The Fighter Staff's building schedule over the next six months accounted for one-third the volume of the entire German building economy in wartime, over 1 billion Reichsmarks, nearly half of which went to Kammler's Führer Staffs, the lion's share for the "great building" complex of Mittelwerk (170 million Reichsmarks alone).[79] Moreover, Kammler's Special Inspections themselves combined civil engineers from the air force, the SS Office Group C, and private industry alike and therefore rested on the kind of mutual trust and admiration that Saur and Milch readily accorded the SS. The guts of this *Behemoth* therefore fed upon cooperation far more than chaotic, internecine strife, even as the Allies were battering the beast into its final submission.

There is also no denying that by the end of the summer of 1944 the Fighter Staff had made good its claims to be a "community of activists hardened by practical experience." Fighter aircraft production had doubled, despite huge disruptions caused by the transfer of production plants underground. We should remember that the Allies landed in Normandy in June and by August were quickly overrunning important production centers in France. American and British forces would soon advance on the industrial regions of Holland and Belgium, while the Red Army was driving hard in the East. By August, the Fighter Staff had achieved its goals despite these and subsequent disruptions to the war economy. Increased production was due mostly to the rationalization of assembly that Saur's production experts forced upon German firms. For the first time, German aircraft works introduced mass assembly lines and applied single-purpose machine tools to simplify labor operations and cut down on training time. These measures were partly a response to the increasing employment of forced laborers, be they eastern European civilians or concentration camp prisoners. By 1944 workers under some kind of compulsion made up almost 20 percent of the German work force. They enjoyed no rights of citizenship, and many did not even speak German. Germany's industrial managers had to simplify and rationalize tasks in order to make any headway.[80] The conclusion is difficult to escape that the Fighter Staff succeeded not merely in rationalizing aircraft production, but also in realizing its members' ideals of racial supremacy in its slave-labor factories—also its conscious objective. Men like Saur and Kammler set out to wed their beliefs in National Socialism to their organizations, and they unquestionably did so. Only the Allied victory put an end to this hour of the engineer.

TOTAL WAR AND THE
END IN RUBBLE

Just as total war completely reoriented the SS's construction corps, it also greatly changed the German Commercial Operations (DWB); but whereas Hans Kammler drew ever more important projects into his ambit during the Third Reich's hour of the engineer, the SS companies began to dissolve. Their increasing marginalization was not for lack of resources or effort. The SS intended to put the DWB at the service of the war economy. After Heinrich Himmler had toured the Reich in September 1942, he had ordered Oswald Pohl to begin "fabrication of window and door frames" as well as "roofing tiles" to repair bombed-out cities. "The fabrication must start in 10 days," Himmler added with typical impatience.[1] In pursuit of Himmler's directive, that very month Pohl pressed Speer to place building supply orders with the DWB, and Speer had quickly agreed.[2]

Himmler's new order marked the diversion of the SS corporations' own slaves to the war effort. After the amalgamation of the Inspectorate of Concentration Camps with the newly consolidated Business Administration Main Office (WVHA) in early March, Pohl had already diverted some prisoners to armaments plants, but the few, paltry initiatives to lease prisoners or concentration camp factory halls to private industrialists (Gustloff, VW, IG Farben–Auschwitz, Steyr-Daimler-Puch) still involved an insignificant proportion of the SS's total slave-labor force. Most working inmates still toiled under the bludgeon of the camp SS within the SS's own industrial empire. Well into 1942 Himmler and Pohl had tried to preserve the DWB for the "peace building program." Now reorienting the German Commercial Operations for total war became one of Pohl's top priorities alongside the leasing of slaves to private industry.

It is worth recalling that the systemic problems that had long preoccupied the DWB throughout 1942 were now becoming general. The whole of German industry was coming to rely on ever larger pools of forced labor and heightened brutality. Speer began taking this reorganization of the war economy in stride; his offices "modernized" factories, introduced high-tech innovations, and simultaneously managed the low-skill levels typical of forced labor. The Allies, who generally assumed that Germany had reached the apex of its industrial capacity in 1939 or 1940, watched with a mixture of dismay and consternation as production levels rose steadily in 1942, 1943, and even through mid-1944. As shown, SS managers had long shared the same *enthusiasm* for modernization as Speer and sought to apply it to industry in the camps, but they had also consistently lacked the key components to make it work, namely shared managerial consensus and the necessary technical competence to turn enthusiasm for modern machines into practical factory systems. The DWB's filials stalled at exactly the point that the larger German war machine accelerated to full throttle. "It will work!" Pohl had declared to Himmler, and he even claimed that gearing up the DWB for total war would lay the foundations of expanded capacity to be used in the coming "peace program," but late 1942 actually marked the beginning of a downward spiral that ended in the ruins of the Third Reich.[3]

"The DAW endeavors to make its own obligatory contribution to victory," claimed one manager. "The entire capacity of all plants of the German Equipment Works have been mobilized for armaments and essential war work. . . . we are taking in important repair work and producing goods as subcontractors." He continued, "War industry contracts also necessitated the acquisition of a number of new machines and equipment and the construction of a new building."[4] His statements show several things. First, typically, SS subsidiaries sought to keep up and running as subcontractors to essential war suppliers— but rarely as war suppliers themselves. Furthermore, SS managers were well aware that their wartime contribution represented an emergency conversion, not the standard tack of SS business development. What holds for all companies caught up in any total war held for the German Commercial Operations as well: producing war goods was an obligatory but temporary adjustment. The DWB was not primarily an armaments producer and never had been. In peacetime, as Himmler repeatedly stated, the SS hoped to return to its settlement dreams. (Even when the German army began its last, full retreat in 1943, Himmler nevertheless desperately tried to organize "Aryan" settlements in Eastern Galicia for the "ethnic Germans" driven out by the Red Army further to the east.) The WVHA sought to use subcontracts for war goods to protect its industrial capacity and thus survive for an ever more imaginary "peace." In addition, as the allusion to subcontracting indicates, the DWB's adjustment took place in cooperation with private firms and industrial planning authorities alike. Of necessity, this also meant that the SS worked within networks of firms, not as a "state within a state." Last, no matter how trivial the contributions of the DWB, SS officers indulged the delusion that each and every one was key to the war effort. "Decisive for victory" became a new watchword, even if the DAW mostly produced petty articles like metal ammunition cases and skis for winter troops. German Noble Furniture switched production lines from "settlement furniture" to bunks, tables, beds for barracks, and casings for airplane radios. These things were hardly "decisive." Allach Porcelain Manufacture made perhaps the most absurd statement, when its top manager, Rudolf Dippe, reported, "The production of war-essential dishes and crockery in the year 1943 corresponds to 52.1 percent of total output."[5]

SS industry had begun with large installments of self-delusion, hypocrisy, and poorly managed investments. This could hardly change overnight in the Third Reich's final hours. Despite talk of "decisive contributions," actual manufacture began to disintegrate, even as the Office Group D and Office Group C expanded their influence. This is not to say that the Office Group W did not expand, but a quick comparison to 1941 demonstrates the contrast. Back then, the 3,000 inmates at IG Farben–Auschwitz represented the largest number of prisoners assigned by the Main Office for Budgets and Building to any single armaments plant. The DWB, by contrast, had employed about five times this number. When, after 1942, the IKL started to become an integral part of the war effort, the number of prisoners available leapt from the tens of thousands into the hundreds of thousands; from around 21,000 when war broke out in September 1939 to about 70,000 in mid-1942, and then 110,000 in September of that year.[6] Reflecting this increase, the DAW more than doubled its work force from 3,650 to 7,402 prisoners. Yet allocations to private industry now dwarfed these numbers. For example, Buchenwald held 65,000 working prisoners in 1944; only 400 worked directly for the SS's own companies. The vast bulk fed into a network of sixty-six external labor camps on lease to private armaments firms through Gerhard Maurer's Office D2. SS managers wished to maintain their missionary spirit, yet they had only rarely translated high rhetoric into concrete systems of production. The war now took away whatever margin of error that was left to them, and the DWB quickly began to fracture along the fault lines of its own internal inconsistencies.[7]

Kurt May's German Noble Furniture demonstrates the way in which the DWB tried to contribute to war production and remain true to its idealism but also how, ultimately, the DWB was falling apart. Of note, its aims for the peace building program never receded from the foreground. May consistently declared that his workers were "engaged in important work for the SS, especially for the eastern settlements and the development of the SS and Police

Strongholds."[8] Indeed his workers might have remained so, but one of German Noble Furniture's factories, D. Drucker in Brünn, burned to the ground under mysterious circumstances on the night of 14 January 1942. One of May's closest associates in midlevel management, the civilian Paul Bauer, subsequently killed himself. Some believed Bauer had committed arson. Whatever the case may have been, in March 1942 the SS Criminal Police opened a full investigation of May.

D. Drucker AG and other furniture factories under May had rivaled the TexLed as pearls of SS entrepreneurship. In 1941 D. Drucker was still running in the red, but May had managed to cut its debts by a third within the year (from 3.4 million to 2.2 million Reichsmarks). The factory had introduced mass production techniques into the furniture industry of the Protektorat, and in 1941 its turnover rose to 30,461,494 Reichsmarks. With the D. Drucker factory in ashes, however, German Noble Furniture confronted a situation similar to that of the DESt in 1939. It had spent money on new machines that now lay in ruin. Now resources were much scarcer, and only armaments contracts and cooperative relations with clients, not utopian missions, might have enabled the WVHA to rally the support of Reich authorities in order to rebuild the Drucker factory. May had held some minor contracts with the aircraft manufacturer Messerschmitt since 1940 for the subassembly of wooden steering elements as well as radio casings. The Reich Air Ministry, which oversaw the company's Messerschmitt contracts, was impressed by Drucker's past performance and helped secure the materials to reconstruct the plant. The aura of the best, latest, most scientific manufacture that May had nurtured contributed to this approval, and after an influx of armaments money, German Noble Furniture could report by June, "We have succeeded in our restless reinstallation of a modern assembly line."[9] Thus the DWB was able to protect this filial by finding a niche in the war economy, although it is important to note once again that "supplying the Waffen SS" or "gaining control over the economy" had nothing to do with these contracts.

Despite apparent success, the scrutiny of the SS's own Criminal Police continued. Pohl supported the investigation, and May had to leave SS top management. The reasons given for his dismissal had to do with a mixture of racial supremacy and anticapitalism. The Criminal Police discovered that May had been unusually generous, in their eyes, to the firm's former Jewish owner during the deals of 1940. This carried the tinge of racial turpitude. Needless to say, in the years before total war when May had closed the deal for Drucker, German businessmen still formally bought out Jewish owners. It was still common practice to maintain some sham of legal and financial respect, and May had acted in this tradition. He had paid Drucker for his factory and arranged for the man to emigrate to New Zealand, something he claimed made him a "rescuer" after the war. This had been no altruistic deed, however; May had simply followed the norm of business practice at the time. By 1942, on the other hand, Jewish policies had radicalized far beyond disenfranchisement to genocide and the Criminal Police judged May's actions soft on Jews. Pohl seems to have been offended too, as he was wont to be when any subordinate did not abide by the highest Nazi purity. Just as Pohl removed Kommandanten for misbehavior (as opposed to incompetence—a very different criterion), he removed May from top management of the German Noble Furniture in October, even though he was competent.[10]

An official statement shortly after May's departure forcefully underscored the reason for his replacement: "The main task of the corporation is, first of all, to weed out all Jewish traces of the operation. And through the installation of new, modern machines as well as through the restoration of the buildings to working order, the corporation must be reorganized in such a way that corresponds to all the demands of today's time."[11] Pohl and his managers were convinced that the machines of modernity were part of the WVHA's mission and, further, that rooting out all Jewish "traces" constituted a necessary step toward the implementation of those means for "today's time." The WVHA set racial cleansing as a precondition for modernization. By seeming, even in retrospect, to have "protected Jews," May had disqualified himself as a trustworthy modernizer in Pohl's eyes. May's case stands in contrast to others in which SS managers were caught making incompetent or even criminal deals but convinced superiors that they truly believed in SS principles. Such officers were often allowed to continue in the DWB.[12] Only the fact that ideology mattered and had consequences within SS corporate bureaucracy can account for this managerial shakedown.

It was also a sign of growing chaos within SS industries, which, however, never prevented Pohl from launching new initiatives. The WVHA continued the frenetic acquisition of new holdings when the chance presented itself: sawmills, health food processors, settlement cooperatives, and additional publishing concerns. Pohl also indulged the SS's charlatan fascination for "inventions of all kinds" by supporting a man named Lumbeck who had patented a special process for bookbinding. His company was pompously titled the Lumbeck Corporation for the German Book Industry, and it proposed to preserve German culture for all posterity by making a binding that would last forever. Pohl invested in the company when the transition to total war threatened Lumbeck with liquidation. In keeping with the same missionary spirit with which the old Office III–Business had begun, most of the

WVHA's new acquisitions, like Lumbeck, continued to complement various aspects of the SS's cultural crusade.[13]

At the end of 1942, the Office Group W also initiated plans to take over some of the petty industries that Odilo Globocnik had organized around the district of Lublin (where Globocnik was SS and Police Führer [SSPF] under the Higher SS and Police Führer [HSSPF] of the General Government, Friedrich-Wilhelm Krüger). By moving to take over Jewish work camps under Globocnik, Pohl was trying to enforce the WVHA's monopoly over all industrial policy within Himmler's organization, but, as shown in chapter 5, Pohl's attempts to assert this authority among the Higher SS and Police Führer were far from successful. HSSPF officials proved jealous of their fiefdoms of petty perquisites just as the Kommandant staffs before them had guarded camp workshops. Various camps where Jews were forced to labor for their captors persisted in the Lublin district, even long after the main ghetto of the city of Lublin had been liquidated. In June 1943 the total number of concentration camp prisoners numbered around 200,000, but in the General Government there were between 300 and 400 additional work camps for Jews. This semi-official network of slavery run by various Higher SS and Police Führer held somewhere between 120,000 to 200,000 Jews. The imprecise estimates of post-war historians are doubtless a reflection of the amateur management of the SS and Police Führer who ran them, for they tended to neglect precise records. Few of these work camps represented systematic efforts at industrial production, and they went by many names, including Labor Ghettos, Work Camps, and Julags (for *Judenlager*, Jewish Labor Camps). They reached the height of their expansion in the spring of 1943 at the very time that the SS liquidated almost all other ghettos in the General Government.[14]

More so even than within the IKL camps, however, the effort to take over Globocnik's industries brought the German Commercial Operations into head-on collision with the senselessness of genocide. Globocnik had played a key role in initiating the extermination of the European Jews in late 1941 and continued to play a key role thereafter. He had brought medical experts to the Lublin district who had formerly invented the gas chambers for Hitler's euthanasia campaign. He also initiated experiments to test carbon monoxide gas against prussic acid as the most efficient poison for murder. The fact that Globocnik was erecting industrial plants both to destroy the European Jews and to exploit their labor at the same time is only one indication of the SS's fundamental lack of coherent labor policy, and in the Lublin district the WVHA would prove no more capable of addressing the contradictions than it had in the past.

Lublin's work camps nevertheless warrant special attention because Max Horn, the WVHA officer assigned to Globocnik's industries, was one of the few SS executives who quickly comprehended the obvious connection be-tween the mistreatment of prisoners and poor production. Yet when he tried to raise the issue, he received no support from the central offices of the WVHA in Berlin. Without backing he could not halt genocide even if only to drive his "work Jews" as slaves. In the end, Globocnik's superior officer, HSSPF Friedrich-Wilhelm Krüger, ordered the liquidation of those few Jews reserved for SS industry in spite of Horn's vocal protests, an example of the failure of SS industry in its bloodiest dimensions. This was also a genuine struggle for power born of two competing visions of what it was to serve as an SS man. Yet as polycratic as even the SS could be, as multifaceted its ideals and motiva-tions, racial supremacy and the impulse to murder trumped all others and always found the most violent reinforcement. They also met the most pa-thetic, most ineffectual resistance.[15] Such was Maximillian Horn.

It did not help Horn's cause that the work camps and labor ghettos in Lublin were a mess. Globocnik had long planned to integrate slave labor, light industry, and agrarian settlements as part of his SS strongholds but had succeeded in collecting only a hodgepodge of unrelated companies. By 1943 he had gathered together a glassworks at Wolomin; a peat-cutting operation at Dorohucza; a furrier in Trawniki; textiles, basket-weaving, and carpentry workshops in Radom; and a manufacturer of iron fittings, earth- and stone-works, a pharmaceutical firm, and a brush factory in Lublin. Just as Globoc-nik's oversight of building schedules had failed to evolve into systematic man-agement (see chapter 4), so too these ventures fared no better. Horn was supposed to consolidate them in an East Industries GmbH (OSTI) as a subsid-iary of the DWB, but Pohl's vague directions to OSTI merely stated: "[The] main task . . . consists in definitive utilization of Jewish manpower for the interests of the Reich."[16] What purpose OSTI was to serve in the DWB was therefore unclear from the outset, and the SSPF of Lublin continued to main-tain one work camp under his own control for "self-supply."[17]

To have any hope of accomplishing his goal, Pohl would have had to define the WVHA's industrial relationship to the Higher SS and Police Führer of the General Government, Krüger, but the WVHA's SS Economic Officer under Krüger, Erich Schellin, would prove no help. Compounding difficulties, Pohl had already helped push aside the mercurial and grudging Globocnik in SS construction back in 1941. Into these Byzantine rivalries stepped Horn, a newly recruited business administrator. Unlike Schellin, the longtime crony of the HSSPF, Horn did come intent on wringing productive labor from the SS's victims. He was a young, spry entrepreneur and had received formal, systematic training in the methods of modern accounting at the Polytechnic University Stuttgart. Like Hans Hohberg, he had earned a doctorate as a certified public accountant.

Horn's own ideological commitment is difficult to judge. He came to the WVHA only after enrolling in the Waffen SS in 1942, and the WVHA assigned him to the management of SS companies instead of the front. He probably saw service initially as a duty, but he performed it with effort and dedication. He was clearly anti-Semitic, and, typical of the SS's productivism, he speciously identified sound management with the superiority of Aryans. Conversely, he blamed the poor management of the work camps upon the inferiority of Jews. "The factory organization is, measured by German relations, not modern," he wrote of industries in Litzmannstadt. He then stated why: "The labor organization that prevails in the ghetto lacks Aryan factory supervisors (Jews as factory supervisors!). . . . The Jew takes charge of orders and raw materials and delivers the finished product. Under way from raw materials to finished product each and every control is neglected!"[18] Despite this anti-Semitism, no evidence suggests that National Socialism stirred Horn to activism. He never, for example, dedicated free time to work for the SS's various causes like many of Pohl's early managers.[19]

Pohl left Horn to decide details of OSTI's operations directly with Globocnik, who became the firm's codirector. OSTI's charter further granted Krüger himself a spot on its board of trustees, later to be replaced by Schellin. Effectively, this meant that Horn had to face the entire HSSPF apparatus. Globocnik was an SS general, a hard-minded man who expected to get his way. He was wont to act alone, to act audaciously, and to seek approval after the fact. By contrast, Horn was an exemplary specimen of the kind of SS man that officers like Globocnik despised as pencil-pushing, sidelined do-nothings, and his staff in Lublin shared a simmering contempt for Horn that it often let him feel. One of Globocnik's associates remarked, "East Industries! I feel nauseated when I just hear the word industry!"[20] In the face of such open hostility Horn also had to struggle against the WVHA's generally murky purpose in forced-labor enterprise, a point of friction on which the WVHA itself could never generate consistent policy.

Furthermore, the SS's crossed purposes at OSTI quickly confused Horn himself. One of his reports asked: "Must this mandate be regarded primarily from a political-police perspective or from an economic point of view?" If the OSTI's intentions were "primarily of a political-police nature, political considerations (concentration of the Jews) have to rank foremost" and "economic considerations have to remain in the background." However, he continued, if the WVHA really wished to achieve industrial output, "economic considerations must predominate in the matter of the concentration of the Jews."[21] That is, the OSTI must arrange the organization of Jewish work camps flexibly according to the convenience of ongoing, steady production and not the primacy of policing. Horn also pinpointed the exact problem: the Jews were starving. "The total output . . . in the ghetto is catastrophically low," he noted; therefore, "the first order of business is the sustenance of the ghetto [*Deckung des Gettobedarfs*] and then may come the first contributions to the war economy."[22] If the SS wished to exploit prisoners, Horn stated bluntly, some modicum of preservation of life had to take precedence over the denigration of the work Jews and—with the genocide in full swing—over their wholesale murder.

Horn sought to force Pohl to choose either the primacy of production or the primacy of policing, but the WVHA chief merely marked "Both" in the margin of his report. This was not wholly absurd in the murderous political economy of Nazi Germany: the SS's engineering corps had achieved "both" by developing techniques of extermination through work. When faced with similar matters, Kammler's engineers had been able to take direct control by exerting their technical expertise. Thus they had been able to manage the SS's ideological goals within complex technological systems. Had Horn possessed real technical skills capable of converting the Jewish work camps into indispensable industries for the war effort, as Kammler had done with his Building Brigades, his position might have been different. But he was only a certified public accountant. The technical staff supposed to back him up within the DWB was neither competent nor dedicated. He could not take charge of factory operations the way that Kammler's Office Group C commanded its construction sites. When Horn petitioned for the preservation of the labor-ghetto Jews, if not as human beings, then as a factory resource, he found these sentences crossed out of his reports.[23]

Exactly what "political" goal the HSSPF staff had in mind became clear by November 1943. The previous year, at about the exact time that Pohl sent Horn to put the OSTI on a sound business footing, Globocnik had taken charge of what was code-named the Operation Reinhard. The operation was twofold. First, Globocnik intended to rid the East once and for all of Jews. Second, he designed the operation to extract all possible remaining wealth from the murdered Jews for transfer to Reich accounts. At first there seemed no contradiction between industry and genocide, and the WVHA participated in these plans from the beginning by fencing the stolen property. To grasp the scale of the operation, Globocnik shipped over 1,000 boxcar loads of used clothing. Even the fillings of the teeth from the dead Jews were laundered through SS accounts, and Pohl planned to plow back the capital derived from the Operation Reinhard into the SS companies, especially as investment in the modernization of OSTI. By the summer of 1943, Globocnik had murdered most of the Jews who had managed to survive in one or another of the SSPF work camps, but some few lingered. Only when these last also became marked for extermination did the Operation Reinhard begin to run counter to the WVHA's own interests in running slave-labor camps.[24]

OSTI had planned to use the capital of hundreds of thousands of dead Jews to run industries that employed no more than 16,000. The total robbed assets from dental gold and other valuables from Operation Reinhard came to between 7 and 23 million Reichsmarks. Estimates vary so wildly because the heterogeneous nature of the loot—everything from gold watches and memorabilia to minted coins—makes it difficult to judge absolute worth. This is a horrendous sum to contemplate when one imagines how many gold teeth or scant valuables sewn up into the victims' pockets it took to constitute it. And it is easy to understand the common belief that the Operation Reinhard served to finance the German war effort. It was the last, bloodiest phase of Nazi Germany's longstanding effort to exploit the Jews before their ultimate destruction.

Total confiscated Jewish assets due to disenfranchisement, insurance fraud, and countless other criminal activities seem to have amounted to something over 400 million Reichsmarks. This was a significant sum, especially in an economy strapped for foreign exchange, but not enough to finance a very large part of the war economy. The WVHA had taken little part in these earlier efforts, which proceeded largely due to the cooperation of the Reich Security Main Office with private banks, in and outside of Germany, and civilian administration. For example, the Vichy regime extended disenfranchisement policies in the nonoccupied areas of France before they were forced to do so by the German occupation.[25] To return to the WVHA's role in the final phase of looting, however, it should be remembered that the Jewish communities of eastern Europe, the primary victims of Operation Reinhard, were not outstandingly wealthy. Even the highest estimates of the gold taken from them, 23 million Reichsmarks, did not match the total capitalization of some individual subsidiaries among the many SS companies. This was small change in a war economy in need of billions, and if the Operation Reinhard booty had truly been enough to make a difference to the war effort, the numerous agencies that cooperated or had knowledge of Pohl's operations would have no doubt prevented the WVHA from reserving it for the wasteful SS companies.[26] It was a dismal example nonetheless of "polycratic" organization in which cooperation was far more important than conflict. The German National Bank held an open account for the funds; the Finance Ministry and the Economics Ministry also knew of the operation.

Pohl's own intention to reinvest the Jewish gold in OSTI leaves little doubt that the WVHA originally wished—with Himmler's support—to reserve some Jews who were "fit to work." This was also in keeping with Himmler's continual orders to preserve some few Jews from wholesale genocide to provide slave labor for core SS companies and the labor brigades of Kammler's civil engineers. In November, Horn was still planning future capital expenditures in the district's work camps, an undeniable indication that he fully expected to rely on a minority of Jews as a steady work force *and* the capital inflow from the murder of the vast majority. Globocnik, however, showed unusual zeal for the SS's cultural dream of an ethnically "cleansed" Europe and put this goal first. Nothing changed in September 1943 when Himmler transferred Globocnik to Trieste. Whereas Globocnik had tried jealously to preserve some kind of influence over the firm as a source of personal slush funds, his successor, Jakob Sporrenberg, showed almost no interest in having "Jewish industries" at all.[27]

As the new SS and Police Führer of Lublin, Sporrenberg also took command as the Red Army was advancing in the East. By August the whole German army was retreating after being lured into a trap at Kursk, and by October the Soviets had pushed them back to the Dnepr River. The German retreat would not end until the Red Army reached Berlin less than two years later. The district of Lublin had formerly been far to the rear but now lay close to Germany's easternmost front. "[Himmler] has newly received another order," Krüger remarked to his staff officers as early as April. "In a very short time the cleansing of the Jews must be finished." "The Jewish camps represent a great danger in the General Government," another police official remarked.[28] In a state of heightened paranoia, the SS staff began to look upon even the emaciated Jews of the labor camps as potential "partisans." Desperate to survive and perhaps sensing the coming Nazi defeat, throughout 1943 what was left of the devastated Jewish population in the ghettos and concentration camps of the East did begin to rise in revolt: in April at Warsaw, in August at Treblinka, and again in October at Sobibor. Thus, in October, Himmler gave the order to Krüger to liquidate the Jewish work camps in the eastern parts of the General Government. In three separate locations in the Lublin district Sporrenberg had prisoners dig zigzag "air-raid" trenches. On 2 November about 3,000 SS men, including a detail from Auschwitz, arrived for what the SS named, with malignant irony, "Harvest Festival." Over the next two days they filled the trenches with 42,000 to 43,000 corpses, almost all the remaining work Jews in the district, including the labor force of OSTI.[29]

The murder of his own workers bitterly surprised Horn and even incensed him.

Despite the continuous difficulties that OSTI had to overcome during construction of its plants and factories it was possible to take over and improve these industries until 3 Nov. 1943. On 3 Nov. 1943, 70 German supervisors, 1,000 Poles, and 16,000 Jews were employed. On 3 Nov. 1943 all Jews were removed. . . . The construction and completion of work done so far became completely valueless through the withdrawal of the Jewish labor.[30]

Horn went on to condemn Globocnik's and Krüger's obstruction of sound management both before and after the WVHA's takeover.

Pohl's attitude toward the final liquidation remains unclear. He and Horn undoubtedly knew that Jews were being murdered by the hundreds of thousands—they expected extra capital for OSTI through Operation Reinhard—but they did not know that Sporrenberg would murder every last worker in their factories, nor was this even consistent with Himmler's official policy of reserving some small minority of Jews for the SS's own companies and construction crews. In fact, in other areas of the occupied East, the SS did not eliminate every last work Jew. Over the course of 1943 and early 1944, the WVHA took over the remaining Jewish ghettos of the Reichskommissar for the East, the Gestapo prison of Warsaw, and a Jewish labor camp in Kraków. These became new concentration camps—Riga, Kaunas, Vaivara, Warsaw, and Plaszow—and began to transfer Jews into war work. Why should Lublin have been different? By all indications Pohl did not get along well with Krüger or Globocnik, and most likely Sporrenberg, following the example of his predecessor, took initiative on "Harvest Festival" without consulting the WVHA, but after the fact Pohl must have consented, for no records show any outrage or irritation on his part—as records clearly do show in Horn's case.[31]

When the barbarity of the Operation Reinhard finally broke in upon his factory world at Lublin, only then did Horn rise in peevish anger at the Holocaust. Suddenly in November he wished to argue that Jews were a valuable resource, but earlier he had argued that they were an inferior work force. He now complained, "Eight more Germans who were not trained book-keepers have to be employed to take up the work, which is completely unknown to them and which was partly destroyed during the action of 3 November 1943."[32] The glaring contradiction in his own anti-Semitic beliefs and his later complaints seems never to have occurred to Horn. All he had to offer at that point were complaints: "Deprived of Jewish clerks, the accounts of the OSTI at first lost 'the ground underneath their feet' in the true sense of the phrase."[33] This did not stop him, however, from wrangling with Sporrenberg over the property confiscated from the Jews of OSTI. Once he had finished complaining, however, he was concerned only that he escape all accountability for the disruption caused to SS industry and nothing more.

Ultimately, Max Horn must be classed among the "normal bureaucrats" studied closely by Christopher Browning. He was not an "activist," like Globocnik, but conformed to what Browning calls "accommodators." They "waited for the signal from above" but then did their level best to carry out the Third Reich's policies, even radical policies of murder.[34] Horn was exceptional only in that he actually did less than "accommodate" and more than just wait. He actually lobbied his superior officer, Pohl, to halt killing in the name of productivity. Given that industrial killing of the Jews in gas chambers had been proceeding in the Lublin district since early 1942, Horn's objections carried little weight. Yet even if he actively opposed killing (of his workers only), in one respect he was exactly like Browning's "accommodators." Browning has stressed the importance of local consensus between institutions as well as across the boundaries of hierarchy within them. OSTI demonstrates that specific issues of consensus mattered no less when conflict erupted. The vast majority of Nazi midlevel bureaucrats had largely bought into the ideological presuppositions that defined Jews as a "problem," as impediments to "modern" operations. Although Horn opposed Globocnik and Sporrenberg, all SS men at Lublin seem to have been of one mind on this issue. Horn had therefore worked out few cogent arguments for the preservation of the Jews. This, in turn, was due at OSTI at least in part to Horn's and the SS's general racial supremacy and anti-Semitism. By associating "modern" management and business with the Aryan "race," Horn neglected the real technological skills and organization it would have taken to convert the OSTI into something truly useful to the German war effort. Racial supremacy mattered no less simply because Globocnik and Horn fought over how to exploit the Jews of OSTI. OSTI was the final meaning of carrying out both "economic" and "political" concerns without acknowledging the differences between them. If Horn saw his work Jews as "unmodern," as inferior, how could he then argue coherently that his factories needed them?

Modern Management and Its Discontents

Horn's inability to assert the WVHA's mandate for production against the Higher SS and Police leadership renders the most harrowing example of the consequences of the DWB's fractured managerial community. Although the WVHA professed a crusade, it could not formulate even simple goals or build them into factory systems. Besides the obvious contradictions of racism, other disputes were also tearing apart the DWB. A fascination for "sweet machines" ran counter to the SS's mission to force prisoners to work for the Reich, and anticapitalism existed alongside deep capital investment. Now, in the midst of total war, various key managers began to split with Pohl. Nevertheless, how they split and over what issues demonstrated the enduring consequences of ideological commitments in SS top management.

For example, Pohl's engineer Erduin Schondorff had shown no interest in adapting the DESt's "modern" brickworks to the SS's mandate for "modern" penal institutions for slave labor. He consequently formed no bridge to industrial managers—the kind of relationship that the engineers of the Office Group C or Maurer in Office D2 built so quickly. Schondorff did get involved in calculating labor rents, as private industry began to object that the WVHA charged a higher "wage" to outside corporations than it did to its own subsidiaries. Up to 1942 SS companies had booked 30 Pfennigs a day per worker to the Reich, a fee supposed to match a fixed compensation for the prisoners' daily room and board. The DWB argued that it should pay no more than this extraordinarily low fee (approximately 5 percent of going wages, even for unskilled labor) because SS companies deserved special status due to their "communal" nature. Of note, the figure was nothing more than a whimsical dictate. Reich ministries had not arrived at the rounded figure of "30 Pfennigs" after calculating real expenses for shelter and provisions; they had decided arbitrarily what prisoners should be forced to subsist upon.[35]

By the end of 1942, when the WVHA started to serve German industry as a labor lord, the Office Group D was determined to prevent capitalists from profiteering and began a thoroughgoing reorganization of labor rents. A compromise was quickly reached in which both the DWB and outside industries agreed to peg labor fees to actual productivity, but carrying through accurate comparisons of prisoners' productivity with those of civilian workers would have demanded the willing cooperation of SS managers with civilian factory engineers in order to draw up tables that mirrored actual factory floor conditions. When pressed to help in this effort, Diploma Engineer Schondorff simply stated that the task was pointless. He reported, reasonably enough, that skill levels among inmates varied widely and that "the masses of concentration camp inmates . . . almost never contain skilled brickworkers."[36] He seems never to have considered innovating to adapt to a low-skilled work force (as had the Textile and Leather Utilization GmbH), nor did he ever demand that the IKL raise subsistence rations for his workers (as had civilian engineers at Dora-Mittelbau). He did identify some of the same factors as Max Horn:

The provisions and shelter, the disciplinary cordon of guards, and the many other special measures that affect the prisoner's existence allow no chance to achieve the productivity of civilian workers. Their low physical energy, which is sapped already under these conditions, is further burdened by the psychological weight of their long incarceration. . . . Their complete disinterest in the care of the valuable machines and tools that are entrusted to them and the attendant high rate of depreciation and increased waste of raw materials also make their contribution to higher factory costs.[37]

Even in the warm summer month of August, when the weather did not exact a toll on prisoners as it did in December and January, the IKL reported 194 deaths at Sachsenhausen out of 26,500 prisoners, 118 at Buchenwald out of 17,600, and 150 at Neuengamme out of 9,800. The DESt brick factories under Schondorff's control were located at these camps. Truly addressing such problems would have meant seeing the factory system as something more than an aggregate of modern machines, something that included both machinery and the human dynamics of labor organization. As for the mass of prisoners, Schondorff regretted that "whereas a private firm [in the metalworking industries] can use an assembly line which sets the tempo of the work among civilian laborers, in the brickworks that is very different. Here there is nothing that sets a calculable, regulated pace of work for the workers."[38] In actuality even this was false. As shown, the camp SS could and did regulate work tempo in order to crush the will of inmates, at times in ways that were eerily similar to the rational methods of Tayloristic time-motion studies. In addition, technology like the ring oven could be designed to set the pace of production. Schondorff's complaints actually showed his ignorance of operations in his own factories. Horn had at least tried to intervene and preserve prisoners at East Industries, if only as slaves, but the engineer did nothing but complain as his workers weakened and died.

It is tempting to see Schondorff as a kind of stubborn "pragmatist" cleaving fast to the cold, hard facts of slave-labor inefficiency, but nothing was "practical" about his calculations. He claimed to have calculated viable labor rents based on a 50 percent rate of productivity, but he had actually pegged fixed rents arbitrarily at 50 percent of those of civilian wages and then computed backward to determine what the output-per-prisoner would have to be to justify this figure. He lifted the "civilian wages" he used from the metalworking trades, not the building supply industries, and the comparison was groundless. Schondorff's "modern" management proved just as arbitrary as the IKL's original calculation of 30 Pfennigs a day for labor rents.[39]

Schondorff had never really cared about the reality of factory floor operations of any kind, but had come to the SS in pursuit of "sweet machines" and
the ideals of his elite "school culture" of engineering professionalization. He
withdrew more and more into research, engineering science, and laboratory
work. Solving the DESt's real dilemmas would have involved something more
than the simple installation of automated equipment, with which Schondorff
was much more at home. As he wrote his reports, the Heinkel Works of
Oranienburg and other factories were achieving efficiency rates comparable
with that of civilian labor forces with concentration camp prisoners. And,
after all, Germany's huge spurt of growth in war production since Albert
Speer had taken office rested on compulsory labor, which proceeded hand in
glove with the forced "modernization" of German industry during total war.
Therefore Schondorff's statements do not evince a "practical" nature, or an
austere value-free "rationality" supposedly natural to engineers. If he had
truly occupied himself with "pragmatic" questions, the most practical thing to
do would have been to see the DESt as a technological system in its entirety,
to adapt management and machinery to slave labor, or to exert some effort to
preserve prisoners (as had rocket engineers at Mittelwerk). He might have
used his technical expertise to wrest control of production away from the
camp SS (as had Kammler), but they continued to drive inmates to mortal
exhaustion in DESt factories.[40]

As Schondorff's increasing captiousness betrayed, some who had helped
the SS companies in the early years began to withdraw their cooperation, and
this was also clearly the case with Hans Hohberg. Like Schondorff, Hohberg
did not find himself alienated as a pure "pragmatist" among "ideologues." He
had come with an idealistic commitment to the Führer principle, and the
WVHA's "organic" holding company was largely his brainchild. Even after the
war, when he tried to defend his innocence at the Nuremberg War Crimes
Trials, Hohberg insisted that he had struggled to "guarantee the rights of the
people's national community [*Volksgemeinschaft*] in the economy," little more
than a thinly veiled reformulation of the unitary "German will" he had proposed to wed to the SS companies in 1939 and 1940.[41] At first, his ambitions
had overlapped with the interests of men like Walter Salpeter and Pohl, and
the accountant's specialized knowledge had proved invaluable. More so than
Horn at Lublin, top managers like Schondorff and Hohberg had initially
thrown themselves into the activities of the DWB. Nevertheless, they also tried
to bend the SS companies to their own values, and, so doing, they exercised
moral decision. When they could no longer get their way, only then did they
begin to withdraw their managerial skills from the WVHA.

Even in the early years many SS executives who were Nazi fundamentalists resented Hohberg. As early as 1941 one manager of DESt had pleaded with Pohl to block Hohberg's appointments of civilian accountants who were not committed SS men. Karl Mummenthey is an example. He counted among the cadre that had got in on the reorganization of DESt at the beginning, the young corporate lawyers that captured WVHA officers after the war described as "the new, leading managers [who] came in many cases out of idealism to the economic undertakings of the WVHA and did not allow themselves to be led astray from their motives. One believed that one could find something exemplary here."[42] Mummenthey was proud of the quick progress he and Salpeter had made in getting rid of Arthur Ahrens and participated in the decisions surrounding the recruitment of Schondorff. As early as 1941, however, Mummenthey began to express discomfort over the influx of managers whose dedication to the SS did not match his own. After some newly recruited DESt managers had gone on vacation, he wrote to Pohl personally, "Is there any consideration that [the replacement] is a civilian?"

> Up till now the Main Office Chief has been of the opinion that only SS officers of the Main Office may be assigned as chief executive officers and official representatives [Prokuristen]. At the Cooperative Dwellings and Homesteads GmbH a representative was just denied an assignment for precisely this reason.[43]

The lawyer reserved his particular irritation for Hohberg, who was not an SS man and who was demanding the appointment of handpicked civilians. Mummenthey harbored suspicions about these newcomers. The point of conflict was one of differing levels of commitment and, consequently, trust in an impersonal hierarchy. Mummenthey often went out of his way to evaluate the personalities of the SS managers placed within his divisions and recognized that an influx of managers with no regard for the SS's goals threatened projects that were dear to him. This should not be construed merely as professional jealousy, although that certainly could have played some role. Mummenthey was already well established and hardly had reason to feel insecure about his own position. Hohberg did not put Mummenthey's doubts to rest by tersely informing him that Pohl had personally approved all non-SS appointments. Meanwhile Pohl had also written to Mummenthey directly, giving a flat no to his queries.[44] SS officers would not run SS companies.

By 1943 that early distrust had turned to animosity and with good reason, for Hohberg actively objected to his SS peers' self-conscious neglect of profit in favor of a "cultural" calling. To Hohberg, the SS's productivism was nonsense. He believed that financial administration meant sound capital accounting for fat profit margins and solubility. As a member of the board of trustees of the East German Building Supply Works GmbH (ODBS) Hohberg acted to block takeovers of bankrupt factories and quickly clashed with the ODBS's chief executive Hanns Bobermin. Bobermin continually pushed for the acquisition of modern works, whether they were profitable or not. The cause or the crusade, not profitability, justified the investments. One report submitted as far back as October 1940 enthusiastically trumpeted the potential of a cement factory at Golleschau, southeast of Auschwitz, and urged a new investment of millions in order to purchase modern machines, even if "in the foreseeable future this does not yield any noteworthy savings."[45] To Bobermin and other gung-ho SS managers, Golleschau represented an opportunity to modernize. Hohberg, on the other hand, stressed the obsolescence of the machine park and turned Bobermin's optimistic advocacy of new investment into pessimistic dissuasion of any new investment at all.[46] Profits, to Hohberg, remained essential to corporate modernity, while the SS had been fighting to associate modern business with productivism—an effort to make factories serve national culture and identity rather than "mammon." Although Hohberg was himself an ardent advocate of modern business organization, and expected the Führer principle to enhance sound corporate structure, in the end he did not wish to overthrow capitalism root and branch.

Hohberg himself became increasingly disgruntled, for Pohl occasionally dressed him down. Earlier Pohl had backed him over loyal SS managers like Mummenthey, but now the WVHA chief began to mock the accountant. "Your reports arrive so late that they have little or absolutely no worth for me," Pohl told Hohberg.[47] Pohl also began to scribble scorn on Hohberg's memorandums and reports. When Hohberg urged a system of wage reform in late 1942, Pohl penciled "so what" in the margins.[48] The barbs must have rankled the officious accountant, who was wont to believe himself the sole competent man among bumblers. Not unlike the lazy and stupid Hans Baier, Hohberg—a much more capable man—measured his colleagues by the extent to which they followed his directions, thought like he did, and promoted his concepts. Now that Pohl began to veto his initiatives, Hohberg reciprocated by complaining to others of the incompetence of the WVHA chief.

He found much worthy of criticism. Despite four years of constant effort, many DWB executives had continued to neglect their books as they had always done. Most filials undertook their first comprehensive audit only in 1943, although Pohl had ordered this at Hohberg's suggestion back in 1940. Nor had the DWB ever fully cleared its legal status with the Reich Economics Ministry. Even Hohberg's "organic" holding company had become a half-baked enterprise. "Neither the auditors nor the assessors have dedicated serious thought to the status of the legal relationship of the WVHA Office W4 [the DAW] to the Reich or to the Nazi Party or to the German Equipment Works as a corporation," he wrote to the Economics Ministry at the beginning of 1943. "And they have not thought at all whether and, if so, how the SS as an organ of the party—without the right under public law to incorporate—itself can even legally own property."[49] These criticisms actually applied equally to Hohberg, who sat on numerous boards of trustees, something certified public accountants were forbidden to do by German business law (at least without forfeiting their accreditation). He had also become, de facto, the most important business leader under Pohl but had no official status as such. So he too participated in a long tradition of unselfconscious hypocrisy in SS industry. Here he was exposing the incompetence of an institution that he had played an undeniably central role in building, but he blamed the DWB's ongoing shambles on everyone but himself. Thus, it should be no surprise that authorities like the Economics Ministry began to turn a dour eye on the querulous accountant at the same time that SS managers also started to feel betrayed by him.[50]

Nevertheless, the Economics Ministry took action on Hohberg's information. It demanded a proper audit, and many within the WVHA suspected a new influx of uncommitted outsiders. In preparation, Pohl reshuffled the top management of the DWB one last time in order to remove Hohberg. In September 1943 he created a new Staff W. But this last reorganization betrayed Pohl's flagging initiative more than it marked anything truly new. Maybe Pohl was already succumbing to the growing detachment from reality that set in generally in the last months of war, for he chose to replace Hohberg with his personal friend, the foolish but devout Hans Baier (who had remained up to this point a "management professor" at the SS officer school at Bad Tölz).

The German Commercial Operations subsequently began a bickering self-destruction. Since 1940, the question of loyalty had kindled resentment against Hohberg and his civilian accountants. Many blamed the muddle on Pohl's abandoned promise never to trust SS business to outsiders. Top administrators wanted to maintain control over their political and cultural mission, believing as many of them did that the SS should act as a core vanguard of Hitler's new Germany. That meant avoiding the risk of dilution by other interests, especially capitalist, business interests. "The chief of the Main Office [Pohl] has been warned from many different quarters," wrote another corporate lawyer, also from the days of Walter Salpeter. "Not least, many outside organizations may gain a deep insight into the most intimate affairs of the SS and perhaps even in the political affairs of the SS."[51] This officer was none other than Leo Volk, who wrote a manifesto on SS corporate ideals, criticized "pure capitalist thinking," and urged SS recruits to overthrow "the era of the liberal economic system."[52] He wanted the SS kept from the gaze of outsiders: "We have our own public accountants who can examine those corporations."[53]

It is tempting to see the intrusion of the Economics Ministry into SS corporate affairs as a kind of struggle between competing agents of power, but Pohl and the ministry quickly acted to quell conflict by selecting individuals whom both organizations believed would take mutual interest to heart. The SS appealed to Fritz Kranfuß. He was a high-ranking SS man in constant contact with Himmler's Personal Staff, but he also served in the Deutsche Revision und Treuhand AG, an accounting firm long established by the state (before the Nazi period) to perform public audits. The director of the Revision und Treuhand, Richard Karoli, had originally intended to supervise the audit personally, but Karoli was not an SS man. Although Pohl was willing to defer to him, the Staff W openly objected. Pohl and Kranfuß then arranged a compromise with the director's brother, Hermann Karoli. Unlike Richard, Hermann was an SS man, and he quickly mollified the DWB managers, who agreed to cede him a supervisory role in the sweeping audits. Even in activities as supposedly objective and mundane as auditing, the sense of trust built on tokens of political and ideological conviction in modern organizations made the difference between acceptable and unacceptable organization men, between cooperation and bureaucratic impasse. That the SS could find a mutually palatable SS man among the ranks of a potentially antagonistic Reich ministry is also further evidence of how polycratic structure created many different venues for cooperation and overlapping networks of contacts. In this case it averted conflict and internecine strife. By the end of 1944, Pohl expressed his warm personal gratitude to all involved and expressed hope that the coming audit would clear the way for the future growth of SS industry.[54]

This importance of trust, even of "community," in SS corporate enterprise should come as no surprise, and precisely because it mattered to the SS's daily management of slave labor and genocide, this book has made it a central theme. Despite Hans Mommsen's claim, among others, that the Nazi "bureaucratic machine . . . functioned practically automatically," in reality no dynamic modern bureaucratic structures operate that way.[55] Ideals and a sense of community mattered even in the most banal of SS affairs, and therefore the content of those ideals also mattered. Understanding this is all the more relevant because of the recent phenomenon of Daniel Goldhagen's *Hitler's Willing Executioners*. Academic historians have repeatedly criticized Goldhagen for the shallow nature of his argument, namely that the Holocaust can be explained only by what he alternatively calls "eliminationist anti-Semitism," "hallucinatory anti-Semitism," and, finally, "orgiastic anti-Semitism."[56] But because Goldhagen has offered a powerful argument that ideology matters, the broad interest in his book is perhaps understandable. If many historians have rightly attacked Goldhagen's simplicity, established scholars have hardly been innocent in this regard. For instance, one respected sociologist proposes that the popularity of the Nazi Party can be explained only with reference to "material self-interest." Another prize-winning book also quite typically explains that the concentration camps can be explained only as an expression of "absolute power" and that, therefore, any historical pursuit of ideological explanation for their development is "superstition."[57] Many would have us believe that not only did the Holocaust's victims come like sheep to the slaughter, but their Nazi butchers did so as well.

It should be obvious by now that WVHA officers did not get up early in the morning due to rabid anti-Semitism alone, although many were rabid anti-Semites. Their plexus of motivations was far broader even than this hateful prejudice. But people like Pohl, Mummenthey, and Hermann Karoli—and supposed "pragmatists" like Hohberg or Schondorff—were not sleepwalkers, and ideals did indeed matter to them precisely because they are embedded in the very nature of dynamic bureaucratic operations. These institutions were and are intended to render local, individual experiences and even physical artifacts (like the bricks and stone of the DESt) fungible, amenable to collation, interchangeability, and abstract transfer. When even the driest statistics arrived from the bottom rungs of an impersonal bureaucracy, any top manager had to believe that this abstract information represented what he would have reported if it were possible for him to make observations personally (increasingly a question of specialized knowledge as well as time, speed, and distance). Those modern institutions that run well are continually concerned with what it is now fashionable to call "corporate culture"; likewise, those that run badly often do so because their middle echelons drift without any durable consensus. All the codified techniques necessary to create the fungibility of experience—down to the most banal methods of accounting—never were autonomous forces within an iron cage of bureaucracy but depended on the maintenance, input, and cooperation of people, the new masses of midlevel managers, engineers, and white-collar workers. From among these "new men" the WVHA had recruited its staff.[58]

The End

There was something surreal in Pohl's final reorganization and the ascendance of Baier, a failed school instructor, to be the chief of the new Staff W. One wonders why Pohl even bothered. Baier reported in 1944, "The result of our audits is clear; the SS companies are insolvent."[59] This was essentially where Pohl had started in 1939. Five years had served to change nothing except the scale and scope of mismanagement in the SS companies. Germany's prospects in the war were deteriorating so rapidly that Hermann Karoli never seriously took to sorting out profits and losses amid the amateurish morass of the SS's "organic" holding company.

Nevertheless, Pohl's "fury for work" was becoming an absolute value in and of itself. His organization had, from the beginning, tended to mistake resolve and energetic action for purpose and direction. Now this tendency ended in a kind of eerie Nazi existentialism: whatever one did, one called it "decisive" for the war—regardless how trivial, far-fetched, or outright ridiculous. To face reality would have meant facing, at the same time, that the unified German "will," as the Nazis had conceived it, had proved such a paltry thing and had achieved nothing but the ruin of a continent and that the hated subhuman Slavs, the Red Peril itself, and the multiracial, cultureless, Wrigley's Spearmint Gum–chewing Americans had now come hammering at the gate.

Pohl launched several initiatives in the last months of the war. In January and February 1944, the WVHA founded a corporation to advise, protect, and remunerate inventors. Here the SS maintained its idealistic posturing to the end. Leo Volk took pains to declare that the Technical-Economic Development GmbH, as it was christened, "does not deal with a mere private business enterprise, but rather . . . its purpose is to utilize patents in the interest of the general good."[60] There is no record that it actually developed anything. There was also a research and development institute, Phrix Research GmbH, which sought patents for miracle nutrition drinks, supposedly to sustain the ever more embattled and poorly provisioned German troops in the field. So the SS remained true to its blinkered technological romanticism. Himmler expected that "explosive devices will appear all of a sudden [*urplötzlich*], due to the progress of technology. The effectiveness and speed of these devices will put our newest explosives for the Vengeance Weapons [i.e., the V-1 and V-2 rockets] in the shade."[61] Himmler knew about German nuclear reactor development, but whether he knew much about the atom bomb is an open question; however, no doubt remains that the Germans had absolutely no chance to produce an atomic weapon, and any somber soul in the project knew this as well.

In May 1944 the SS founded a corporation for the production of shale oil (Schieferöl GmbH) and tried to integrate vertically by acquiring shale- and peat-mining operations. This effort came at exactly the time that Hitler ordered a newly appointed "General Kommissar for Immediate Measures" (Edmund Geilenberg) to raise synthetic oil and fuel production. The general

Kommissar supported the SS's effort to contribute some kind of miracle breakthrough. Himmler made Pohl "deputy for the extraction of shale oil," and all invested high expectations in a new invention, namely a charcoal-pile process (*Meilerverfahren*). The project quickly turned into a disaster: "A catastrophic rainstorm has caused extreme damage, so that, for example, a landslide engulfed Work 4, where both large collectors have crushed together. . . . Pathways and the grounds are a morass."[62] At this time the Allies were bombing steadily, and one raid killed an SS executive at the works. As 1945 arrived, with the Allies perched on the frontiers of the German Reich itself, the chaos intensified, but the WVHA stuck it out to the end.[63]

The sales representative of the corporation wrote in 1945, "*Oil is flowing*" (emphasis in original).[64] He included a rather personal commentary to his friend in the Staff W. "Well now, I still firmly believe in a good outcome to the war, no matter how bad it might seem now," but then he went on in a tone that showed even he had admitted the end.[65] He described the plight of his family. His parents, at seventy, were hiding out in a stranger's house, but at least they were in safety. They would not be harmed in the invasion. His wife too, was in hiding; she was expecting their fourth child: "No one could have foreseen this development, for our belief in Germany was too unassailable. We could not believe in an unhappy turn. This belief we carry with us still, and we do not give it up. Germany will need children and ever more children after this struggle."[66] He proudly bore a vial of oil to Berlin as proof that he had fulfilled his duty to the last. It was the end of March 1945.

Not all of the WVHA's efforts at the end of the war met with such pathetic disaster, and in those cases in which WVHA Office Groups could both deploy skilled management and maintain close cooperative relations with other state ministries and private corporations, they waxed as ever stronger players in the Third Reich. Thus Maurer's importance among top industrial managers only increased. By late 1944 the working population under the Office Group D2 rose to over 500,000 prisoners as the German forces retreating from the former occupied territories undertook to incarcerate as many able-bodied civilians as they could. Hungarian Jews also made up a significant portion of the new prisoners at this time, diverted from the genocide to various industrial projects. Most prisoners went to work on fighter aircraft production within the Fighter Staff, usually on the construction of underground production factories.[67]

Numerous letters praised Maurer for the many services he rendered and petitioned him for more. A letter from the head of German aircraft production is just one example: "It has been reported to me that we owe the smooth operation of this action [the SS Labor Action] essentially to your competence and cooperation."[68] "In the last weeks I have had to rely upon the support of SS Colonel Maurer . . . almost daily," another manager, Paul Budin, wrote to Pohl. Speer had placed Budin in charge of a special antitank weapon, and Budin elaborated that his company, HASAG, "already works today with over 10,000 prisoners and is more than satisfied with their productivity and their attitude. I feel it is my duty to communicate to you that without the before-mentioned support the full operations could never have been mastered at their current level, not even close."[69] This relationship to the SS was all too typical of industries seeking labor, not exceptional.

Those who worked with the Office D2 began to achieve levels of efficiency comparable with those of civilian laborers. Women prisoners at Henschel Aircraft Works in Oranienburg, for example, achieved rates of "100 percent efficiency" with "American methods," as one Henschel manager proudly proclaimed.[70] The concentration camp prisoners were, of course, paid nothing; they were undernourished; and each worked an average of sixty hours more every week than their civilian counterparts. Henschel had imposed a kind of managerial arrangement similar to that of the Mittelwerk GmbH. It elicited the willing participation of its captive work force by providing some modicum of sustenance, by maintaining direct technical control over production, and by holding the brutality of the camp SS at bay (at least relatively). Henschel assembly lines were less deadly than, say, the tunneling sites under Kammler. This still meant that Henschel managers used the violence of the SS as a managerial resource, and all inmates knew that failing at the assembly line meant almost certain transfer and death. It says much for the normalcy with which the contempt for freedom was viewed in the Third Reich that civilian managers had come to depend on terror and the threat of murder to do their job.

In October 1944 Speer succeeded in gaining hold of the entire allocation of labor throughout the Reich, a belated coup seen by many as the final defeat of the SS's aspirations to "become independent in armaments." In reality, the SS and the Ministry of Armaments and War Production had never worked in closer partnership. Speer continued to rely on Maurer's office. Germany's work force was now a mishmash of German civilians and various categories of forced laborers: eastern civilians, prisoners of war, and, at the very bottom of the ladder, concentration camp inmates and Jews. Speer wanted to eliminate confusion and waste but hardly wished to eliminate Himmler's contribution. The Office D2 now simply received requests channeled through Speer instead of haphazard petitions from industry. Even in January 1945, when Maurer left Oranienburg to join an elite unit that fired the V-2 rockets, he continued to manage labor allocations from the field. In the midst of administrative chaos, the authority of the SS proved most durable in exactly those areas where it had generated effective managerial bureaucracy and supported bonds of cooperative trust both within and outside the SS. The Nazi Party Gauleiters even began to speak of a Speer-Himmler axis in the war economy, but a troika would have been a more appropriate metaphor: one drawn by the team of Speer, Kammler, and Maurer.[71]

Kammler's Special Inspections also received a steady stream of praise. His own end is worth examining in some detail because he became increasingly involved in the one serious attempt on the part of Himmler to take whole armaments plants under the wing of the SS. Himmler had sought since August 1943 to command the whole V-2 rocket project. He did so because he truly believed that the SS could push the project to prompt completion. Speer and Himmler's own subordinates had simply ignored him, but now he became more and more persistent. Himmler was becoming obsessed with Mittelwerk and came to believe that the relentless energy of men like Kammler and their miracle technology represented the one, last hope for the German war effort. Frustrated in attempts to extend the SS's direct grip over rocket development, Himmler stepped up secret-police surveillance of the civilian managers at Mittelwerk, and in March 1944, as Göring, Saur, and Speer ordered the expansion of the Harz caverns that housed Mittelwerk, he made his most drastic move. He arrested Wernher von Braun and several other leading engineers.[72]

Curiously most evidence of von Braun's actual arrest, secret meetings, and conspiratorial machinations stems from postwar memoirs of those German engineers that the United States recruited to build up its own rocket program. There is nothing unusual about missing Gestapo files, for the vast majority were destroyed either purposefully or through bombing raids. But the SS was a huge bureaucratic organization. Conspiracy to "take over" the rocket program in this fashion should have left numerous traces in organizations like the WVHA or the Special Staff–Kammler. Why is the evidence so meager? Alfred Jodl's record of the evidence held by the Security Police mentions mostly concerns for the political purity of commitment among the rocket team. Von Braun and members close to him had supposedly talked about the rockets as mere spaceships, not wonder weapons that would win the war; another had made comments about the war turning out badly. These grounds for arrests differed not at all from those for others undertaken by the SS against leading industrialists at the time, most of which were prompted by volunteer informants, not infiltrators. In fact, volunteer informants had prompted earlier SS arrests within the Peenemünde community.[73]

There is further reason to doubt the tales by men like Walter Dornberger, Arthur Rudolph, and Wernher von Braun told after the war. Postwar testimony is not inherently suspect, yet the fact is that these men lied about their relationship to the SS and did so in a methodical way—to conceal any record of their own complicity. The rocket team's systematic mendacity has always caused the indignity of survivors, who had to bear the brunt of the truth, and even that of some German civilians who worked at Mittelwerk. These witnesses are hardly ever brought to light, certainly not as often as the rocket team's celebratory sagas about itself, which one civilian drafter described as "a totally disgusting scandal [*Schweinerei*]."[74] Men like Dornberger or von Braun were quick to mention adversarial brushes with the Gestapo that made them seem like "resisters." Had they explained exactly how Himmler had first come to Mittelwerk, namely at their bidding, they would have had to reveal their own sordid role. True enough, as with so many of the SS's relationships, once Himmler arrived as an invited guest he had become increasingly hostile and unruly and had refused to leave the rocket team to its work, but half-truths of a grand SS conspiracy to "infiltrate" the V-2 rocket program serve only to reinforce stereotypes of the SS as a marauding agency bent on total control. It is therefore well to consider how and exactly over whom Himmler came to exert authority at Mittelwerk and what actually changed in consequence.

Von Braun's arrest in March never led to an SS "takeover" in any case. The transition came months later after an unrelated chain of events. On 20 July Wehrmacht officers planted a bomb in the Führer's headquarters and staged an abortive coup. The now frightened Führer ordered the liquidation of all involved. The ill-fated assassination attempt thoroughly discredited the Wehrmacht, which had controlled the rocket project until then. In addition, the Wehrmacht's reputation had already suffered on account of the Allied landing at Normandy. By 9 July the Allies had occupied the French town of Caen, and by the end of the month they had completely broken through the German defensive—not to mention that the Red Army had launched its own offensive on 22 June, three years to the day after Hitler's invasion of the Soviet Union. Shortly thereafter the German armed forces in the East were decimated. Thus the assassination attempt of 20 July came toward the end of an intense string of disasters for the Führer and the Wehrmacht alike.

Hitler quickly turned to the ever-shrinking circle of his true paladins and put the Reichsführer SS in charge of Germany's military forces on the home front (the "Replacement Army," roughly equivalent to the U.S. National Guard). This included authority over the Army Weapons Development Office and, by extension, the V-2. Frederick Ordway and Mitchell Sharpe inform their readers that the "ambitious officer" Kammler "was already in control of the V-1 and V-2 production" and that Himmler had been champing at the bit to take charge of the miracle weapons projects.[75] Hans Mommsen seems at times to confuse hopelessly the Mittelwerk GmbH with the SS's own industrial property.[76] Even as careful and knowledgeable a historian as Richard Overy has been taken in by the postwar reconstruction of this drama, claiming that Himmler took control of the rockets in August 1943. In fact, Himmler did not order Kammler to take over until August 1944, and then only as a combat commander of the Waffen SS under Himmler as Commander in Chief of the Home Front. The Allies were closing in from east, south, and west. The home front was quickly becoming the only front, and only under this somewhat mundane transfer of authority did the SS take charge of the rockets at all. When Himmler did take hold of the miracle weapons, Mittelwerk's civilian production management had already long been in the ascendance over von Braun's team (as might be expected in any modern research-and-development project in transition to full-scale operations). Among the production staff, few found the SS irksome at all, and, in the end, the SS never had to resort to any strong-arm methods.[77]

An otherwise insignificant correspondence within Himmler's Personal Staff shows both the excitement, desperation, and dilettantism of these last months. Since the end of 1943 Himmler had been pushing the General Plenipotentiary for Chemical Production to transfer factories beneath sandstone formations along the Elbe. Like all industrialists who had more pressing dealings with more competent industrialists, the plenipotentiary had brushed Himmler off. Now a year later, the "General Kommissar for Immediate Measures," Edmund Geilenberg (who, like Gerhard Degenkolb, never hesitated to rely upon the SS), had ordered the transfer of synthetic-fuel factories underground at the very same location. "With special satisfaction," reads a short report to Himmler, "I wish to bring to your [Himmler's] attention that now the enactment of exactly that transfer has come about that you urged in December of last year. . . . Once more the bitterest experiences were necessary in order to put into motion the necessary decision."[78] "This case is just one more proof that the Reichsführer is always right," Himmler's adjutant replied.[79] Himmler's star was rising with a pace matched only by the Third Reich's decent into annihilation.

Acting upon his specific convictions of what "was always right," Himmler transferred the V-2 rockets to Kammler, his most capable engineer. All agreed that research and development must be subordinated to serial production and combat operations, and Himmler ordered Kammler to do everything possible to deploy these weapons. Of note, Kammler neither interfered with production itself in any significant way nor disrupted its civilian management. If Himmler really had wished the SS to control Mittelwerk as an independent SS factory, or, conversely, if all civilian managers on the project felt threatened by this "takeover," neither Kammler's initiatives nor the cooperation he enjoyed among Mittelwerk's midlevel managers evinced this in the least. His only notable intervention was his callback of the civilian Gerhard Degenkolb, who had been removed for voicing his opinion that Hitler had grown incompetent. Although Degenkolb was never known to hold his tongue, Kammler knew and trusted his commitments from the days when they had drafted the Mittelwerk GmbH's operating guides. The hardheaded industrial manager returned to the fold. Supervision of the assembly line, subassembly, and supply remained under the energetic leadership of Albin Sawatzki and Arthur Rudolph. The SS never even tried to appoint its own officers to these positions, nor is there any evidence that the SS attempted to acquire the Mittelwerk GmbH as an SS corporation. At the end of 1944 Mittelwerk hit peak levels of production with a work force composed overwhelmingly of slaves. In August 374 rockets had rolled from the line; by September that number had climbed to 629; and production did not fall below 600 a month until the utter collapse that set in after March 1945.[80]

Kammler's Special Inspections continued to expand tunnels in the Harz, adding a liquid oxygen plant to the east. His only new duty lay in the leadership of special battalions, one made up of SS and two of Wehrmacht soldiers, in charge of firing the rockets. The SS Building Brigades also built secret launch sites in western Germany. Among other military installations ordered by Himmler in these last desperate months, one received the comic-book-villain-sounding name "Undertaking X." In November, Kammler received command of the German antiaircraft rocket, code-named Enzian, a project that was stillborn. In January 1945 Hitler also removed the V-1 rocket from the control of the German air force. Joseph Goebbels wrote in his diary, "In the deployment of these new weapons one must recognize that it is always only one individual who has really achieved a great and unique thing . . . Kammler."[81] But Kammler accomplished these "great" deeds only in the "shadow empire of skeleton organizations, false hopes, and self-delusion," as Michael Neufeld has accurately described it, one built on fantasies of the Third Reich's salvation through the most modern high-tech weapons.[82]

It was, however, an empire of delusions shared by many of Hitler's industrial elite, not only by the SS. That winter Speer took the opportunity to visit Kammler's combat division in the field:

On my last trip to the Ruhr region I visited the SS Division at Special Disposal under SS General Kammler. During the visits five shots were delivered to distant targets.

The division made an outstanding impression on me. The leadership personnel consists of young, excited officers, who are engaged in this thing with all their hearts. I believe, therefore, that it would be right, if the combat mobilization of the Flak Rocket (Enzian) were also to be given over to this division, and a onetime Technical Combat Troop [Einsatztruppe] should be formed from the core of this division.

It seems that SS General Kammler possesses an extremely unusual gift for such technical units that we should exploit.[83]

As Speer's enthusiasm shows, his ministry's embrace of the SS never slackened. The Armaments Ministry continued to coordinate industrial matters. Meanwhile Kammler and Himmler issued combat orders for the weapons that they still earnestly believed would carry Hitler's Germany to final victory.

Himmler's intervention in these "miracle weapons" coincided with the fading confidence of some magnates within German industry. By late 1944 top managers began to see Germany's industrial capacity threatened with extinction, and some began to withdraw their active cooperation for the first time. They usually did so covertly, attempting to delay underground transfers in order to protect their machines. Often, ironically, they took refuge in bureaucratic structure in order to resist the Third Reich's final compulsory orders; they claimed that this or that directive was impossible due to technical difficulties, transport, or labor shortages. In one case, Kammler reacted by having an executive of BMW arrested and interrogated by the Gestapo.[84] Speer's own deputy, Karl Otto Saur, cut off all appeals for moderation: "We are dealing with a case in which we have been consciously cheated out of 6 weeks' time, and BMW is still not moved into the underground positions reserved for it."[85] Ministerial officials and German air force engineers who shared the SS's anti-capitalism and fanatic devotion supported Kammler's call for the arrest. They also initiated such strong-arm tactics on their own. When some industrialists had begun to balk, this served only to confirm the prejudices long held by Nazi fundamentalists like Saur. They sought out and backed up the SS when "mere private capitalists" came under attack.

Saur's outburst was nonetheless just one indication that the Third Reich had begun to cannibalize itself. A growing tide of police brutality washed over the armaments sector. Long ago industrialists had turned to the SS for help—to "Aryanize" the German economy or to keep forced laborers in line. Now some in turn became the victims of the very murderers they had invited into their midst. Two members of the board of directors of the Deutsche Bank were executed in 1944 for expressing doubts about Hitler. They were, however, turned in by their own peers, not SS "infiltrators." One was denounced by another Deutsche Bank director. Such brutality marked the growing desperation of an inescapable ruin. Hitler issued a spasm of orders dictating new general plenipotentiaries, the empowerment of special committees, and deputies at special disposal. All sought to copy the model of the Fighter Staff and expected the same results. The titles of these institutions began to sound as grandiloquent and vague as the destruction of Germany was becoming absolute. At the same time Hitler still demanded his impregnable "great buildings." When Speer repeatedly sought to warn that these fantasy projects would drain the building economy (over one-third of which was already tied up in the Fighter Staff program) dry, Hitler put the great buildings in the hands of Speer's subordinate Xavier Dorsch, who eagerly promised to build six of them. Kammler remained in charge of the Mittelbau complex, itself designated a great building site, and arguably the most important.[86]

The most pressing shortage in the straining war economy—as it had been since 1936—remained human labor. "We must fish through every penal institution in all of Germany," one Fighter Staff member declared in exasperation.[87] Yet by mid-1944 there were no prisons that had not already been fished out. Even Kammler's system of extermination through work began to break down as he exhausted the last reserves of concentration camp prisoners. By the end of 1944 the civil engineer could no longer boast of having completed all his projects in record time. In July he had reported 90 percent of his A Projects as finished (i.e., those that required new installations and renovation of existing underground tunnels), but by year's end he had problems completing the remaining 10 percent. He had planned to have the B Projects (those requiring complete new tunnels) done by the New Year of 1945. He could not keep his word. His success had depended on the constant influx of fresh prisoners, and there were simply no more to be had. They were either already dead or working under the aegis of some other authority.[88]

Beginning in 1945 a cascade of lofty-sounding titles fell upon the chief of SS engineers: "General of the Army Corps at Special Disposal," "Deputy of the Führer for Breaking the Air-Raid Terror," and, finally, "General Plenipotentiary of the Führer for the Turbo-Jet Fighter." In his hands, Kammler held the V-1 (Buzz Bomb), the V-2, antiaircraft rockets, and the development of the first combat jet plane (the Messerschmitt 262), among others. He ordered the centralization of production for all "miracle weapons" at Mittelwerk, which he now code-named the "R-Sector," for reasons that are hard to discern. In the meantime, he paused to write to Himmler's adjutant that he looked forward to the future, when the SS could concentrate on the settlements of the New Order and the monumental Führer buildings. His telegrams and briefs crossed Himmler's desk from all corners of the shrinking Reich, hinting at a frenetic, insomniac rush. In one, Kammler submitted a graph detailing an exponential growth in SS construction far into 1946. He was obviously losing touch with all reality, for the reality was that all he had served and lived for now lay in rubble. At Mittelwerk itself, where all began to sense the end, work ground to a halt. Rumors after the war spread that Kammler and other SS officers wanted the prisoners herded into one of the tunnels, at which point they would order the entrances and exits dynamited. Nothing of the sort happened. The American Third Armored Division rolled into Nordhausen on 11 April. Kammler fled eastward, unlike the majority of the rocket team, which fled west. He seems to have taken part in fighting in Prague, and evidence suggests that he eventually killed himself or perhaps had his own adjutant shoot him on 9 May. His personality was consistent with such an end. Kammler believed in the Reich and sought to serve it with all his might until its very extinction. When the end came, he could no longer imagine his future and chose not to face responsibility for his past.[89]

The end was approaching fast. A few days before Dachau was liberated, Pohl gathered what men remained to him and held a banquet there. The prisoners outside were starving as they had always done under the WVHA, but this was no concern of the celebrants. They had fled south to be captured by the Americans, by whose hands they expected more lenient treatment than by those of the Red Army. And who could deny that they were right? Pohl, it is true, was hanged in 1951. But he was the only one of the WVHA top managers executed by the Americans. The engineers of the Central Building Directorate of Auschwitz, for example, almost all escaped justice. With his wife, Richard Glücks committed suicide and thus seems to have met an end similar to that of Hans Kammler. So the laziest and the most capable administrator of murder in the WVHA likely shared the same fate.[90]

EPILOGUE

Seventeen WVHA officers, including the disgruntled Hans Hohberg who had been kicked out of the DWB in 1943, found themselves in the dock at the Nuremberg War Crimes Trials. Here they pathetically pleaded their innocence. Hans Baier claimed ignominiously—if somewhat honestly—that he had been too incompetent to have masterminded industrial killing and slave labor. The fact that this was a separate issue from having participated in murder seems not to have occurred to him. Others like August Frank—who had helped launder Jewish gold in the Operation Reinhard—bombastically claimed to have been too competent; that is, their bureaucratic expertise had placed them in such demand that they had had no time to concern themselves with the details of all WVHA affairs.

Ironically, one of the pillars of these men's defense was their claim to be "modernizers" of German industry. Pohl told his judges that the SS drive for modern production demonstrated his innocence because, had the SS wanted to exploit slave labor, "it would have been cheaper not to go too far into modernization."[1] A civilian worker for the German Earth and Stone Works rendered perhaps the most poetic statement in this regard: "If I mention the term 'modern equipment' here, I understand by that a perfect blending of the various individual factors, of the most up-to-date machinery, and technical perfection, so that manual work was limited to insignificant proportions."[2] Remember, however, this "perfect" system was precisely what the WVHA never succeeded in producing. Eugen Kogon, a survivor of Buchenwald, dryly remarked about the SS's beloved machines (in this case "modern" pumps): "I can have pumps and still sink into the water up to my stomach when it is bad weather."[3]

Of course, Pohl did not invent "modernization" theory, and the stock phrases he and his officers used were neither original nor eloquent. Some of the most powerful expressions had already been formulated before the war's close. Robert K. Merton and Karl Popper, to name but two examples, argued that an inherent link tied science, and the technology derived from science, to Western, liberal-capitalist democracy and, further, that this was due to the value-free, nonideological character of such knowledge and its artifacts.[4] When WVHA officers tried to argue that their "modern" technology had been inherently liberating, they failed to convince anyone at Nuremberg. But they nevertheless sought deftly to associate themselves with a new ideology of "value-neutrality" emerging in Western modernization theory in order to put distance between themselves and Nazi fundamentalism.

Yet when Pohl and his enthusiastic young officers had written their earlier manifestos, they had wished to invest their institutions with anything but "neutrality." Whether midlevel managers, certified public accountants, lawyers, or engineers, all of them representatives of new, modern professions, WVHA officers had promoted slave labor, the crassest racism, and the New Order in the occupied East as similarly modern; in short, they had seen no contradiction in counting as modern the very barbaric ideals that intellectual historians would later decry as "antimodern." After the war, only the wrecked machinery and abandoned organizational charts of the WVHA remained, the dead skeleton where once had been the animated limbs of modern institutions and technological systems. Only after the war, when, literally and figuratively, the grisly boneyards of these systems were all that remained, did the SS men under Pohl desperately try to claim that managerial rationality and the machines of modern industry somehow existed as a humane force apart from, even opposed to, Nazi Germany, as if they had not worked for close to a decade in order to make their systems embody Nazi ideals every step along the way. Most perversely they had helped integrate modern organization and technology into the factory of extermination at Auschwitz-Birkenau.[5] But we can understand the SS's modernization drive only if we do not conflate "modernity" with "rationality" and "pure" technocratic instrumentalism, or insist that modernization necessarily leads to a democratic polity, or the full-flowering of the Enlightenment. Such was, however, the gist of much modernization theory of the 1950s and 1960s, and it sometimes persists into the present.

The Nazis' vision of modernity was only one among many that demonstrate how irrational belief in modernization can be.[6] In a sense a straight line runs from Heinrich Himmler's first preoccupation with "German inventors" in the early 1930s, through the SS brickworks, to the V-2 rockets of the war's end. In all cases the SS failed due to a hyperventilating fascination with such gadgets that they called "modern." Through these they hoped to express the highest values of Hitler's Germany. In the first case, the German inventors, Himmler backed frauds who sold themselves as modern men. In the case of the SS's industries, the SS did so again in the person of Arthur Ahrens and also chose technology that almost ensured that its slave-labor factories would fail in the most embarrassing way. Finally, the SS *and* others such as Albert Speer backed the V-2 rocket and its eerie underground mass-production factory in the Harz Mountains. This example is important because it shows that the SS was not necessarily incapable of mastering modern production systems. Nevertheless, the wonder weapons had almost no strategic value. The V-2 delivered a warhead of 1,000 kilograms of explosives and was not as accurate as the SCUD missiles deployed by Iraq in the Gulf War. On the other hand, each V-2 used up to six times the resources of more practical military hardware like fighter planes.[7]

What explains this pattern other than an almost blind faith in salvation through "modern" machines, machines that Himmler and other prominent National Socialists associated with the German "will" or the German "inventive spirit"?[8] These systems failed *because* of the SS's ideological drive for modernity. We can understand this only by conceiving of "modern industry," defined here as vertically integrated mass-production factories and centralized, bureaucratic command and control, as an ideology. It was not nor has it ever been the end phase of a rational process of historical development or the pinnacle of industrialization.[9]

Naturally the SS's own "modernization" project did not conform to the many claims made by postwar historians and sociologists in the name of a "modern" society. Pohl's and Himmler's vision was partial and cannot be taken as an overarching definition of "the modern world." Here, as elsewhere, National Socialism opened itself to multiple interpretations, as polysemous in its meanings as it was polycratic in organization. A heterogeneous set of institutions, let alone social norms, has continually accompanied the rise of industrial, technological society. Unsurprisingly, the Nazis' approach was selective, but National Socialists have hardly monopolized selective approaches to "modernity." Fritz Stern, George Mosse, and Ralf Dahrendorf have all associated liberal democracy with "modernization," a characterization that is hardly less arbitrary, as much as we sympathize with their political goals more than Hitler's. Others like Zygmunt Bauman often wish to insist that National Socialism is *the* definitive moment of modernity, which they set equal to "rationality" and "technocracy." This is no less selective. The history of the WVHA does show, however, that the Nazis sought the benefits of modernization not by reacting against any single, predictable set of cultural changes that supposedly accompany it but by trying to make these changes conform to their own dogma. Everyone engaged in "modernization" attempts to fix a distinct set of norms and political systems to industrial change by insisting that their vision represents the one destiny open to humankind, that their modern technology will accomplish ideal social changes to which the only alternative is backwardness. Nazism's participation in this effort makes it "modern."

The WVHA officers at Nuremberg received sentences that ranged between ten years and life imprisonment. Pohl and three others were condemned to death. Some measure of justice might have been confirmed in these sentences, but in the 1950s almost all of the officers were pardoned or had their sentences reduced. It is an open question whether Pohl, if he had not been hanged in 1951 after appealing repeatedly for stays of execution, might have succeeded just a year later when the wave of pardons began. Those with "lifelong" sentences served only fifteen years; those condemned to death, but who escaped the gallows, were released after twenty. Few of the rest sat out their prison terms for more than a handful of years.[10] "Modernity" played a small role in these waves of pardons. The 1950s also witnessed the first ossification of the Cold War, and the West perceived the need for an economically strong and "modern" Germany as a bulwark against communism in the East. Thus the Western Allies joined West German leaders of the 1950s, who saw no reason to let Nazi industrialists languish in jail.[11]

And nowhere was the protean nature of "modernity" more pronounced than in the Anglo-American vision that quickly emerged after World War II. At that time, Walt Whitman Rostow was just one voice among many that began to promote economic "modernization" throughout the world in order to encourage "viable, energetic, and confident democratic societies." Of note, he coupled this to an argument that modern technology and managerial acumen were "neutral with respect to the political issues that rouse men's passions."[12] Such associations formed the kernel of an Anglo-American consensus and constituted what might be defined as an ideology of technocratic modernity. Because the value-neutral nature of managerial and technological systems had also supposedly confined the ambit of moral sensibility within industry and modern organization, many assumed that "ordinary" men in managerial institutions had never really been cognizant of Nazism's ideals. It was no accident that the WVHA engineers escaped justice more often and more completely than other professionals of the WVHA.[13]

For Gerhard Maurer and others tried in Poland, justice was swifter but equally unsure. For Polish communists, it did not prove sufficient to show that Maurer had aided in murder and driven hundreds of thousands as slaves. What mattered was that he be tried as a puppet of finance capitalism. The strategy backfired, however, for Maurer, like almost all committed SS business executives, had always conceived of the WVHA as a vanguard of Nazi productivism. In his hands business was to manufacture German culture and not merely earn profits, and he resented private capitalism and consumer society for their cultural anarchy. His judges therefore easily convinced him that the entire disastrous war had been caused by pursy, cigar-smoking men coveting their bankrolls in dark rooms; it had been the financiers, the bourgeoisie, that had plotted ruin for Germany's honest, working man. If the Americans promoted a new vision of "modernity" under the banner of liberal capitalism, which quickly displaced the Nazis' vision of modernity, the satellites of the Soviet Union seem to have settled for substituting an international capitalist conspiracy for the Nazis' "World Jewish Conspiracy." "I did not know," cried Maurer, "that capitalists, big businessmen, bankers, and large landholders already supported the Führer of the NSDAP and helped him into power before 1933 in order to exploit the party for their own greedy, profit-grubbing interests."[14] In the shadow of the gallows he wrote three such declarations, each increasingly febrile and repetitive. At first his Polish attorneys were excited. They transcribed the first confession into German typescript and translated it into Polish, presumably for distribution; the second, they only transcribed into type; the third, they left in Maurer's cramped handwriting. Even they recognized that the man was ranting. He was hanged in 1948, much more swiftly than Pohl, but after a trial that had abandoned universal criteria of justice for a bogus historical narrative and was therefore as much a travesty as the Nuremberg pardons.

It would be somehow comforting to consider all the SS men involved in this murderous chaos as ignoramuses, idiots, or retrograde cretins, but Maurer and Kammler were capable men. They were neither venal nor corrupt in any mundane sense. They were creative, interventionist managers and, in this limited sense, "rational." Convinced of their purpose, they victimized millions of men and women. Karin Orth estimates the total number of concentration camp prisoners killed during the tenure of the WVHA's control over the IKL at 1.9 million.[15] What competent administration the WVHA created and maintained had thus increased the scale of killing. Kammler's modern technical oversight made productive labor and the destruction of life compatible and its scope greater still. These managers had engineered the Holocaust and served as its daily supervisors.

The temptation to represent them as "one-dimensional men" is great. That so many people took part in the Holocaust and that so many found it more or less a worthy activity boggles the imagination. In fact, Fritz Stern notes that a man otherwise so wide-ranging in his contemplation as Albert Einstein could never come to terms with the fact that National Socialism represented a real social movement fueled by belief and genuine motivation rather than greed or careerism.[16] Not without reason have so many declared it incomprehensible, a black hole in the understanding of humankind.

Perhaps this is also why the perpetrators of the Holocaust are so often presented as "cogs" caught in the wheels of a murderous system—because imagining that they committed such monstrous deeds out of conscious human passion is so difficult. The bureaucratic "cog" is one of the longest-standing metaphors of modern bureaucracy in histories of National Socialism. In a trial in the 1950s against Gerhard Peters (the chief executive officer of the Degesch chemical firm, which supplied gas and advice to the SS for the extermination of human beings) judges used exactly the words "cog in the wheels of a giant machine."[17] Arendt's famous book on totalitarianism refers to Hitler's administrators as "being just a number and functioning only as a cog."[18] Even a careful historian of the Holocaust like Christopher Browning uses exactly this language: "In the Third Reich, specialists whose expertise normally had nothing to do with mass murder suddenly found themselves minor cogs in the machinery of destruction."[19] Now Browning (this is from 1985) has gone on to stress local initiative and the dependency of central authority in Berlin on the willing participation of "little men" in the Holocaust; in this very article he also stresses the radicalism of leading figures such as department chiefs. Nevertheless, Zygmunt Bauman picked up exactly this passage—of all things—and cited it in full in *Modernity and the Holocaust* (1989). Here he asks us to believe that it was not the conscious initiative of perpetrators and their first principles of racial supremacy that "caused" the Holocaust but some kind of global mentality of science and technology spawned by the Enlightenment.[20] Such arguments divide German society from its institutions, such that Germans do not act through institutions but institutions act upon Germans.

Overemphasis on the polycratic structure of National Socialist rule might be considered another facet of this same fundamental impulse. When it is applied in tandem with sensitivity to motives (as its authors intended), it can yield an accurate picture of the dynamic interaction and radicalization of Nazi Germany. When it degenerates into a focus on mere "power struggles," we are instead led to believe that Nazis pursued naked, internecine strife as if for its own sake. Again, as with the "cog in bureaucratic machinery," what is missing is how all too human the Nazis were. Naturally the National Socialist state was "polycratic," and the SS was so even within the boundaries of its own offices and departments. Himmler urged his underlings to follow the Führer principle. He told them that their jobs were "what an SS man made of them" and that they had better take initiative with the authority vested in them. In this sense he spurred them on to the very independent, wild initiatives written about in Franz Neumann's *Behemoth*. But Himmler's statements differ hardly at all from similar clichés that, say, chief executive officers of modern bureaucratic corporations today or at any time in the past fifty years have doled out to their vice presidents and office chiefs.

No doubt, disputes and different factions within the SS *do* demonstrate "polycracy" to some extent, yet what large institutions differ in this respect? And was polycracy, so often defined as internecine strife, the definitive factor in daily operations? What is more, is polycracy anything unusual or particularly National Socialist? Despite fractious disputes in the atom bomb project, for example, the young men assigned to the thankless, tedious task of computing equations for the physicist Richard Feynman worked harder when Feynman disobeyed direct orders and disclosed the purpose of their work to them.[21] It is not difficult to imagine that the same was true within the WVHA. The most hardworking individuals were those who knew and believed in what they did. They were also the most likely to be sought out by Speer's officials or private industry for help. By contrast, where we find mere venality, as in the case of Hans Baier or Erich Schellin, the WVHA was the most incompetent and least capable of any mutual effort.

There is no denying that conflicts occurred, even between cooperating institutions. They were all the more pronounced when all shared the same overarching goals but disagreed passionately about their implementation. They were even more bitter when competing ideological programs collided and could not be reconciled. Nevertheless, we should not be too quick to overemphasize dissent over consensus. This is not to underestimate the fundamental contributions of those who have brought the polycratic structure of National Socialism to our attention. For instance, Franz Neumann finished his *Behemoth* just before the German invasion of the Soviet Union. At the time, the Holocaust had yet to begin in its full dimensions. Common understanding of National Socialism was impartial at best. Many then viewed Hitler's Germany as a monolithic edifice in which the Führer gave orders and all followed in lockstep. Neumann stressed instead the decentralized nature of control in the Third Reich. Peter Hüttenberger later picked up on the multifarious authority under Hitler's rule and developed the theory of polycracy in one of the single most influential essays about Nazi Germany.[22] (Neumann had used the term earlier but only to refer to the proliferation of extraparliamentary institutions in what he called the "era of monopoly capitalism" before the war; that is, Neumann had not considered polycracy to be anything unique to the structure of National Socialism.)[23] Raul Hilberg was also deeply influenced by Neumann, and *The Destruction of the European Jews* still stands more or less unchallenged as the definitive text on the Holocaust. As did his teacher, Hilberg emphasized the importance of "organizational innovations" and the constant transformation of National Socialist institutions.

Nevertheless, Hilberg tended toward judgments of modern bureaucracy extremely common in the 1950s and 1960s. For example, in his first edition he wrote that "German bureaucracy was so sensitive a mechanism that in the right climate it began to function almost by itself."[24] In Hilberg's day, the judgments of Arendt (her *Origins of Totalitarianism* appeared a decade before Hilberg's book and her *Eichmann* two years after) or those of sociologists like C. Wright Mills were predominant. Mills, for example, wrote that midlevel administrators inhabited a world of the "cheerful robots," and he dedicated a section to this theme in his influential *White Collar*.[25] Hilberg would have had to reinvent the wheel in order to overturn such interpretations of bureaucracy in 1961; in all likelihood he agreed with them as did most intellectuals of his day. How could it be otherwise? It was not his primary purpose to understand modern organization; rather it was his secondary purpose so that he could understand the Holocaust. Major advances in organizational theory concerning modern bureaucracy and technology (Alfred Chandler, Oliver Williamson, Thomas Parke Hughes, Douglass North, James Beniger, Jürgen Kocka) did not start to appear until after Hilberg had written *The Destruction*, in most cases fully one generation later.

Perhaps this is partly why, over time, Hilberg began to distance himself from mechanistic metaphors of the Holocaust, especially after the questionable spin such interpretations were given by prominent German historians (although not exclusively these). It is therefore interesting to note that Hilberg omitted entirely his famous introductory paragraphs on self-operating bureaucracy when *The Destruction* was translated into German in 1982. Yet exactly these passages about mechanistic bureaucracy have lived a life of their own in later authors' work, and Hans Mommsen and Wolfgang Sofsky still argue this. Since the end of the 1960s both apologists like Albert Speer and left-leaning historians like Mommsen have posed as the moral conscience of their nation by condemning the "structures" of bureaucracy and the "mentality" of technocracy while dodging any principled confrontation with the vast, willing complicity in the Holocaust. Here a "self-acting" bureaucratic machine caused the Holocaust. Others have struggled to put back ideology ever since.

Why did so many of the "best and the brightest" find the murder of the European Jews the right thing to do? With some exceptions, Holocaust studies too seldom pose this question, but rather, Why did no one resist? and the answer provided by Mommsen and so many others: bureaucracy and large-scale technological systems had confined and morally deadened ordinary Germans. In fact, fully twenty years after the myth of Albert Speer as a "technocrat" has been thoroughly debunked, Mommsen has used exactly Speer's arguments to present Ferdinand Porsche—who, remember, was one of the first industrialists to solicit slave labor from the SS—as a pure technocrat. Porsche was supposedly a man with passion only for technical designs, and thus oblivious to the crimes perpetrated around him. It is a powerful argument, for people like Porsche can be condemned and let off the hook at the same time.[26]

Hilberg's original interpretation actually opposed this. Prefiguring later theorists of modern organization and technology, Hilberg stressed the many faces of agency as well as the necessary cooperation of institutions. The organized homicide of an entire segment of Europe's population "must ultimately feed upon the resources of the entire organized community."[27] Perhaps Hilberg subtly distanced himself from metaphors of a "self-operating" Holocaust because he always sought to stress initiative and motivation *as well as* reliance upon institutional structure. All along he had discovered that, contrary to the current wisdom of his day, ideology and conscious dedication mattered in large-scale institutions.

Furthermore, when Hilberg spoke of "organizational innovation," he usually meant the erection of a new institution. He did not generally pursue the collective biographies of their inhabitants or their daily work world. He was, and again understandably so, mostly concerned with the inputs and outputs of organizations, not their "guts." When he spoke of the dynamism of modern bureaucracy he mostly meant his powerful narrative model which identified the progression of (1) definition of victims, (2) their disenfranchisement, (3) their concentration, (4) their exploitation, and (5) their destruction. What this book tries to show in no way contradicts Hilberg's narrative model, but I do think the difference is important and complementary. I have tried to show how the dynamism of modern bureaucracies stems from the active identification of the bureaucrats themselves. These institutions operate through the steady flow of information up and down well-defined, impersonal hierarchies. In daily operations ideological consensus serves a clear function: it promotes trust and identification, which enables bureaucrats to rely upon information gathered by others whom they cannot know personally or even supervise consistently. Thus organization and ideology mutually reinforce each other, and the former is mere dead structure without the latter. Conversely, modern institutions are quite vulnerable to dissent. When this occurs, they quickly become stagnant or dysfunctional.

There is perhaps one last point upon which this book might complement the work of those who have studied modern organization in the Nazi genocide. The bureaucrats who concentrated, processed, and ultimately sent the victims to their death, remarks Hilberg, "could dip into a vast reservoir of administrative experience, a reservoir which church and state had filled in fifteen hundred years of destructive activity."[28] Yet it is interesting to note how many core institutions in Hitler's Germany *lacked* ancient tradition. It was not the venerable German army or the post office that continually pushed forward the "final solution to the Jewish question" or slave labor; nor was it the staid civil servants of the Reich Chancellery. There is little question that these were complicitous, but they did not form the avant-garde of genocide. The SS took that position as did the Nazi Party Chancellery and the young men of Speer's Ministry of Armaments and War Production—that is, new, dynamic organs of the fascist state.

At its final destruction, the SS could look back on a history of, at most, twenty years. Even that was stretching it. Its leadership never changed hands from the founder generation. Likewise young or newly founded state industries, those that owed their existence to the Third Reich, blazed the trail of slave labor in cooperation with the concentration camps: Volkswagen, the Gustloff Works, the Hermann Göring Works, Heinkel. Venerable corporations with a continuous lineage of entrepreneurship dating back generations were the exception rather than the rule. The Fighter Staff, responsible for some of the worst crimes of slave labor, existed for less than six months. In short, these were all modern rather than traditional bureaucracies. As a rule, they were conscious of this fact.

Several factors divide modern from traditional bureaucracy. First, as Robert Musil parodies in *Mann ohne Eigenschaften*, civil servants are supported by state power, whereas the new professions must support themselves by what they know.[29] Partly in consequence, traditional bureaucracies seldom relied on statistics because they could rely on social prestige. This is confirmed by Ted Porter in his investigation of statistical thinking. Statistics arise most often where diverse accountability creates the need to respond quickly and transparently. Modern bureaucracies use statistics rather than appeal to social authority. Numbers are the medium of fungible information, and modern bureaucracies use them not merely to record and chronicle but to command and control. The model for modern bureaucracy never was the traditional European civil service but the modern, American corporation new to the late nineteenth and early twentieth centuries.

Throughout the Third Reich the practices typical of modern and traditional forms sometimes lay side by side in the same institutions. A good example might be the "Deathbooks" of Auschwitz, which represent a more traditional chronicling of information whose most typical form was the list or, in traditional business organization, the bound ledger. The Deathbooks of Auschwitz recorded date of death, name, and prisoner number, but also the place of origin for people whose origins the SS was trying to erase; they also recorded the names of the father and mother of the deceased. Their main purpose was descriptive rather than command and control, and those in positions of decision generally made scant reference to them. It is unclear what purpose they were ever supposed to serve, except to keep information safe for the eyes of the state: here was the 1,500-year-old tradition of bureaucracy. The form of such records differed little from the records of birth, death, and marriage kept for centuries by churches or synagogues.[30]

The Deathbooks proved useless to the SS Labor Action, and in early 1943, at the very time that Gerhard Maurer was increasingly extending his modern statistical control, the Political Office of the camps ceased keeping them at all. When Maurer insisted upon the standardized flow of information things began to change. At one level he undertook a gross reduction of the data collected in the Deathbooks. For example, many of his forms dispensed with prisoners' names, place of origin, or fathers' and mothers' names. Aggregate numbers sufficed instead. But at another level, he accelerated the speed with which information was gathered and sent to the WVHA managers who made allocations. The forms collected regularly by camp doctors and orderlies yielded up mortality rates at a glance at graphs and charts; they recorded not just information but the crucial element of time and changes over time. The Office Group D could then respond to fluctuation and the wishes of modern industry. It would be an exaggeration to suggest that Maurer's system worked according to a flawless rational plan, and he never overcame the resistance to modern management among all Kommandanten. But he did make information flow through the Office D2 in a dynamic fashion, enough to make centralized command and control over labor possible for the first time.[31]

Like the ambitious men in many of the emerging disciplines and new professions of the 1920s and 1930s, Maurer and the vast majority of those who worked for the WVHA were young men. More often than not, they were not the German "mandarins" of tradition-bound institutions. In some cases such institutions were closed to them.[32] Such "new men" tended to seek out experimental organizations with the air of impending revolution about them. Those who held doctorates as public accountants under Hohberg or the Diploma Engineers of the Office Group C had earned degrees that counted as recent innovations in the world of German education. More traditional, humanities-trained professionals did not necessarily view them as prestigious. Nevertheless, such up-and-coming professionals worked in "interdisciplinary" fields and hoped to apply their special knowledge to social reform. They believed they could solve Germany's pressing problems if only they got the chance and the power to do so.

The three most important departments of the WVHA, the Office Group D–Inspectorate of Concentration Camps (IKL), Office Group W–Business, and Office Group C–Construction, yield three distinct patterns of interaction between ideology, the new managerial professionals, and modern bureaucratic and technological systems.

Traditional interpretations that place ideology—often taken to be synonymous with unrestrained irrationality—and pragmatic, rational organization at opposite ends of a spectrum could not find a better illustration than the Office Group D–Inspectorate of Concentration Camps. Nevertheless, it was the content, rather than the mere presence, of ideology in the IKL that proved so detrimental to modern industry and organization. The officers of the concentration camps were distinguished by a strong esprit de corps. Their consensus can be characterized by what I have called the primacy of policing.[33] In this context, the purpose of labor was to provide a predictable method for controlling and punishing prisoners, even unto death. As self-styled political soldiers, the most typical officers of the IKL were also hostile to modern bureaucratic administration, which they defined as the work of pencil-pushing do-nothings. This meant that the skills necessary to operate modern industry were not only unknown to most of them but actively despised. It should be no surprise that modern industry failed in the concentration camps and that the results were disastrous for the prisoners, especially because neither Pohl nor Maurer ever strictly removed the management of labor sites from the ultimate authority of the Kommandanten.

The Office Group W–Business provides a case in which some knowledge of modern industry and technology was clearly present—in contrast to the IKL. Yet here any unifying sense of purpose was lacking. After initial and disastrous technological failures at the German Earth and Stone Works (DESt), Pohl recruited experts like the certified public accountant Hans Hohberg and the brickworks engineer Erduin Schondorff in order to bring skill into the SS companies from the outside. But these men (and others) never found common cause with the existing cadre of young SS officers committed to the SS's Nazi fundamentalism. On the other hand, most dedicated SS businessmen had a background in either corporate law or sales. Engineers were not necessarily foreign to them, but engineering was. Thus, in exactly those places where Pohl's German Commercial Operations mustered the most technical skill—among Schondorff's engineering pupils—it failed to enroll their commitment. Where Pohl could muster the most commitment (Salpeter, Mummenthey, Bobermin, Volk), on the other hand, his officers lacked the necessary knowledge to syncretize SS goals in technological and organizational systems.

No one in the Office Group W ever engineered slave labor into a viable system as Hans Kammler's Building Brigades and Construction Directorates did in the Office Group C. Instead, by the end of the war, Erduin Schondorff had withdrawn increasingly into a world of almost pointless academic research. Failure and chaos reigned. The only notable exception among SS companies, the Textile and Leather Utilization GmbH, proves the rule. Unlike other filials of the German Commercial Operations, here technical and financial management was unified and dedicated to the SS's goals; here, too, a prevalent ethos of "women's work" throughout the textile and garment industry had yielded machines that were adapted to low-skilled and unmotivated workers.

Last, in Office Group C–Construction, Kammler recruited a dedicated officer corps that was neither riven by internal dissent like the Office Group W nor lacking in organizational or engineering talent like the IKL. Here, just as the camp SS had held Theodor Eicke in awe in the early days of Dachau, so too did Kammler's subordinates admire his enormous capacity for work, his breadth of knowledge, his dedication, and his achievements. Some may not have liked him, but he impressed everyone. Unlike the IKL, however, the collective identity of the SS civil engineers included a commitment to sound technical competence and the mastery of modern organization, something commonly associated in the SS with racial supremacy and National Socialism. Whereas the industrial systems of the German Commercial Operations truly developed like a Behemoth, expanding in all directions without systematic planning, Kammler engineered solutions that harmonized otherwise contradictory SS intentions. His civil engineers did not measure productivity in terms of yield per prisoner, but in terms of the speed of construction. Unlike Erduin Schondorff, Kammler was more than ready to substitute techniques of labor-intensive work for labor-saving machines when the latter were scarce to come by, and he used modern managerial bureaucracy to do so. Last, these engineers repeatedly found common ground for cooperation with engineers among the polycratic and overlapping offices of state ministries and industry, some of whom—Karl Otto Saur and Gerhard Degenkolb come to mind—saw in SS officers like Kammler kindred spirits.

In the IKL, ideological consensus obstructed the function of modern institutions; in the Office Group C–Construction, by contrast, ideology reinforced and enhanced function. This also underscores how ideology could be "functional" in the Third Reich and in modern institutions in general. First of all, as Robert Gellately has pointed out, ideological goals did not operate as independent variables that mechanically determined organizational outputs.[34] Rather ideology lent otherwise impersonal institutions an air of common identity and purpose. In doing so, it engendered trust. It created a milieu in which individuals who were otherwise strangers could believe that each was essentially interested in the same thing. Once individuals came to identify with the organization, the organization could more easily mobilize their collective action, sometimes in pursuit of completely different ends. None of Kammler's engineers knew in early 1941 that they would be engineering the Holocaust or managing extermination through work; many of them probably could not even have conceived this at the time. But they did come to share in the collective identity of the SS civil engineering corps, its racial supremacy, its utopian sense of mission. When it came time to design the gas chambers, they interpreted this as merely another part of that larger mission.

•

Perhaps most disturbing for me, the new interdisciplinary professionals who rallied to National Socialism included historians. Many of the senior professors who taught Germany's leading generation of social historians of the 1960s had in fact pursued what was called "eastern research" in the 1930s and 1940s. They involved themselves deeply and directly in Himmler's plans for a New Order, and hoped to synthesize the tools of geography, sociology, and history to create a science of the "Volk." Arguably the most famous living German historian, Hans-Ulrich Wehler, completed his dissertation under the tutelage of one such former "eastern researcher," Theodor Schieder. In the winter of 1941–42 Schieder wrote of the assimilated Jews of Bialystok, "This pure outer whitewash [of assimilated culture] has only made it better for the Jewish element, whose pure race has proved itself just as true as ever before, to occupy important key positions in business."[35] Schieder advocated what we would now call ethnic cleansing of Poles as well.

In February 2000, I had the great fortune to see Wehler give a public lecture in Munich, and to my mind, what was most remarkable was the discontinuity with this ugly past. Wehler's owl-like brows rose and fell as his eyes roamed the lecture hall and twinkled with curiosity. He was so obviously committed to a democratic Germany based upon civil liberties that there can be no doubt that, to him, "eastern research" was a deep embarrassment. He does share in common with his teachers the basic proposition that deep structures in society should be the basic unit of historical research. Like them, he has abandoned simple histories of "great men," "glorious national deeds," and traditional history of ideas. He and his school have also sought to work with statistical methods and have often organized large, interdisciplinary research projects. This too was characteristic of "eastern research." But Wehler has never entertained categories of "race" or "biological destiny"—so important to the New Order. Whatever the purely structural similarities, his first principles distinguish him clearly from his forebear.

Curiously, Wehler related that, as a young historian, the question always arose with American colleagues, Why did no one ever speak out against

National Socialism? From this, he said, he had drawn the lesson that "we had always to speak out, even at risk of putting one's foot in one's mouth [*selbst wenn man ins Fettnäpchen tritt*], rather than react too late"—that is, rather than allow democracy to go under as it had with a whimper in the Weimar Republic.[36] Over the course of a long career he and other German social historians have continually gone out of their way to do exactly this.

Nevertheless, I could not help feeling that the question American professors had continually pressed upon Wehler was the wrong one, for the fact is that so many of the young, the most capable, the most intelligent—men like Hans Kammler or Gerhard Maurer or the "eastern researchers"—*had* spoken up. Not their silence or inability to resist but their willingness to serve and the content of what they said fed the dynamism of National Socialism. Not all were "orgiastic anti-Semites," but so many of them found enough of themselves and their ambitions fulfilled by one or another myriad theme of National Socialism that they were indeed Hitler's willing executioners. As I listened to this still spry historian, the more pertinent question seemed to me, not why they *did not* speak out, but why they did.

NOTES

A number of documents cited in the notes can be found in multiple archives, particularly those related to the Nuremberg Trials, and thus are not cited to a particular archive below. I have used the standard designation of Nuremberg documents (NO, NG, NI, PS, etc.) when it was known to me. References to Defense Document Books, Prosecution Document Books, and Protocol are to Trial IV, vs. Oswald Pohl et al., held at Nuremberg. I have carried out research in these various document collections in different locations at different times. In any case where an English title is given for a German document, I have relied upon Nuremberg translations. I have used documents at the National Archives, Washington, D.C.; the Imperial War Museum, London; the Institut für Zeitgeschichte, Munich; and the Bundesarchiv Potsdam and Koblenz, now located at the Bundesarchiv Lichtefelde, Berlin.

ABBREVIATIONS

In addition to the abbreviations found in the text, the following abbreviations are used in the notes.

APMM	Archiwum Panstwowego Museum na Majdanku (Archive of the State Museum of Majdanek)
BA MA	Bundesarchiv-Militärarchiv Freiburg
BAK	Bundesarchiv Koblenz
BAP	Bundesarchiv Potsdam
BDC	Berlin Document Center
CIOS	Combined Intelligence Objectives Sub-Committee
FHA	Führungshauptamt (Waffen SS General Staff)
GBA	Generalbevollmächtigter für den Arbeitseinsatz (General Plenipotentiary for the Labor Action)
HTO	Haupttreuhandstelle Ost
IfZ	Institut für Zeitgeschichte, Munich
IMT	International Military Tribunal
KL	Konzentrationslager (concentration camp)
KZ	Konzentrationslager (concentration camp)
NG	Nazi Government (Nuremberg Trials Documents)
NI	Nazi Institutions (Nuremberg Trials Documents)
NO	Nazi Organization (Nuremberg Trials Documents)
OKW	Oberkommando der Wehrmacht (German Armed Forces General Staff)
OT	Organization Todt
PHA	Personal Hauptamt (Personnel Main Office)
PMO	Panstwowe Muzeum w Oswiecimiu (State Museum Auschwitz/Birkenau)

RG	Record Group
RJM	Reichsjustizministerium (Reich Justice Ministry)
RLM	Reichsluftfahrtministerium (Reich Air Ministry)
RuSHA	Rasse- und Siedlungshauptamt (Race and Settlement Main Office)
RWM	Reichswirtschaftsministerium (Reich Economics Ministry)
Súa	Státní ústrední archiv, Prague
USHMM	United States Holocaust Memorial Museum, Washington, D.C.
Vha	Vojensky historicky archiv, Prague
W-SS	Waffen SS
ZSL	Zentrale Stelle der Landesjustizverwaltung, Ludwigsburg

INTRODUCTION

1. Herbert, *Fremdarbeiter*, 270, gives the exact statistic of foreign workers in the overall German work force as 26.5 percent in August 1944. Regarding the loyalty of German midlevel management, see Prinz, *Vom Mittelstand zum Volksgenossen*.

2. Sketch, sometime in 1944, T-976/18.

3. Mumford, *Technics and Civilization*, 177.

4. Arendt, *A Report on the Banality of Evil*, 26 and 49 respectively; compare Allen, "The Banality of Evil Reconsidered," 253–94.

5. Arendt, *A Report on the Banality of Evil*, 54.

6. Ibid., 46; compare Safrian, *Die Eichmann Männer*, 43, and 23–67 in general.

7. Beniger, *The Control Revolution*, 279.

8. Weber, *Gesammelte Aufsätze zur Religionssoziologie*, 203–5.

9. Giddens, *The Consequences of Modernity*, esp. 137–44; Bourdieu, *Outline of a Theory of Practice*, 72–95; Habermas, *Theorie des kommunikativen Handelns*, 1:209–368 and esp. 2:459–61; Zunz, *Making America Corporate*, esp. 55–58, for the careers of engineers as midlevel managers; Zeitlin and Tolliday, "Employers and Industrial Relations between Theory and History," 1–34. Discussion of the role (or its absence) that ideology played in the genocide has erupted again in the wake of Daniel Goldhagen's *Hitler's Willing Executioners*. Although Goldhagen's critics are right to point out the book's methodological and substantive flaws, there can be little doubt that Goldhagen has gained such wide acclaim because he addressed the role of ideology, which many historians have neglected. Many recent works have begun to put ideology back into the picture, among them Bartov, *Hitler's Army*; "Passing into History," 162–88; and *Murder in Our Midst*; Herbert, *Best*; Orth, *Die Konzentrationslager-SS*.

10. Mommsen, "Die Realizierung des Utopischen," 186.

11. Lifton, *The Nazi Doctors*, argues that SS men were forced to lead double lives to overcome their repulsion for their work, a quandary that in fact few ever faced.

12. Safrian, *Die Eichmann Männer*.

13. My prosopographical studies of the WVHA officer-managerial corps are based primarily on names gathered from the "SS Führer des Wirtschaftsverwaltungshauptamt," BDC Hängeordner 1206–1313. My study includes 164 officers who at one point in their SS careers served in significant positions in the WVHA or its predecessors. In a complete list from 1944, out of a total of 138 administrative officers, 85 worked as administrative officers in the service of other offices and not directly in building the WVHA.

Only those above the rank of *Hauptsturmführer* are listed with their administrative function. This leaves around 60 top-ranking officers when one adds 5 above the rank of general (not included in the document). I excluded, for instance, officers who led the Truppenwirtschaftslager of the W-SS. These officers were much more subordinate to the FHA than the WVHA. I also excluded officers who worked in the administrative offices of other SS Hauptämter for similar reasons. Extrapolating the same ratios for *Hauptsturmführer* included in the totality of all administrative officers in the SS in 1944, 390, yields an estimated activated staff of around 121 officers in the service of the WVHA. My data are, however, significantly more expansive than this, although by no means exhaustive. They include officers who left the WVHA's forerunners in its early days for various reasons and are not included in the "SS Führer des Wirtschafts-Verwaltungsdienst."

14. For examples of scholarly work, see Dawidowicz, *The Holocaust and the Historians*, 20; Mommsen, "Die Realizierung des Utopischen," 217; Friedländer's introduction, xlv. More generally, see Evans, *In Hitler's Shadow*. The intellectual historian Ernst Nolte also alluded to this "uniqueness" in his article that incited the "Historikerstreit"; see Nolte, "Vergangenheit die nicht vergehen will." Appleby, Hunt, and Jacob, *Telling the Truth about History*, 7, deploys the same rhetoric in an American *Historikerstreit*. Not the precision of these statements but their ubiquity is interesting: that a historian as far left as Mommsen or as far right as Nolte should share the same opinion about technology and Nazism, although none have ever really studied it.

15. Kammler, end of Dec. 1941, "Bericht des Amtes II–Bauten über die Arbeiten im Jahre 1941," USHMM RG-11.001M.03: 19 (502-1-13).

16. Hayes, "Polycracy and Policy in the Third Reich," 190–210.

17. Hüttenberger, "Nationalsozialistische Polykratie," 417–42, quotation from 421. Neumann prefigured Hüttenberger's theory in *Behemoth*. Allen, "The Banality of Evil Reconsidered."

18. Kafka, *The Trial*, 86; Koselleck, *Preußen zwischen Reform und Revolution*, esp. 663–71; Caplan, *Government without Administration*; and Mommsen, *Beamtentum im Dritten Reich*. Hilberg, *Destruction of the European Jews*, long ago stressed the importance of "organizational innovations."

19. Allen, "Ideology Counts." A recent exaggeration of demonic Nazi efficiency can be found in Black, *IBM and the Holocaust*.

20. Smelser, *Robert Ley*, 155, and on DAF ideology, 174–79. Mommsen and Grieger, *Das Volkswagenwerk*, is discussed extensively in n. 8, pp. 322–23, and n. 3, p. 331.

21. Quotation from Wisselinck, "Ernährung und Ausbildung der Lehrlinge: Deutsche Erde- und Steinwerke Granitwerk Gross-Rosen," 2–7 Mar. 1944, T-976/18. See also Wisselinck, "Aktenvermerk Granitwerk Gross-Rosen," 30 Mar. 1944, T-976/18, and another similar case, Opperbeck to Chef W Baier, "Bericht des Hauptbetriebobmannes SS-Ostuf. Wisselinck vom 13 Sep. 44 über den Besuch im Werk Butschowitz am 31 Aug. 44," 3 Oct. 1944, T-976/18. By contrast Wisselinck praised SS companies that withheld rations from their prisoners and made efforts to secure extra rations for civilians (see Wisselinck to Chef W Baier, "Granitwerk Gross Rosen," 29 Feb. 1944, T-976/18, and Wisselinck's reports throughout this collection). Jaskot, "The Architectural Policy of the SS," 100–159, contains excellent analysis of the SS's stone quarries. See also NO-2132, Dr. Claussen, directive of Reichsminister für Ernährung und Landwirtschaft, 7 Apr. 1942.

22. Karny, "'Vernichtung durch Arbeit.' Sterblichkeit in den NS-Konzentrations-lagern," 140–47.

23. Pohl to Obersturmf. Ketterer, Betriebsobmann der Granitwerk Gross-Rosen, "Gross-Rosen/Striegau," 24 Mar. 1944, T-976/18.

24. At around this time, correspondence begins to refer to Wisselinck as the Sonder-Reichstreuhänder der Arbeit für sämtliche SS-Wirtschaftsbetriebe (Special Reich Trustee of Labor for all SS Business Operations). He was obviously being promoted for excelling at his job. This title appears on various reports in T-976/18.

25. Wisselinck to Stab W, "Vorschläge zur weitgehendsten Ausschaltung von Ver-untreuungen in der Gemeinschaftsverpflegung," unspecified, sometime in March 1944, T-976/18. Compare a similar case at Jonastal/Ohrdruf labor camp, Remdt and Wer-musch, *Rätsel Jonastal*, 52.

26. All quotations from Wisselinck, "Die SS-Siedlungen bzw. die Werksiedlungen in den wirtschaftlichen Unternehmungen der Schutzstaffel," 17 May 1944, T-976/18.

27. Exemplary would be, respectively, Mayer, *Why Did the Heavens Not Darken?*; Turner, "Big Business and the Rise of Adolf Hitler," 89–108; Goldhagen, *Hitler's Willing Executioners*; and Mosse, *The Crisis of German Ideology*.

28. Evans, *In Hitler's Shadow*, 36.

29. Wisselinck, "Die SS-Siedlungen."

30. Beniger, *The Control Revolution*; Zunz, *Making America Corporate*; Chandler, *The Visible Hand*, and, for a comparative perspective that includes Germany, see *Scale and Scope*. Regarding the German case in particular, see Kocka, *Die Angestellten*, and "Scale and Scope," 711–16. Regarding the Weimar and Nazi period, see Prinz, *Vom Mittelstand zum Volksgenossen*.

31. Gellately in particular has warned against rigid causal explanations that invoke ideology; see "'A Monstrous Uneasiness,'" 180. Allen, "Technocrats of Extermination."

32. Wisselinck, "Die SS-Siedlungen."

33. Smelser, *Robert Ley*, 19.

34. Nolan, *Visions of Modernity*; Overy, *The Nazi Economic Recovery*; "Mobilization for Total War in Germany, 1939–1941"; "Germany, 'Domestic Crisis' and War in 1939"; and "'Blitzkriegswirtschaft'?"

35. Peukert, *Inside Nazi Germany*, 193.

36. Advertisement, *Der Vierjahresplan: Zeitschrift für Nationalsozialistische Wirt-schaftspolitik* 3 (1939): 235. This was a special issue and the advertisements were extremely glossy. Compare Overy, "'Blitzkriegswirtschaft'?," with Ritschl, "Die NS-Wirtschafts-ideologie," 48–70, and Mason, "Domestic Dynamics of Nazi Conquests: A Response to Critics," 161–89; "Labour in the Third Reich, 1933–39"; and "Some Origins of the Second World War," 67–87.

37. Quoted after Ackermann, *Heinrich Himmler als Ideologe*, 37.

38. See Herf, *Reactionary Modernism*. MIT's engineering faculty in the interwar period was just as enamored of reactionary "antimodern" architecture as were the worst Nazi hacks. See Sinclair, "Inventing a Genteel Tradition: MIT Crosses the River," 1–18. Compare Alder, "Innovation and Amnesia," esp. 300, and *Engineering the Revolution*. On the case of German engineering during the Enlightenment, see Brose, *The Politics of Technological Change in Prussia*. Recently David Lindenfeld has pointed out that "inter-

pretation of the Holocaust rests on a double caricature—of the Third Reich on the one hand, and of the legacy of the Enlightenment on the other," in "The Prevalence of Irrational Thinking in the Third Reich," 368. The history of technology can make a significant contribution to Herf's interesting beginnings precisely by departing from these caricatures. See also Latour, *We Have Never Been Modern*, esp. 67–70; Dietz, Fessner, and Maier, " 'Der Kulturwert der Technik,' " 24; and Zilt, " 'Reactionary Modernism' in der westdeutschen Stahlindustrie?," 191–202.

39. Jünger, *Der Arbeiter*, 318. Regarding Jünger, who was not himself a fervent Nazi, and National Socialism, see Herf, *Reactionary Modernism*, 70–108. See also Nolan, *Visions of Modernity*.

40. Abrahamson, *Against Silence*, 42.

41. What could be more indicative than Henry Ford's own autobiography with Samuel Crowther, *My Life and Work* (Garden City, N.Y.: Garden City Publishers, 1922). Translated, *Mein Leben und Werk* was a best-seller in Germany. Its popularity was not due primarily to Ford's own anti-Semitism; nevertheless, some Nazis promoted Fordist production as commensurable with their racial aims. See Hughes, *American Genesis*, 284–94.

42. For a review, see Peukert, "Alltag und Barbarei," 51–61, esp. 55–57. The classic text in this field is probably Kershaw, *Popular Opinion and Political Dissent*. For an interesting case of the Nazi government's retreat from its ideology in industrial policy, see the discussion of women in the work force in Winkler, *Frauenarbeit im Dritten Reich*, 44–46. Histories of everyday life have recently begun to emphasize popular support; see Gellately, *Backing Hitler*.

43. Some very sophisticated historians of Nazism believe the exceptionalism of National Socialism should always be highlighted above its common roots in Western culture; for example, Friedländer, "West Germany and the Burden of the Past"; Maier, *The Unmasterable Past*, 66–99.

44. Hilberg, "Significance of the Holocaust," 101.

CHAPTER ONE

1. Naasner, *Neue Machtzentren*, 198–99, cites one of Adolf Hitler's monologues. See also Koehl, *The Black Corps*, 10–26; Buchheim, "Die SS—das Herrschaftsinstrument," 30–33.

2. Kissenkoetter, *Gregor Straßer*, 22–27; Corni, "Die Agrarpolitik des Faschismus," 391–423, and *Hitler and the Peasants*, 18–33.

3. Koehl, *The Black Corps*, 26–30.

4. James, *The German Slump*, 148. The Nazis, he argues, capitalized on despair and directionlessness among German businessmen by peddling a sense of the future at a time when organized capitalism could no longer provide one. Regarding engineers and this same sentiment, see Hortleder, *Das Gesellschaftsbild des Ingenieurs*, 111. See also Trommler, "Between Normality and Resistance"; Herbert Ziegler, *Nazi Germany's New Aristocracy*, 49–51, 74–78, 93–124; Koehl, *The Black Corps*, 34, 40–55. Compare Kater, "Zum gegenseitigen Verhältnis von SA und SS," 339–79.

5. Compare Naasner, *Neue Machtzentren*, 242–50. Koehl, *The Black Corps*, 79, 92, 287ff.

6. Segev, *Die Soldaten des Bösen*, 119–52; Boehnert, "The Third Reich and the Problem of 'Social Revolution,'" 203–17.

7. Letter, 21 May 1933, submitted to Himmler by Pohl upon his application to the SS, BDC SS Personal-Akten Pohl, Band 1. See also Koehl, *The Black Corps*, 109–13.

8. Pohl to Himmler, 24 May 1933, BDC SS Personal-Akten Pohl, Band 1. Compare NO-1205, Affidavit of Oswald Pohl.

9. Testimony of Oswald Pohl, Protocol: 1258, 1254–60 in general; Lebenslauf from ca. 1932, BDC Personal-Akte, and Lebenslauf, BDC RuSHA Akte Oswald Pohl; NO-2343, NO-2618, Affidavits of Oswald Pohl. See also Wolf Lubbe to Hans Baier (both Navy buddies of Pohl's), 18 June 1942, BDC SS Personal-Akte Hans Baier.

10. Kracauer, *Die Angestellten*, 7. In my prosopography I have counted the *Abitur* as elite education, sufficient to prepare one for a business career. I did not count *Volksschulen* and apprenticeships as modern occupational training. Expanding educational institutions between the wars provided valuable modern occupational training, which was, however, not considered elite—for example, the *Fachschulen, Technikum, technische Lehranstalten, Handelsschulen*, or military training programs. I am indebted to Kees Gispen, who helped me to clarify some of these categories. Scanning some of the SS managers' dissertation titles should leave little doubt about their acquaintance with modern management: Hanns Bobermin, "Die Rationalisierung des kaufmännischen Büros im industriellen Grossbetrieb und ihre Wirkungen auf Angestellte" (Wirtschaftswissenschaft diss., University of Rostock, 1930); Walther Salpeter, "Verbotene und unsittliche Geschäfte im Steuerrecht" (Jura. diss., University of Halle-Wittenberg, 1933); Max Horn, "Zum Bewertungsproblem in der Jahresbilanz der Unternehmung" (Leipzig, 1935); Leo Volk, "Die Übertragung des Anwartschaftsrechts aus bedingter Übereignung" (Mosel, 1936); Hans Kammler, "Zur Bewertung von Geländeerschliessungen für die grossstädtische Siedlung" (engineering diss., Hanover, 1931). Compare Herbert Ziegler, *Nazi Germany's New Aristocracy*, table on 104, 113–17, 147.

11. NO-1224, Oswald Pohl, 24 June 1932, "Why Am I a National Socialist and Why an SA Man?" See also NO-1225 and NO-2343, Affidavits of Oswald Pohl. Pohl had three children and made sure that his only son attended an NSDAP school at Plön: 12 Feb. 1944, "Ergänzungsbogen und Veränderungsmeldungen"; Lebenslauf submitted to Himmler in May 1933, BDC SS Personal-Akte Oswald Pohl. On the Führer principle after the war: Protocol: 1258.

12. Quotations from Pohl, 9 Feb. 1942, "Befehl Nr. 3," T-976/36. Coyner, "Class Consciousness and Consumption," 310–31. The interpersonal relations between the secretaries of Himmler, Pohl, and other high-ranking SS officers is well presented by Koch, *Himmlers graue Eminenz—Oswald Pohl*.

13. NO-2571, Affidavit of Oswald Pohl. NO-1574, 1 Oct. 1935, "Verwaltungschef der SS," shows the organizational structure of this early period. Georg, *Die wirtschaftlichen Unternehmungen der SS*, 27; Tuchel, *Konzentrationslager*, 245–48, 259; Naasner, *Neue Machtzentren*, 248. Forty elite officers started in the low-level, paramilitary administrative system created by Pohl. Only five of these had been unemployed. Of eighty-seven officers already in SS service at this time and who then later joined Pohl, nineteen, or almost a quarter, had suffered some kind of unemployment. Nevertheless, these statistics are significantly lower than those for the concentration camps.

14. Himmler to Pohl and various SS Hauptämter, 1 June 1935, "Personalverfügung Nr. 20"; Pohl's telegram to Karl Wolff, 3 June 1935; and Wolff's answer of the same day, BDC SS Personal-Akten Oswald Pohl.

15. Wegner, *Hitlers politische Soldaten*, 105–8; Tuchel, *Konzentrationslager*, 245–48; Naasner, *Neue Machtzentren*, 252–53; Sydnor, *Soldiers of Destruction*, 3–36.

16. Eicke cooperated closely with top officials in the Security Main Office, who already had secured budget lines in both regional and national interior ministries. Tuchel, *Konzentrationslager*, 211–18, 248.

17. Ibid., 247–64.

18. August Frank to Hans Baier, 26 May 1940, and Baier to Frank, 24 May 1940, BDC SS Personal-Akten Hans Baier; Testimony of August Frank, Protocol: 2249, 2326–28, and 2248–50; NO-1576, Affidavit of August Frank. Later in the war he would also reorganize the payroll system for the OKW, a sign of the cooperation between military and SS.

19. Trischler, "Führer-Ideologie im Vergleich," 45–88.

20. Koehl, *The Black Corps*, 36–59.

21. NO-542, Salpeter, undated [mid-1939], "Tasks, Organization and Finance Plan of Office III–W of the RFSS"; Oswald Pohl, Protocol: 1666–68.

22. NO-542.

23. Georg, *Die wirtschaftlichen Unternehmungen der SS*, 15.

24. Ibid.

25. NO-1016, Leo Volk, "Organization und Aufgaben der Amtsgruppe W." In the trial Protocol: 58 prosecutor Robins stated that this report was written for Leo Volk by Heinz Fanslau. NO-1016 includes a cover letter by Fanslau who sent the manifesto in Volk's absence, but it is unclear that Fanslau wrote the document alone. Likely the two men (or more) collaborated.

26. Anonymous, 22 Nov. 1937, "Der technische Fortschritt als Ausgleich mangelnden Lebensraumes," BAK NS3/110; compare Herf, *Reactionary Modernism*, 156–61. See also Kocka, *Die Angestellten*, 142–70; Buchheim, "Befehl und Gehorsam," 215–318.

27. NO-542, Salpeter, undated [mid-1939], "Tasks, Organization and Finance Plan of Office III–W of the RFSS"; Georg, *Die wirtschaftlichen Unternehmungen der SS*, 18–19.

28. Compare Fairbairn, "History from the Ecological Perspective: Gaia Theory and the Problem of Cooperatives in Turn-of-the-Century Germany," 1203–39. See report on Dachauer Bauentwicklung, 1 Mar. 1938, BAK R2/28350.

29. PS-1992, after Georg, *Die wirtschaftlichen Unternehmungen der SS*, 21.

30. Petropoulos, *Art as Politics in the Third Reich*, 99.

31. Knoll, "Die Porzellanmanufaktur München-Allach," 116–19.

32. Testimony of Karl Wolff, Protocol: 2130. SS figurines are reproduced in Heskett, "Design in Interwar Germany," 281. See also Huber, *Die Porzellan-Manufaktur Allach-München GmbH*, 50–54.

33. Unsigned reports and balance sheets on Anton Loibl GmbH, covers 1936–41, T-976/7: 255–85; NO-542, Salpeter, undated [mid-1939], "Tasks, Organization and Finance Plan of Office III–W of the RFSS."

34. Georg, *Die wirtschaftlichen Unternehmungen der SS*, 12–24.

35. Ibid. See also Schulte, "Verwaltung des Terrors," 99–102.

36. See Pohl to SS Personalkanzlei, 3 Sept. 1937, BDC Personal-Akte Walter Salpeter. See also BDC Personal-Akte Heinz Schwarz.

37. Naasner, *Neue Machtzentren*, 386. Wegner, *Hitlers politische Soldaten*, 107, table 4, gives the exact budget for the *Verfügungstruppen* from 1935 to 1938 at 479.8 million Reichsmarks.

38. NO-1574, 1 Oct. 1935, "Verwaltungschef der SS," lists four functional divisions besides Pohl's *Adjutantur*: Budget, Accounting, Clothing and Equipment, and Accommodation. "Mitteilung für die Waffen-SS," 18 Apr. 1940, BDC Wirtschaftsverwaltungsanordnungen, lists eight functional divisions: Budget, Payroll, Clothing, Construction and Accommodation, Quartermaster and Supply, Legal Office and Property, Accounting, and Spargemeinschaft der SS. NO-4007, Affidavit Hubert Karl.

39. The term "total institution" used in the section heading comes from Goffman, *Asylums*.

40. Rudolf Höss, "Meine Psyche, Werden, Leben, u. Erleben," IfZ F 13/2: p. 26. See also Orth, *Das System der nationalsozialistischen Konzentrationslager*, 28.

41. Broszat, "Nationalsozialistische Konzentrationslager," 37–73; Tuchel, *Konzentrationslager*, 121–58; Richardi, *Schule der Gewalt*, 48–87, 119–54. Compare Drobisch and Wieland, *System der NS-Konzentrationslager*, 43–46.

42. Segev, *Die Soldaten des Bösen*, 120; Tuchel, *Konzentrationslager*, 160–66; Orth, *System der nationalsozialistischen Konzentrationslager*, 28, 51.

43. In general, Tuchel, *Konzentrationslager*, 128–40, 159–204; quotation from 219.

44. Quoted from Pingel, *Häftlinge unter SS-Herrschaft*, 62. Hans Speier has pointed out the appeal of "warrior" administration to German *Angestellten* in *German White-Collar Workers*, 103–8. Compare Buchheim, "Befehl und Gehorsam"; Orth, *Das System der nationalsozialistischen Konzentrationslager*, 36–37; Drobisch and Wieland, *System der NS-Konzentrationslager*, 188–93; Segev, *Die Soldaten des Bösen*, 119–52.

45. Höss, "Autobiography," 29; Richardi, *Schule der Gewalt*, 119–54; Tuchel, *Konzentrationslager*, 121–58. Friedlander, "The Nazi Concentration Camps," 33–69, also gives biographical sketches of the Dachau staff.

46. Segev, *Die Soldaten des Bösen*, 120–26, 151; Orth, *Die Konzentrationslager-SS*.

47. Segev, *Die Soldaten des Bösen*, 133.

48. Theodor Eicke, 1 Oct. 1933, "Disziplinar- u. Strafordnung für das Gefangenenlager (Dachau)," and "Dienstvorschriften für die Begleitpersonen und Gefangenenbewachung," BAP, PL5: 42053; Richardi, *Schule der Gewalt*, 119–54.

49. Eicke, "Disziplinar- u. Strafordnung," and "Dienstvorschriften." Compare Richardi, *Schule der Gewalt*, 119–54.

50. Segev, *Die Soldaten des Bösen*, 127. See also Testimony of Oswald Pohl, Protocol: 1327–29.

51. Cited after Segev, *Die Soldaten des Bösen*, 145. See also Drobisch and Wieland, *System der NS-Konzentrationslager*, 250–61.

52. Tuchel, *Konzentrationslager*, 282–85; Drobisch and Wieland, *System der NS-Konzentrationslager*, 435; Order of Heinrich Himmler, 14 Dec. 1938, "Sanitätswesen der bewaffneten Teile der SS und der KL," SF-01/16334 Militärgeschichtliches Archiv Potsdam.

53. NO-4007, Affidavit of Hubert Karl. Compare Testimony of Oswald Pohl, Pro-

tocol: 1631–33. Dr. Stuckart to Pohl, 7 Aug. 1938, "SS-Totenkopfverbände und Konzentrationslager Vollmacht," T-175/37, mentions a previous Vollmacht from 13 May 1937, showing that Pohl controlled the budgets of Prussian KLs from at least mid-1937. Compare Tuchel, *Konzentrationslager*, 258–78.

54. Tuchel, *Konzentrationslager*, 238–39; Wegner, *Hitlers politische Soldaten*, 95–112.

55. Eicke to RFSS, 20 Oct. 1938, BDC Personal-Akte Robert Riedl; Testimony of Pohl, Protocol: 1766.

56. Eicke also stridently opposed Reinhard Heydrich, who argued in 1934 that the mismanagement of the camps was grounds for their incorporation into the Sicherheitsdienst. Segev, *Die Soldaten des Bösen*, 142; Wegner, *Hitlers politische Soldaten*, 100–103. Power struggles should not be exaggerated, however. At other times Eicke and the Gestapo cooperated with little friction. See Tuchel, *Konzentrationslager*, 206–7.

57. Testimony of Karl Wolff, Protocol: 2193.

58. Eicke to RFSS, 5 Oct. 1934, BDC SS Personal-Akte Karl Möckel; NO-4007, Affidavit of Hubert Karl. For Eicke's later praise, see BDC Personal-Akte Anton Kaindl and BDC Personal-Akte Robert Riedl, various evaluations. Tuchel, *Konzentrationslager*, 275–77.

59. Staatssekretär des RJM-Berlin, Dr. Jur. Roland Freisler, "Arbeitseinsatz im Strafvollzug" (title article for 13 Sept. 1940), *Deutsche Justiz* 102 (1940): 1021–25. This is exactly the function of labor identified by Goffman, *Asylums*, 141ff. Incidence of its practice is described clearly by Richardi, *Schule der Gewalt*, 83–87, 150–54, and Drobisch and Wieland, *System der NS-Konzentrationslager*, 13, 76–81. Tuchel, " 'Arbeit' in den Konzentrationslagern," 456–59.

60. These were the rules codified by Eicke's direct predecessor, Hilmar Wäckerle, PS-1216, "Sonderbestimmungen (Lagerordnung)—Dienstvorschriften. Vermerkung über Todesfälle-Besprechung," 29 May 1933 (one month before Eicke), submitted by Dr. Wintersberger to the Staatsanwaltschaft bei dem Landgericht München II. Compare Richardi, *Schule der Gewalt*, 69–72, 88, where he notes that 10 percent of the prisoners at the original Dachau camp were Jewish. See also Pingel, *Häftlinge unter SS-Herrschaft*, 37; Tuchel, *Konzentrationslager*, 153–56.

61. Theodor Eicke, 1 Oct. 1933, "Disziplinar- u. Strafordnung für das Gefangenenlager (Dachau)," and "Dienstvorschriften für die Begleitpersonen und Gefangenenbewachung," BAP, PL5: 42053.

62. From Béon, *Planet Dora*, xxxiii. Freund, *Arbeitslager Zement*, 168–72, contains an excellent treatment of typical social structure among camp prisoners.

63. Bartel et al., *Buchenwald*, 222. See also Drobisch and Wieland, *System der NS-Konzentrationslager*, 435.

64. Kaienburg, *Vernichtung durch Arbeit*, 200–201.

65. Testimony of Fritz Schmidt, who survived Aussenlager Zwieberge, in Bartel et al., *Buchenwald*, 269. Compare Frederick Taylor, *Principles of Scientific Management* (New York: Harper, 1911), 57–62.

66. Goldhagen, *Hitler's Willing Executioners*.

67. Compare Richardi, *Schule der Gewalt*, 143–46; Drobisch and Wieland, *System der NS-Konzentrationslager*, 76–81, 119–22, 207–10.

68. Tuchel, *Konzentrationslager*, 52–62, 143, 192–202; Kaienburg, *Vernichtung durch*

Arbeit, 56–63; PS-1469, Oswald Pohl to Himmler, 30 Sept. 1943, "Todesfälle in den KL"; Karny, "'Vernichtung durch Arbeit.' Sterblichkeit in den NS-Konzentrationslagern," 140–45. In the Amtsgruppe D-IKL in Oranienburg I found only one officer who possessed technical training, a *Meistermechaniker*, about the rank of foreman. It is not clear what his function was, however. BDC SS Personal-Akte Ernst Schulz. A typical profile for the camp SS officers might be that of Heinrich Schwarz, the *Arbeitseinsatzführer* at Auschwitz. He had trained as an *Elektrotechniker* but had long given up the trade by the time he took over industrial management at Auschwitz III–Monowitz. During the Weimar Republic he had worked as a technician in photo studios. BDC Personal-Akte Heinrich Schwarz; Pingel, *Häftlinge unter SS-Herrschaft*, 37–39; PS-1216, "Sonderbestimmungen (Lagerordnung)—Dienstvorschriften. Vermerkung über Todesfälle-Besprechung," 29 May 1933 (one month before Eicke), submitted by Dr. Wintersberger to the Staatsanwaltschaft bei dem Landgericht München II; Theodor Eicke, 1 Oct. 1933, "Disziplinar- u. Strafordnung für das Gefangenenlager (Dachau)," and "Dienstvorschriften für die Begleitpersonen und Gefangenenbewachung," BAP, PL5: 42053. See also "Lagerordnung für die Konzentrationslager," with organizational schematic, BAP, PL5: 42053, dated by Höss from 1936.

69. Cited after Boehnert, "The Third Reich and the Problem of 'Social Revolution,'" 203.

70. Oswald Pohl to Hans Baier, 6 Oct. 1937, "Nachwuchs für SS-Verwaltungsführer," BDC SS Personal-Akten Hans Baier.

71. After Segev, *Die Soldaten des Bösen*, 145.

72. Oswald Pohl, 8 Sept. 1942, "Befehl Nr. 31," T-967/36; Nuremberg Trial document R-129, Pohl to Himmler, 30 Apr. 1942, "Incorporation of the Inspectorate of CC into the WVHA"; Pohl to Amt D, 30 Apr. 1942, T-967/36: 994-5. The need to manage rather than drive labor has never been self-evident. Compare Hobsbawm, "Custom, Wages, and Workload in Nineteenth-Century Industry," 232–55.

73. Testimony of Pohl, Protocol: 1713–16; Kaienburg, *Vernichtung durch Arbeit*, 90–97, 180–220, 262–79; Allen, "The Puzzle of Nazi Modernity," 527–71.

74. NO-1574, 1 Oct. 1935, "Verwaltungschef der SS"; NO-4007, Affidavit of Hubert Karl; NO-2613 and NO-4406, Affidavits of Franz Eirenschmalz. Pohl's predecessor, the SS treasurer, had always maintained a department for "shelter and accommodation" (*Unterkunft*). As early as 1935 Pohl expanded this office and added a subdivision for construction (*Bauten*) and quickly raised the subdivision to the level of an independent department within the Personal Staff. See also Bolenz, *Vom Baubeamten*, 122–24; Slaton and Abbate, "The Hidden Lives of Standards."

75. Führerstellenbesetzungsplan für den Stab des IKL, 1 June 1940, IfZ Fa 183.

76. See, for example, NO-2197, Affidavit of Ernst Krone, who strung the electric fences at Sachsenhausen.

77. NO-2615, Hans Kammler and Weigel to Pohl and Himmler, 29 July 1944, "Bericht über den Einsatz der 5. und 1. SS Baubrigade." See also the very detailed descriptions of brigade operations in Schmidt-Klevenow, Sept. 1943 (date partially burned off document), "Einleitung eines Ermittlungsverfahrens gegen Max List," BDC SS Personal-Akte Max List.

78. Gispen, *New Profession, Old Order*, 224–31; Ludwig, *Technik und Ingenieure*, 17–20. Of the thirty-three SS engineers who served in the SS before 1939 (in the camps but

also elsewhere in the SS) for whom I have found data, twenty-three held less than elite education; in other words, like Weigel they had trained at vocational schools or in apprenticeships but not at the polytechnic universities.

79. BDC SS Personal-Akten and RuSHA Akte Gerhard Weigel, esp. Lebenslauf. In all, only nine of fifty-six engineering officers had suffered such unemployment, seven of them from this early cadre.

80. Hortleder, *Das Gesellschaftsbild des Ingenieurs*, 109–11; Ludwig, *Technik und Ingenieure*, 281.

81. "Lebenslauf," BDC SS Personal-Akte Gerhard Weigel. Compare Tuchel, *Konzentrationslager*, 197, 393–94.

82. Sloppy management of the KL Bauleitungen is mentioned in NO-4007, Karl's Affidavit. See also Karl Bischoff to Bauinspektion Ost, 16 Apr. 1942, "Baupolizeiliche Behandlung," USHMM RG-11.001M.03: 23 (502-1-82); likewise, Kammler, 22 Aug. 1941, "Mittelbewilligungen," USHMM RG-11.001M.03: 18 (502-1-1); 4 Jan. 1939, "Vermerk" and Greifelt to Chef des Pers. Stabes, 5 May 1939, "Vierjahresplan," NS19/1084; Frank to Hauptabteilung V4, 21 Apr. 1938, BAK NS4/59; Pohl to Eicke, 31 Aug. 1938, BAK NS4/59; Pohl to SS PHA, 10 June 1939, BDC SS Personal-Akte Robert Riedl; Eicke to RFSS, 20 Oct. 1938, BDC SS Personal-Akte Robert Riedl. Tuchel, *Konzentrationslager*, 276, states that after Riedl's entry KL construction was "finally led totally by Pohl," which is surely an exaggeration. Nevertheless, he is right that Pohl's competent representatives were making headway at this time.

83. Hans Gerlach, "Die Kameradschaftssiedlung der SS am Vierling in Berlin-Zehlendorf," *Siedlung und Wirtschaft* 19 (1937): 708.

84. Here and earlier, ibid., 705.

85. Ibid., 699–700.

86. Van Pelt and Dwork, *Auschwitz, 1270 to the Present*, 265–66, on camp architecture in general, and 229–31 for illustrations; Jaskot, *The Architecture of Oppression*, 114–39.

87. Dr.-Ing. Edgar Hotz and Dr.-Ing. Hans Kammler, *Grundlagen der Kostenrechnung und Organization eines Baubetriebs für den Wohnungs- und Siedlungsbau in Stadt und Land* (Berlin: Verlagsgesellschaft R. Müller, 1934), 1.

88. Durth, *Deutsche Architekten*, 23–40, 65–86, 102–3; Ludwig, *Technik und Ingenieure*, 20–35, 105–9; Wengenroth, "Zwischen Aufruhr und Diktatur," 215–23; Gispen, *New Profession, Old Order*, 48–51; Bolenz, *Vom Baubeamten*, 33–39, and regarding housing, 85–86.

89. Durth, *Deutsche Architekten*, 152.

90. NO-1922, Affidavit of Max Kiefer. Compare Hounshell, *From the American System to Mass Production*, 311–13.

CHAPTER TWO

1. Tuchel, *Konzentrationslager*, 225–30; Wegner, *Hitlers politische Soldaten*, 105–8. Regarding Pohl's influence on the IKL: Testimony of Oswald Pohl, Protocol: 1631–33, 1764–66. Evaluation by Pohl of the head of the Prüfungsabteilung, 12 Mar. 1936, BDC Personal-Akte Karl Möckel. Hohberg to Pohl, 18 Sept. 1940, "Übernahme der wirtschaftlichen Betriebe der SS in Dachau in die Deutschen Ausrüstungswerke," and order Pohl to Salpeter, 31 Jan. 1940, BAP, PL5: 42055.

2. James, *The German Slump*, 360–83, and 343–419 in general; Hayes, "Big Business and 'Aryanization,'" 255–71; Overy, *Goering the Iron Man*; Gillingham, *Industry and Politics in the Third Reich*.

3. Thüringen's regional Innenminister, Hillmuth Gommlich, from 24 Apr. 1937, quoted after Jaskot, *The Architecture of Oppression*, 21. Hayes, "Big Business and 'Aryanization,'" 265–68. See also Karl Mummenthey's "Monatsberichte" for DESt, 1941–42, NS3/1346; "Aufstellung über die in den Jahren 1938 und 1939 eingesetzten Häftlinge und Bewertung derselben," ca. June 1940, NS3/1345. Also for the Deutsche Ausrüstungswerke: NO-557, Gerhard Maurer to Pohl, 15 Apr. 1941, "Monatsbericht für Monat März, 1941"; Georg, *Die wirtschaftlichen Unternehmungen der SS*, 40–69.

4. Jaskot, *The Architecture of Oppression*, 87–88, 100–113; Van der Vat, *The Good Nazi*, 83–104.

5. Reichsschatzmeister, 17 June 1938, "Kreditbeschaffung für Zwecke der Klinkerwerke Buchenwald bei Weimar," IfZ, Fa 183. See Jaskot, "Architectural Policy," 106–13, 117–21, and *The Architecture of Oppression*, 58–61. Compare Lane, *Architecture and Politics in Germany*.

6. Van der Vat, *The Good Nazi*, 352. *Der Spiegel* published articles on this family feud in the early 1970s. Georg, *Die wirtschaftlichen Unternehmungen der SS*, 44–45. For sound analysis, see Jaskot, "Architectural Policy," 43ff. and 47. Brenner, "Der 'Arbeitseinsatz' der KZ-Häftlinge in den Außenlagern Flossenbürg," 682–706. See also Mummenthey and Salpeter, 31 July 1941, "Geschäftsbericht der DESt 1940," T-976/26: 1–516.

7. Compare Schulte, "Verwaltung des Terrors," 103–4, and "Stellenbesetzungsplan des Verwaltungsamtes-SS vom 1. Mai 1934," reprinted in ibid., 50. In my opinion there is insufficient evidence to assert that professional control over these small-scale industries was wrested from Eicke's control. Pohl had to reintegrate them in any case in 1938–39.

8. Quotation from Lebenslauf, BDC RuSHA-Akte Arthur Ahrens. See also Reichsschatzmeister, 17 June 1938, "Kreditbeschaffung für Zwecke der Klinkerwerke Buchenwald bei Weimar," IfZ, Fa 183, which documents Pohl's visit to Ahrens on 15 June. Jaskot, *Architecture of Oppression*, 104–5. Ahrens brickworks is mentioned in Mummenthey and Dr. Erduin Schondorff to DWB, 7 Feb. 1944, T-976/25: 400–850, but lists the works in "Lebus." See also Ahrens, 15 Sept. 1938, in BDC Personal-Akte Walter Salpeter, in which Ahrens, not Pohl, appoints Salpeter as deputy *Geschäftsführer* of the DESt. Likewise, "Bericht über die Prüfung der DESt GmbH," 6 July 1939, NS3/880, and undated report [1940], "DESt," NS3/625, list Ahrens as the first *Geschäftsführer* of DESt. Salpeter did not receive discretionary authority until 15 September 1938, when Ahrens wrote him out a *Vollmacht*. Salpeter's discovery of incompetence dates from after this point. The DESt was a GmbH, and all GmbH are required by law to have at least two cofounders. Each had to contribute a minimum of 10,000 Reichsmarks to the starting capital. Salpeter was probably chosen as a "silent partner." Ahrens also had a hand in the bike-light company Loibl as well. See NO-542, Salpeter, undated [between Apr. and June 1939], "Tasks, Organization and Finance Plan of Office III (W)." This document is undated and unsigned, but its general anticapitalist, antibureaucratic, reform tone, as well as its pseudoacademic format, are consistent with the prose style and ideals of Walther Salpeter, who was also the only top SS financial manager engaged in reorganizing the SS with

Pohl at this time. Salpeter also habitually left his memoranda to Pohl unsigned, the only one to do so. Georg, *Die wirtschaftlichen Unternehmungen der SS*, 50 and n, attributes NO-542 to July 1939. Several textual references date the document even earlier, however. Founding of the Deutsche Ausrüstungs Werke is alluded to as having just occurred (3 May 1939). The document also discusses all SS administrative offices as departments of the VuWHA (founded on 4 April), that is, before the foundation of the Hauptamt Haushalt und Bauten on 20 April. This would place the document between 4 and 20 April. See also unsigned, undated report [from 1943 or 1944], balance sheets of Amt W3 (DAW), T-976/7, showing Ahrens as *Geschäftsführer* with Salpeter from 3 May 1939 to 14 Aug. 1939. Georg (62ff.) states that Pohl initiated the *Versuchsanstalt*, but I have found no evidence to support this. Here too Salpeter uncovered incompetent bungling: NO-1044, unsigned [Salpeter], undated [from 1940], "Bericht über das erste Geschäftsjahr der Deutschen Versuchsanstalt für Ernährung und Verpflegung." See Pohl to DESt Klinkerwerk, 16 June 1939, NS3/880.

9. NO-542, Salpeter, undated [mid-1939], "Tasks, Organization and Finance Plan of Office III–W of the RFSS." See also the record of changes in DESt's top management in undated report [1940], "DESt," NS3/625. It was Salpeter who set out to recruit corporate lawyers late in 1938 and early 1939. For instance, see NO-2523, Affidavit of Karl Mummenthey, in which Mummenthey claims that he answered a classified ad published in a legal journal by Salpeter. Pohl intervened to fire Ahrens in June 1939, and it is likely that Ahrens was expelled from the SS at this time. See Pohl to SS Untersturmführer Clemens Tietjen (DESt *Prokurist* under Ahrens), 14 June 1939, and Pohl to DESt Klinkerwerk, 16 June 1939; and Pohl to the Spengler Maschinenbau GmbH, 16 June 1939, NS3/880. The circumstantial evidence is overwhelming that Pohl did not concern himself directly in SS business affairs until around this date. This point is worth clarifying because earlier initiatives came from multiple sources. Pohl was definitely aware but not in control. Georg, *Die wirtschaftlichen Unternehmungen der SS*, 15, 25, points out that Pohl's *Prüfungsabteilung* was listed as an auditor of SS businesses. Little evidence before 1939 suggests that Pohl's offices actually carried out any comprehensive audits, however. Compare Kaienburg, *Vernichtung durch Arbeit*, 75.

10. NO-1044, unsigned [Salpeter], undated [from 1940], "Bericht über das erste Geschäftsjahr der Deutschen Versuchsanstalt für Ernährung und Verpflegung," and NO-542, Salpeter, undated [mid-1939], "Tasks, Organization and Finance Plan of Office III–W of the RFSS."

11. "SS erschließt Neuland. Wo in Deutschland der Pfeffer wächst," *Das Schwarze Korps* (22 Sept. 1938): 4. On agricultural autarky, see also Herbert Backe, the SS officer who succeeded Walter Darré, "Verbrauchslenkung im Kriege," *Der Vierjahresplan* 4 (1940): 266–68.

12. NO-1044, unsigned [Salpeter], undated [from 1940], "Bericht über das erste Geschäftsjahr der Deutschen Versuchsanstalt für Ernährung und Verpflegung," and NO-542, Salpeter, undated [mid-1939], "Tasks, Organization and Finance Plan of Office III–W of the RFSS." See BAK NS3/1038 generally.

13. Dipl. Landwirt Heinrich Vogel to Hohberg, 17 June 1941, NS3/954. Compare Georg, *Die wirtschaftlichen Unternehmungen der SS*, 62–66.

14. Zeller, "Landschaften des Verkehrs," 323–40; Wolschke-Bulmann, "Biodynami-

scher Gartenbau, Landschaftsarchitektur und Nationalsozialismus." See also Biagioli, "Science, Modernity, and the Final Solution," 185–205.

15. NO-542, Salpeter, undated [mid-1939], "Tasks, Organization and Finance Plan of Office III–W of the RFSS"; Drobisch and Wieland, *System der NS-Konzentrationslager*, 76–81; Richardi, *Schule der Gewalt*, 51; Georg, *Die wirtschaftlichen Unternehmungen*, 59; NO-678, Anonymous, undated [1941], "Deutsche Ausrüstungswerke GmbH 'DAW.' "

16. Hohberg to Mummenthey, 10 Sept. 1940, "Bilanz der DESt zum 31 Dec. 39," T-976/25; NO-542, Salpeter, undated [mid-1939], "Tasks, Organization and Finance Plan of Office III–W of the RFSS"; Pohl to the Spengler Maschinenbau GmbH, 16 June 1939, BAK NS3/880.

17. Fritz Todt, "Regelung der Bauwirtschaft," *Der Vierjahresplan* 3 (1939): 764.

18. Beyerchen, "Rational Means and Irrational Ends," 395. These were the exact sentiments of contemporary engineers.

19. Richard Grün, "Mensch und Technik," *Zement* 30 (1941): 637. On Grün, see Allen, "The Puzzle of Nazi Modernism," 554–55.

20. For the American context, see Edwin Layton, *The Revolt of the Engineers: Social Responsibility and the American Engineering Profession* (Cleveland: Case Western Reserve Press, 1971); Zunz, *Making America Corporate*, 67–90; Siegel and Freyberg, *Industrielle Rationalisierung*, 46–50, 77–90.

21. From January 1939, cited after Burleigh and Wippermann, *The Racial State*, 175.

22. Richardi, *Schule der Gewalt*, 69–72, 88. See also Pingel, *Häftlinge unter SS-Herrschaft*, 37; Tuchel, *Konzentrationslager*, 153–56.

23. Orth, *Das System der nationalsozialistischen Konzentrationslager*, 46–54.

24. Grün, "Mensch und Technik," 634–35. For continuity with the postwar period, see Seigel and von Freyberg, *Industrielle Rationalisierung*, 50.

25. Grün, "Mensch und Technik," 636.

26. Johannes Sändig et al., *Ziegeleilexikon* (Leipzig: Dr. Sändig Verlagsgesellschaft, 1938–54), 820; Johannes Fischer, *Die Ziegelei. Anlage und Betrieb* (Berlin: Paul Parey, 1955), 52–54; Karl Spingler, *Lehrbuch der Ziegeltechnik* (Halle: Verlag von Wilhelm Knapp, 1948), 95–97; J. F. Kesper, "Über neuzeitliche automatische Ziegeltrocken-pressen," *Süddeutsche Ziegelwelt* 67 (1936): 263–64.

27. Fischer, *Die Ziegelei*, 45–48, 52–54; Werner Gräff, *Die Ziegelei* (Berlin: Sanssouci Verlag, 1934), 8–15, 22–25; Kesper, "Über neuzeitliche automatische Ziegeltrocken-pressen," 263–64; Spingler, *Lehrbuch*, 95–97.

28. Speer, *Inside the Third Reich*, 144.

29. Sändig et al., *Ziegeleilexikon*, 141–43, 623–24, 822; Gräff, *Die Ziegelei*, 8–15; Fischer, *Die Ziegelei*, 26–29. Another technique, the tunnel oven, could accommodate mechanized transport but was much more wasteful of energy. It was adopted quickly in the United States but rarely in Germany.

30. Mummenthey, 12 Feb. 1940, "Monatsbericht," BAK NS3/1346. On the other technical disasters, see Salpeter, Schondorff, Mummenthey, and (civilian engineer) Zahn, 17 June 1940, "Neuorganization und Dezentralisierung Werk Oranienburg," BAK NS3/1344. On the losses due to Spengler in general, see Hohberg to Mummenthey, 10 Sept. 1940, "Bilanz der DESt zum 31 Dec. 39," T-976/25; Salpeter to Pohl, "DESt," 29 Nov. 1940, and "Gewinn- und Verlustrechnung der DESt," 27 Dec. 1940, T-976/25.

31. Credit from the Golddiskontbank, a division of the Reichsbank, was arranged by the Reichswirtschaftsministerium. Undated report [1940], "DESt," NS3/625, lists credit from the Golddiskontbank at 8 million Reichsmarks. In May 1941 this would be increased by another 8 million Reichsmarks. "Besprechung des Herrn Reichswirtschaftministers und Reichsbankpräsidenten mit dem Chef des persönlichen Stabes des RFSS," 22 Aug. 1939, BAK NS3/1532. Compare Salpeter, Schondorff, Mummenthey, and (civilian engineer) Zahn, 17 June 1940, "Neuorganization und Dezentralisierung Werk Oranienburg," BAK NS3/1344.

32. BDC Personal-Akte Fritz Lechler and Felix Krug; Boehnert, "The Third Reich and the Problem of 'Social Revolution,'" 203–17; Herbert Ziegler, *Nazi Germany's New Aristocracy*, 93–148.

33. Lebenslauf, BDC RuSHA Personal-Akten Fritz Lechler.

34. See esp. Sachse, *Siemens, der Nationalsozialismus und die moderne Familie*; Siegel and von Freyberg, *Industrielle Rationalisierung*; Allen, "Flexible Production at Concentration Camp Ravensbrück."

35. Affidavit of Georg Lörner, Defense Document Books of Georg Lörner.

36. BDC Personal-Akte Felix Krug, esp. evaluations of Lechler from 1 Jan. 1940–30 June 1941.

37. On modern bureaucratic organization, see Beniger, *The Control Revolution*; Wootton and Wolk, "The Evolution and Acceptance of the Loose-Leaf Accounting System"; Chandler, *The Visible Hand* and *Scale and Scope*, esp. 393–592, which contains his comparisons of managerial structures in Germany with those of the United States; Kocka, *Die Angestellten*; *Unternehmensverwaltung und Angestelltenschaft am Beispiel Siemens*; and "Eisenbahnverwaltung in der industriellen Revolution," 259–77.

38. Initially they toyed with the idea of introducing hand looms but decided against this. Compare Zumpe, "Arbeitsbedingungen und Arbeitsergebnisse in den Textilbetrieben der SS," 23.

39. NO-1918, Georg Lörner and Volk to Reichsjustizministerium, 21 Mar. 1942. Krug's claims are confirmed in prisoners' testimony after the war; see, e.g., Bernadac, *Camp for Women*, 143.

40. NO-1918 mentions nine technical managers who were already working at the TexLed in 1940, that is, immediately after start-up. This is an interesting contrast to the arrival of Schondorff and his students after the fact of incompetent technical organization.

41. "Bericht über die Gesellschaft für Textil- und Lederverwertung mbH 31. Dezember 1942 bis 31. Dezember 1943," T-976/24.

42. NO-1221, trade report of 1940–41, compiled by Lechler. The Textil- und Lederverwertung was originally capitalized with 2.7 million Reichsmarks (1 million of which was instantly paid back when the Reich agreed to carry the fixed costs associated with the construction of barracks and work halls at Ravensbrück). Compare Amtsrat Scheck, "Vermerk über die Gesellschaft für Textil- und Lederverwertung, mbH. Dachau und Ravensbrück," undated [probably from shortly after Mar. 1941]. Report covers developments up to quarter ending 21 Mar. 1941, BAP 23.01 Rechnungshof des Deutschen Reiches: 5636.

43. NO-1221, trade report of 1940–41, compiled by Lechler; "Der Mindener Bericht," 4 Jan. 1947 (written by WVHA officers in Allied captivity in Minden), reprinted in

Naasner, *SS-Wirtschaft und SS-Verwaltung*, 178; Scheck, "Vermerk über die Gesellschaft für Textil- und Lederverwertung, mbH. Dachau und Ravensbrück." BAP 23.01 Rechnungshof des Deutschen Reiches: 5636 and NO-1918, Georg Lörner and Volk to Reichsjustizministerium, 21 Mar. 1942.

44. For an excellent analysis of the gendered nature of production in Great Britain and the United States, see Lewchuk, "Men and Mass Production," 219–42. On the textile industry specifically, Green, *Ready-to-Wear and Ready-to-Work*, 162–70.

45. Siegel and von Freyberg, *Industrielle Rationalisierung*, 132–35; Maier, *In Search of Stability*, 19–69. On this sentiment among German engineers generally, Gispen, *New Profession, Old Order*, 114–29.

46. Piore and Sabel, *The Second Industrial Divide*, 213–16, notes that Italy's flourishing garment industry is based on capital-intensive, labor-saving machines and a high-wage labor market. Compare Green, *Ready-to-Wear*, 31–40.

47. Women workers in Japan represent one example among many of similar developments elsewhere. Partner, *Assembled in Japan*.

48. Cited after Bernadac, *Camp for Women*, 145.

49. Arndt, "Das Frauenkonzentrationslager Ravensbrück," 135, 135–42 in general. See also the promulgation of Himmler's orders for the correct bearing toward prisoners at the Mittelwerk GmbH for the production of V-2 rockets: Förschner, 30 Dec. 1943, Verteiler, "Umgang mit Häftlinge"; Rickhey and Kettler, 22 June 1944, "Bestrafung von Häftlingen durch Gefolgschaftsmitglieder des MW. Umgang mit Häftlinge," BAK NS4 Anhang/3.

50. Tillion, *Ravensbrück*, 62, 142–49 generally. This section includes survivor testimony of Lily Unden, Marie-José Zillhard, and Madeleine Perrin and interviews and commentary made after the liberation of Ravensbrück by Lord Russel of Liverpool.

51. Arndt, "Das Frauenkonzentrationslager Ravensbrück," 149–50. Ravensbrück maintained the lowest death rate of any concentration camp, sinking to around three-tenths of 1 percent in late 1943. Nevertheless, this reckoned per month mortality. One prisoner who worked in the camp's administration reported that 1,725 women had died over the entire course of 1943; thus even low monthly death rates added up to substantial numbers. Death rates were lower in the winter of 1943 for all camps, not just Ravensbrück. Here they had been higher earlier in the year and would again climb in 1944–45. Pingel, *Häftlinge unter SS-Herrschaft*, 61–68, attributes lower death rates to the integration of labor in the industrial court.

52. Amtsrat Scheck, "Vermerk über die Gesellschaft für Textil- und Lederverwertung, mbH. Dachau und Ravensbrück," undated [probably from shortly after Mar. 1941]. Report covers developments up to quarter ending 21 Mar. 1941, BAP 23.01 Rechnungshof des Deutschen Reiches: 5636. On the rationalization of labor rents, NO-1035/NO-1289, Hohberg to Pohl and Pohl to all Wirtschafts offices, 21 July 1942 and 30 Dec. 1942; Lechler to Amtsgruppenchef W, 6 Nov. 1942, T-976/15; Zumpe, "Die Textilbetriebe der SS im Konzentrationslager Ravensbrück," 11–41, and "Arbeitsbedingungen und Arbeitsergebnisse in den Textilbetrieben der SS," 11–51; NO-1221, trade report of 1940–41, compiled by Lechler.

53. BDC SS Personal-Akte and RuSHA-Akte Walter Salpeter; BDC NSDAP Partei Karte Walther Salpeter. His SS Stammkarte shows no party membership. See also Wal-

ther Salpeter, "Verbotene und unsittliche Geschäfte im Steuerrecht" (Jura diss., 1933, Universität Berlin und Halle [published in Berlin, 1934]).

54. Brustein, *The Logic of Evil*.

55. NO-542, Salpeter, undated [mid-1939], "Tasks, Organization and Finance Plan of Office III–W of the RFSS."

56. Aktennotiz, 10 Feb. 1939, initialed by Pohl, NS3/1471; NO-542, Salpeter, undated [mid-1939], "Tasks, Organization and Finance Plan of Office III–W of the RFSS." The party treasury proved one of the few sources of resistance to SS business at this stage. The SS bid for brickworks and stone quarries otherwise met with widespread cooperation from the Reich Justice Ministry and the Four Year Plan. The Reich Economics Ministry extended the DESt state credit. The General Building Inspector, Albert Speer, offered advanced payment of 9.5 million Reichsmark for brick and stone, what amounted to an interest-free loan. The city of Hamburg placed similar orders, and the German Workers Front promised a loan of 700,000 Reichsmark. This was a *Polykratie* of mutual coopera-tion. See Reichsschatzmeister, 17 June 1938, "Kreditbeschaffung für Zwecke der Klinker-werke Buchenwald bei Weimar," IfZ, Fa 183; undated report [1940], "DESt," BAK NS3/625. See also Kaienburg, *Vernichtung durch Arbeit*, 97–111.

57. Salpeter to Reichsfinanzministerium Dr. Asseyar, 5 Feb. 1941, "Steuerpflicht der DEST," T-976/25.

58. NO-542, Salpeter, undated [mid-1939], "Tasks, Organization and Finance Plan of Office III–W of the RFSS."

59. Ibid.

60. Salpeter to Reichsfinanzministerium Dr. Asseyar, 5 Feb. 1941, "Steuerpflicht der DEST," T-976/25.

61. Friedlander, *The Origins of Nazi Genocide*, 81.

62. NO-542, Salpeter, undated [mid-1939], "Tasks, Organization and Finance Plan of Office III–W of the RFSS."

63. NO-1044, unsigned [Salpeter], undated [from 1940], "Bericht über das erste Geschäftsjahr der Deutschen Versuchsanstalt für Ernährung und Verpflegung."

64. NO-542, Salpeter, undated [mid-1939], "Tasks, Organization and Finance Plan of Office III–W of the RFSS."

65. Quotation from Reichsschatzmeister, 17 June 1938, "Kreditbeschaffung für Zwecke der Klinkerwerke Buchenwald bei Weimar," IfZ, Fa 183. See also Aktennotiz, 10 Feb. 1939, initialed by Pohl, NS3/1471; Salpeter to Pohl, Möckel, and Sollmann, in House Memoran-dum, 19 May 1939; Salpeter to Pohl, 16 May 1940, BAK NS3/1471. Also NO-542, Salpeter, undated [mid-1939], "Tasks, Organization and Finance Plan of Office III–W of the RFSS," and NO-1044, unsigned [Salpeter], undated [from 1940], "Bericht über das erste Geschäftsjahr der Deutschen Versuchsanstalt für Ernährung und Verpflegung." See also Reichswirtschaftminister, 29 Aug. 1936, "Vierte Verordnung zur Durchführung der Vorschriften über die Prüfungspflicht der Wirtschaftsbetriebe der öffentlichen Hand," T-976/1. Likewise, later, Hohberg, 17 Sept. 1940, "Aktenvermerk," T-976/25. For the inseparable union of pragmatism and ideological conviction in discussions of *Gemein-nützigkeit*, Chef des Prüfungsamtes to RWM, 30 Mar. and 17 June 1939, T-976/1.

66. Pohl to Reichskommissar für die Preisbildung, 19 Sept. 1941, "Erklärung nach §22 KWVO der dem VuWHA angeschlossenen Gesellschaften," T-976/3. On *Gemeinnützig-*

keit, see Waldemar Moritz, "Beamtentum, Sozialismus und Neuer Staat," *Reichsverwaltungsblatt* 55 (1934): 174.

67. Himmler's decision remains veiled due to scant documentation in the early months of 1939. Nevertheless, later documents suggest that Pohl was instructed to take complete charge of the SS companies just preceding April. NO-1451, Heinrich Himmler, 20 Apr. 1939, "Establishment of a New Main Office"; NO-1045, Salpeter to Pohl, 8 May 1940, "Organization des Amtes III A."

68. "Anordnung des Stellvertreters des Führers vom 30 Mar. 39" mentioned in Himmler, 17 Feb. 1940, "Anordnung: Zuständigkeit des VuW-Hauptamtes," T-175/137.

69. Quotation from NO-542, Salpeter, undated [mid-1939], "Tasks, Organization and Finance Plan of Office III–W of the RFSS." Compare Georg, *Die wirtschaftlichen Unternehmungen der SS,* 27–29. Himmler continually strengthened Pohl's authority over SS enterprise throughout the war; e.g., Himmler, 17 Feb. 1940, "Anordnung: Zuständigkeit des VuW-Hauptamtes," T-175/137. The Reichssicherheitshauptamt would also be formed in 1939, as would the Reichskommissar für die Festigung des deutschen Volkstums. Buchheim, "Die SS—das Herrschaftsinstrument," 66–83; NO-1451, Heinrich Himmler, 20 Apr. 1939, "Establishment of a New Main Office"; Testimony of Pohl, Protocol: 1260– 62 and 1303–5; NO-4007, Affidavit of Hubert Karl. Compare NO-2616, Affidavit of Oswald Pohl; Verwaltungsanordnung Nr. 7, 23 Feb. 1937, T-175/97. Pohl organized the new construction office into four subdivisions, each corresponding to different branches of the SS: V-5 a: general construction affairs of the Command Troops (Verfügungstruppen) and Totenkopfverbände (Eicke's Death's Head Units), i.e., military buildings; V-5 b: buildings of the General SS (the traditional SS Section or *Abschnitte*); V-5 c: construction at Dachau's military training center; V-5 d: settlement construction at Dachau. By 1940 construction was again switched to V-4, now an independent office: "Verwaltungsmitteilungen für die W-SS," 18 Apr. 1940, BDC Wirtschaftsverwaltungsanordnungen: Mitteilungen für die W-SS Verwaltungs-Mitteilungen für die W-SS. This shows, if nothing else, the high degree of flux, the repeated attempts to impose administrative order, and Pohl's persistence.

70. Salpeter, Schondorff, Mummenthey, and (civilian engineer) Zahn, 17 June 1940, "Neuorganization und Dezentralisierung Werk Oranienburg," BAK NS3/1344; Salpeter, 26 Oct. 1940, Aktenvermerk and Salpeter to Pohl, 29 Nov. and 27 Dec. 1940, BAK NS3/1344; Memorandum and Salpeter to Pohl, 16 May 1940, initialed by Pohl on 28 Sept., BAK NS3/1471; and Aktennotiz, 10 Feb. 1939, initialed by Pohl, BAK NS3/1471. Salpeter led the president of the Reichsbank on a tour of several DESt factories in August. "Besprechung des Herrn Reichswirtschaftministers und Reichsbankpräsidenten mit dem Chef des persönlichen Stabes des RFSS," 22 Aug. 1939, BAK NS3/1532; Testimony of Pohl, Protocol: 1260–62. See also Georg, *Die wirtschaftlichen Unternehmungen der SS,* 26. Contemporary documents confirm postwar testimony: Dr. Leo Volk, 4 Sept. 1941, "Stichworte zu den Ausführungen des Gruppenführers für die Ansprache bei der Amtschefssitzung," T-976/35. Compare Tuchel, *Konzentrationslager,* 245–48. The division took place on 20 April. NO-1451, Heinrich Himmler, 20 Apr. 1939, "Establishment of a New Main Office"; Himmler, 17 Feb. 1940, "Anordnung: Zuständigkeit des VuW-Hauptamtes," T-175/137; NO-542, Salpeter, undated [mid-1939], "Tasks, Organization and Finance Plan of Office III–W of the RFSS." After the war, Pohl claimed that the

VuWHA maintained responsibility for the budgets of the General SS, but this is not born out by the documents of early 1939 or any of the later activities of the Office III–W. Salpeter, for one, was supposed to devote most of his time to property rights and eminent domain—tasks of the HAHB—yet his Legal Department remained in the Office III–W (thereafter synonymous with the VuWHA). The Legal Office transferred to the HAHB a year later, but the Amt II continued to work closely with DESt to negotiate building costs with the Reich.

71. On Maurer: BDC SS Personal-Akte and RuSHA-Akte Gerhard Maurer; Affidavit of Gerhard Maurer, Defense Document Books of Oswald Pohl; Affidavits of Gerhard Maurer; Interrogation, 20 Mar. 1947, 3 June 1947, and undated, handwritten Lebenslauf, BAP, PL5: 42063. On Mummenthey: NO-2523, Affidavit of Karl Kurt Andreas Emil Mummenthey; BDC Personal-Akte and RuSHA-Akte Karl Mummenthey. He had been in the SS over a year before he joined Salpeter's Legal Department and began his revealing audits of the DESt's books. On Möckel: BDC SS Personal-Akte and RuSHA-Akte Karl Möckel. On Schwarz: BDC Personal-Akte Heinz Schwarz.

72. Hohberg to Mummenthey, 10 Sept. 1940, "Bilanz der DESt zum 31 Dec. 39," T-976/25; Salpeter to Pohl, 29 Nov. and 27 Dec. 1940, BAK NS3/1344. NO-542, Salpeter, undated [mid-1939], "Tasks, Organization and Finance Plan of Office III–W of the RFSS," notes that the Reich had taken over property acquisitions for Allach. Mummenthey, "Monatsbericht," 1940–42, NS3/1346, shows the Reich subsidizing DESt by taking over construction costs, not to mention the cost of the *Wachmannschaften*.

73. Herbert Ziegler, *Nazi Germany's New Aristocracy*, 104–8.

74. Erduin Schondorff, "Fachschulwesen und das Selbstkostenproblem in den Ziegeleibetrieben," *Ziegel und Zement* 43 (1931): 440.

75. Salpeter, Schondorff, Mummenthey, and (civilian engineer) Zahn, 17 June 1940, "Neuorganization und Dezentralisierung Werk Oranienburg," NS3/1344; Oswald Pohl, Protocol: 1713–16. Schondorff's report is submitted in Salpeter to Pohl, 15 Jan. 1941, "Endgültige Planung des Klinkerwerkes Oranienburg," T-976/25. Regarding the SS's application for new credit: "Besprechung des Herrn Reichswirtschaftministers und Reichsbankpräsidenten mit dem Chef des persönlichen Stabes des RFSS," 22 Aug. 1939, BAK NS3/1532; Hohberg to Mummenthey, 10 Sept. 1940, "Bilanz der DESt zum 31 Dec. 39," T-976/25; Mummenthey, "Monatsbericht," 1940–42, NS3/1346; Mummenthey and Salpeter, 31 July 1941, "Geschäftsbericht der DESt 1940," T-976/26.

76. Saul, *Voltaire's Bastards*, 22. Saul blames the Holocaust on unbridled instrumental reason (73–74).

77. Speer, *Inside the Third Reich*, 112. Compare Hortleder, *Das Gesellschaftsbild des Ingenieurs*, 107–38, regarding National Socialism, and 150–63, regarding ideology in general. Schmidt, *Albert Speer*; Sereny, *Albert Speer*; Van der Vat, *The Good Nazi*.

78. Ludwig, *Technik und Ingenieure*, 160–70.

79. Schondorff, "Fachschulwesen," 441–42.

80. Kaienburg, *Vernichtung durch Arbeit*, 94–95. National Socialist support for modernization in the brick industry can be seen plainly by browsing a trade journal like *Tonindustriezeitung. Ziegel und Zement*. Ludwig, *Technik und Ingenieure*, 17–35.

81. Gispen, "German Engineers and American Social Theory," 550–74, and *New Profession, Old Order*, 26–37.

82. Calvert, *The Mechanical Engineer*; Oswald Pohl, Protocol: 1713–16; Affidavit of Roman Fuerth, Defense Books of Karl Mummenthey; BDC SS Personal-Akte Erduin Schondorff.

83. Schondorff, "Fachschulwesen," 442.

84. Ibid., 441–42. Schondorff, "Trockenfehler bei Hohlziegeln," *Tonindustrie-Zeitung* 57 (1933): 893–94, or "Glasieren, Engobieren, Dämpfen von Ziegeln," *Tonindustrie-Zeitung* (three-part title article) 61 (1937): 585–87, 599–601. The last, concluding section does not appear in the journal, perhaps because shortly thereafter Schondorff became involved with the Hermann-Göring-Werke and the SS companies.

85. "Besprechung des Herrn Reichswirtschaftministers und Reichsbankpräsidenten mit dem Chef des persönlichen Stabes des RFSS," 22 Aug. 1939, BAK NS3/1532; unpublished paper of the late Arno Mietschke (for which I thank him), "Technische Angestellte in der deutschen Großindustrie der Zwischenkriegszeit: Das Beispiel der August-Thyssen-Hütte, Hamborn." Mietschke found that 40 percent of the technical employees of this firm joined the Nazi Party. Many had marched into the nationalized steel firms founded by the Nazi government, greatly irritating the private firm Thyssen.

86. Kaienburg, *Vernichtung durch Arbeit*, 93–97.

87. Undated report [1940], "DESt," BAK NS3/625; Kaienburg, *Vernichtung durch Arbeit*, 94–95 and n.

88. Pohl to RFSS, 25 Jan. 1944, "Lehmbauweise Posen," BAK NS19/3115.

89. Salpeter, Schondorff, Mummenthey, and (civilian engineer) Zahn, 17 June 1940, "Neuorganization und Dezentralisierung Werk Oranienburg," BAK NS3/1344.

90. Schondorff, "Fachschulwesen," 442–43.

91. The SS's involvement in settlements is discussed later. See Naasner, *Neue Machtzentren*, 238–39. This seems to be a popular epithet among researchers of the WVHA in general. Hans Kammler, who was forty-four at the time of his death in Prague, is called a "gray eminence" in Agoston, *Teufel oder Technokrat*. Koch applies the title with a bit more credibility in *Himmlers graue Eminenz—Oswald Pohl*. Georg, *Die wirtschaftlichen Unternehmungen der SS*, 129–31. NO-1924, Affidavit of Hans Karl Hohberg. See also the history of the formation of the VuWHA and Deutsche Wirtschaftsbetriebe in Leo Volk, 22 Apr. 1944, "Vermerk für SS-Oberführer Baier," T-976/1, and the chronology of Hans Baier, 22 May 1944, "Prüfung der wirtschaftlichen Betriebe der Schutzstaffel (historische Entwicklung)," T-976/1.

92. Affidavit of Dr. Max Wolf, Defense Document Books of Hans Hohberg.

93. Koehl, *The Black Corps*, 170. This enthusiasm for modern management among mid- and lower-level executives has been most aptly captured in the American context by Zunz, *Making America Corporate*. Compare Kocka, *Unternehmer in der deutschen Industrialisierung*, 56–57.

94. Berthold Manasse, "Der Wirtschaftsprüfer als Berater der wirtschaftlichen Unternehmungen," *Der Wirtschaftsprüfer* 1 (1932): 8 (address to annual meeting). See similar statements in Walther Schreiber, "Die Hauptstelle für die öffentlich bestellten Wirtschaftsprüfer und die Zulassungs- und Prüfungsstellen"; "Was erwartet die Wirtschaft vom Wirtschaftsprüfer?"; and Bernhard Brockhage, Vorsitzender des Instituts für das Revisions- und Treuhandwesen, "Geleitwort," *Der Wirtschaftsprüfer* 1 (1932): 2, 5, and 1 respectively. Compare Zunz, *Making America Corporate*, 199–203.

95. Pohl to Reichskommissar für die Preisbildung, 19 Sept. 1941, "Erklärung nach §22 KWVO der dem VuWHA angeschlossenen Gesellschaften," T-976/3. Hohberg prepared this document, and he included his initials in the copy submitted to Pohl for signing. Compare Hohberg to Finanzamt Börse, 5 May 1943, "Gesellschaftssteuerbescheid vom 9 Sep. 42," BDC SS Hänge Ordner: 3463. See also Trischler, "Führer-Ideologie im Vergleich," 45–88, and Hundt, *Zur Theoriegeschichte der Betriebswirtschaftslehre*, 93–97.

96. By no means is the only example NO-1040, Dr. Hohberg to Chef des Amtes A II and SS Hauptsturmführer Mellmer, 3 June 1943.

97. DWB, "Richtlinien für die Revisionsabteilung," BAK NS3/933; Hohberg, 17 June 1940, "Entwurf eines Vertrages zur Herstellung eines Organverhältnisses," and Anmerkung Salpeter, 18 June 1940, NS3/1471; Pohl to Salpeter, 27 July 1940, and Salpeter to DWB, 17 Feb. 1941, "Einbringung von Geschäftsanteilen"; "Gründung der Wirtschaftsbetriebe GmbH," T-976/1; Hohberg, 27 Sept. 1940, "Aktenvermerk für Gruppenführer Pohl," BAK NS3/1471. Another clipping instructed SS managers in the statistical surveillance of financial operations in modern corporations.

98. *Der Wirtschaftstreuhänder* and book by Herbert M. Casson, excerpt from chap. 4, "Trotz Allem—Siegen!," from 1928, BAK NS3/934.

99. Hohberg, 27 Sept. 1940, "Aktenvermerk für Gruppenführer Pohl," BAK NS3/1471; Chronology of Hans Baier, 22 May 1944, "Prüfung der wirtschaftlichen Betriebe der Schutzstaffel (historische Entwicklung)," T-976/1.

100. Memorandum and Salpeter to Pohl, 16 May 1940, initialed by Pohl on 28 Sept., BAK NS3/1471; Salpeter to Reichsfinanzministerium Dr. Asseyar, 5 Feb. 1941, "Steuerpflicht der DEST," T-976/25; NO-542, Salpeter, undated [mid-1939], "Tasks, Organization and Finance Plan of Office III–W of the RFSS."

101. Walter Salpeter to Reichsfinanzministerium Dr. Asseyar, 5 Feb. 1941, "Steuerpflicht der DEST," T-976/25.

102. See Dr. Kühl, 17 Apr. 1939, "Die steuerlichen Vor- und Nachteile einer Holding-(Schachtel) Gesellschaft," BAK NS3/1471; Mummenthey to Pohl and Hohberg, 17 Sept. 1941, NS3/1344. Later, but in reference to the earlier days, Leo Volk, 22 Apr. 1944, "Vermerk für SS-Oberführer Baier," T-976/1.

103. NO-1016, Dr. Leo Volk to Dr. Schneider, 11 July 1944.

104. Henke, "Von den Grenzen der SS-Macht," 262.

105. NO-1016, Dr. Leo Volk to Dr. Schneider, 11 July 1944; NO-542, Salpeter, undated [mid-1939], "Tasks, Organization and Finance Plan of Office III–W of the RFSS." The SS would come to control 75 percent of sales in bottled water. Also Pohl to Reichswirtschaftsminister Funk, 4 June 1941, T-976/1: 582–84; Georg, *Die wirtschaftlichen Unternehmungen der SS*, 73–77.

106. NO-542, Salpeter, undated [mid-1939], "Tasks, Organization and Finance Plan of Office III–W of the RFSS." This is an internal memo, and there is no reason to doubt its sincerity or infer motives of self-supply. Compare Schulte, "Rüstungsunternehmen oder Handwerksbetrieb?," 560.

107. NO-542, Salpeter, undated [mid-1939], "Tasks, Organization and Finance Plan of Office III–W of the RFSS"; Georg, *Die wirtschaftlichen Unternehmungen der SS*, 76–77; Oswald Pohl, Protocol: 1536–40; Hohberg, 27 Sept. 1940, "Aktenvermerk für Gruppenführer Pohl," BAK NS3/1471; NO-678, Anonymous, undated [1941], "Deutsche Aus-

rüstungswerke GmbH 'DAW' "; Hohberg to Pohl, 18 Sept. 1940, and Pohl to Salpeter, 31 Jan. 1940, BAP, PL5: 42055; Hohberg to Finanzamt Börse, 5 May 1943, "Gesellschafts-steuerbescheid vom 9 Sep. 42," BDC SS Hänge Ordner: 3463; Aktenvermerk Hohberg and Salpeter to DWB, 17 Feb. 1941, "Einbringung von Geschäftsanteilen," T-976/1; Salpeter to Pohl, 15 Jan. 1941, "Endgültige Planung des Klinkerwerkes Oranienburg," T-976/25; Salpeter to Pohl, 29 Nov. and 27 Dec. 1940, BAK NS3/1344; Hohberg to Salpeter, 28 Oct. 1940, BAK NS3/1344; Pohl to Salpeter, 27 July 1940, and "Gründung der Wirtschafts-betriebe GmbH," T-976/1; Georg, *Die wirtschaftlichen Unternehmungen der SS*, 71 and 70–73 in general; Naasner, *Neue Machtzentren*, 393 and 386–427 in general; Kaienburg, *Vernichtung durch Arbeit*, 87–89.

108. Uncritical investment in technology simply because it was considered "modern" was all too common in the 1920s and 1930s. For example, the case of the French auto industry: Cohen, "The Modernization of Production in the French Automobile Industry."

109. Armanski, *Maschinen des Terrors*, 73.

CHAPTER THREE

1. Grundmann, *Agrarpolitik im Dritten Reich*, 22.

2. Koehl, *RKFDV*; Müller, *Hitlers Ostkrieg*; Aly and Heim, *Vordenker der Vernichtung*; and Aly, *Endlösung*. See also Streit, *Keine Kameraden*, 126; Broszat, *Nationalsozialistische Polenpolitik*, 58–59. On the *Einsatzgruppen*, Wilhelm, *Die Truppe des Weltanschauungs-krieges*. On Eichmann, Safrian, *Die Eichmann Männer*, 23–56. On the Nazi leadership's optimism at this time, see Browning, "Beyond 'Intentionalism' and 'Functionalism,'" 220; Heinemann, "Die Kooperation von SS, Polizei und Zivilverwaltung."

3. Quoted after Wasser, *Himmlers Raumplanung im Osten*, 52. See also Müller, *Hitlers Ostkrieg*, 83–114.

4. Quotation from Breitman, *The Architect of Genocide*, 42.

5. Quotation from Wasser, *Himmlers Raumplanung im Osten*, 110. See also Aly and Heim, *Vordenker der Vernichtung*, 125–46.

6. Müller, *Hitlers Ostkrieg*, 83–114, esp. 105. I thank Mark Spoerer for the conversion rate to today's equivalents.

7. Konrad Meyer-Hettling, quoted after Müller, *Hitlers Ostkrieg*, 91.

8. Hauptabteilung Planung und Boden, Stabshauptamt des Reichskommissars für die Festigung deutschen Volkstums, *Planung und Aufbau im Osten. Erläuterungen und Skizzen zum ländlichen Aufbau in den neuen Ostgebieten* (Berlin: Deutsche Landbuch-handlung, 1941), 11.

9. See the catalog of *Planung und Aufbau im Osten*; Müller, *Hitlers Ostkrieg*, 87; Aly and Heim, *Vordenker der Vernichtung*, 135–45; Wasser, *Himmlers Raumplanung im Os-ten*, 20–45, and on the Main Trustee East, 68. In a long list of firms taken over by the SS and transferred to "settlers" in Böhmen and Mähren by 1943, WVHA acquisitions are also listed. These were sand or stone pits, an occasional building supply industry, but never armaments firms or even any metalworking firms or machine-tool firms—which the SS certainly would have needed to enter the armaments industry. See Brandt to Gieß, 10 Feb. 1942; Obst. Berg, 30 Apr. 1943, "Vermerk für Ostbaf. Dr. Brandt," BDC Sam-melliste 8. Likewise, Dipl. Kaufmann Ernst Feichtinger, 18 May 1942, "Erläuterungen zum Organisationsplan der HA. III"; 16 Apr. 1943, "Erläuterungen zu den Bestandslisten

der beschlagnahmten Gewerbebetriebe. Stand 15 Apr. 43," in BAP 23.01 Rechnungshof des Deutschen Reiches: 5622.

10. NG-1912, Max Winkler, undated [Nov. 1939], "Beschlagnahme-Verfügung," signals the HTO's agreement to Pohl's appointment. The date is given by Rössler, "Area Research," 135. See also NI-11314, Oswald Pohl, 4 Nov. 1941, "Aktenvermerk"; NI-11326, Aktenvermerk Hohbergs, 20 Nov. 1941; Müller, *Hitlers Ostkrieg*, 87ff.; Testimony of Pohl, Protocol: 1707; Affidavit of Joseph Opperbeck, Defense Document Books of Hanns Bobermin. Pohl's first title was apparently "Generaltreuhänder für die Ostziegeleien." NO-1043, undated [1940], "Generaltreuhaender fuer Baustofferzeugungsstaetten im Ostraum im Jahre 1940"; NI-11324, Oswald Pohl to Heinrich Himmler, RFSS, 4 Sept. 1941, "Abgabe von treuhänderisch verwalteten Ziegeleien." For Pohl's authority in relation to Gauleiter in the East, see also NI-11321, Oswald Pohl to Gauleiter und Oberpräsidenten des Gaues Oberschlesien Bracht, 14 Feb. 1942, "Ostziegeleien. Dortiges Schreiben von 9.2.42"; NI-11320, Bracht to Pohl, "Generaltreuhänder für Baustoff-Erzeugungstätten im Ostraum," 9 Feb. 1942, from Kattowitz; NI-11322, Pohl to Winkler, HTO, 14 Feb. 1942; NI-11323, Pohl to Gauleiter und Oberpräsidenten des Gaues Ostpreussen, 14 Feb. 1942; NI-11319, Pohl to Winkler, HTO, 9 Feb. and 14 Apr. 1942, "Ostziegeleien, Mein Schreiben vom 14 Feb. 42." The ODBS factories were originally intended to provide jobs and managerial positions for Frontkämpfer returning to SS settlements. NO-1008, Bobermin to Pohl, copy to Himmler, 3 July 1941, "Entwurf eines Schreibens an den RFSS, betr. Abgabe von treuhänderisch verwalteten Ost-Ziegeleien."

11. Rössler, "Area Research," 135.

12. Pohl to Heinrich Himmler, 14 Jan. 1941, T-976/1.

13. Heinrich Himmler, likely formulated by Konrad Meyer, "Grundsätze für die Gestaltung der Städte," in Richtlinien für die Planung und Gestaltung der Städte in den eingegliederten deutschen Ostgebieten, Allgemeine Anordnung Nr. 13/11, 30 Jan. 1942, BDC Sammelliste 54.

14. Vermerk, Lublin, 21 July 1941, BDC Sammelliste 54; Karl Bestle and Karl Niemann, 30 June 1943, "Protokol der Aufsichtsratssitzung 1943 der Deutsche Edelmöbel Aktiengesellschaft"; Kurt May, 23 June 1941, "Jahresrechnung des Vorstandes der Drucker AG, Dampfsägewerke und Holzwarenfabriken in Brünn zur Jahresrechnung 1940"; and undated, unsigned, "Technische Grundlagen"; and "Wirtschaftliche Verhältnisse" of the Victoria AG of Prague [sometime in 1943], in T-976/22; Georg, *Die wirtschaftlichen Unternehmungen der SS*, 76–77. A complete table set produced by Allach/Victoria AG was displayed at the Münchner Stadtmuseum in its exhibit, "Hauptstadt der Bewegung. München unter dem Nationalsozialismus," in 1994.

15. Amtsrat Scheck, "Vermerk über die Gesellschaft für Textil- und Lederverwertung, mbH. Dachau und Ravensbrück," BAP 23.01 Rechnungshof des Deutschen Reiches: 5636.

16. NO-1918, Georg Lörner and Volk to Reichsjustizministerium, 21 Mar. 1942. Compare a similar statement about the SS Kleiderkasse, which was managed by the same men: "Kleiderkasse der Schutzstaffel," 23 June 1941, BAK NS3/954.

17. Daluege's countersignature can be found on the blueprints: Kammler, 14 Aug. 1941, "Polizeistützpunkt Typ A" and "Typ B," WAP Lublin Inventarz Zentralbauleitung Lublin: 237 and 238. Regarding front soldiers, see NO-1008, Bobermin and Pohl, 3 Mar.

1941, "Entwurf eines Schreibens an den RFSS, betr. Abgabe von treuhänderisch verwalteten Ost-Ziegeleien," and NI-11323, Pohl to Koch Gauleiter und Oberpräsidenten des Gaues Ostpreussen, 14 Feb. 1942. Dipl. Kaufmann Ernst Feichtinger, 18 May 1942, "Erläuterungen zum Organisationsplan der HA. III," and the list of properties confiscated, Feichtinger, 16 Apr. 1943, "Erläuterungen zu den Bestandlisten der beschlagnahmten Gewerbebetriebe. Stand 15 Apr. 43," in BAP 23.01 Rechnungshof des Deutschen Reiches: 5629. Regarding the same process earlier in Untersteiermark, see Rücker-Fmbden [?], Vermerk, 11 May 1939, BDC Hängeordner 1025, and unsigned, from within RKF, 1942, "Verwaltungsmässige Form der Untersteiermark," BAP 23.01 Rechnungshof des Deutschen Reiches: 5620.

18. Himmler's speech on 9 June 1942, cited from Aly, *Endlösung*, 292.

19. Hayes, "Big Business and Aryanization," 267–71; Dieter Ziegler, "Die Verdrängung der Juden aus der Dresdner Bank," 187–216; James, "Die Rolle der Banken im Nationalsozialismus," 25–36.

20. Dresdner Bank, *Volk und Wirtschaft im ehemaligen Polen. Nur zur persönlichen Information!* (Berlin: Dresdner Bank Volkswirtschaftliche Abteilung, 1939), 38. The booklet proudly pointed out, with no small degree of satisfaction, that the Soviet-occupied zone of Poland contained fewer and poorer-quality industries. See also NI-6341, Walter Salpeter to Professor Meyer, Dresdner Bank, 4 Jan. 1940, and NI-7866, August Frank to Oswald Pohl, 2 Sept. 1942.

21. NO-1043, undated [1940], "Generaltreuhaender fuer Baustofferzeugungsstaetten im Ostraum im Jahre 1940." The document is unsigned, but Prosecutor Robins at Nuremberg attributed this report to Volk, who did not dispute the matter.

22. NO-1016, Leo Volk, "Organization und Aufgaben der Amtsgruppe W." See chapter 1.

23. NO-1016, and see also BDC SS Personal-Akte Leo Volk and Affidavit of Gerhard Hoffman, Defense Document Book of Leo Volk.

24. On Bobermin's commitment and pride, see NO-1008, Bobermin to Pohl, copy to Himmler, 3 July 1941, "Entwurf eines Schreibens an den RFSS, betr. Abgabe von treuhänderisch verwalteten Ost-Ziegeleien"; Testimony of Pohl, Protocol: 1707. See also BDC Personal-Akte Hanns Bobermin; NO-1566, Affidavit of Hanns Bobermin. Service to RKF: NI-11327, Bobermin to Hohberg, 22 Dec. 1941, "Tätigkeit und Geschäftsergebnis 1941"; NO-1012, Pohl's order of 28 June 1941. Bobermin's technological enthusiasm and anticapitalism: NO-1006, Bobermin to Baier, 2 Apr. 1944, "Taking over of brickworks Bonarka"; NO-1045, Salpeter to Pohl, 8 May 1940, "Organization des Amtes III A"; NO-1299, Hohberg to Pohl, 26 May 1941, "Limitation of the Sphere of Work of Office III/A." ODBS founded: NI-11315, 1 Apr. 1944, "Beglaubigte Übersicht aus dem Handelsregister Abteilung B"; Bobermin to Göbel, 7 Dec. 1943, "Organverhältnis-Steuerentrichtung," PMO, D-AuIII/Golleschau, Band 2; NO-1021, Bobermin to Hohberg, 23 Nov. 1941, "Russland-Ziegeleien/Bericht an RFSS."

25. NI-11325, Pohl to RFSS, 4 Sept. 1941, "Aufgabe von treuhänderisch verwalteten Ziegeleien"; NO-1043, undated [1940], "Generaltreuhaender fuer Baustofferzeugungsstaetten im Ostraum im Jahre 1940"; NI-11324, Oswald Pohl to Heinrich Himmler, RFSS, 4 Sept. 1941, "Abgabe von treuhänderisch verwalteten Ziegeleien."

26. Fritz Todt, "Regelung der Bauwirtschaft," *Der Vierjahresplan* 3 (1939): 764.

27. NO-1043, undated [1940], "Generaltreuhaender fuer Baustofferzeugungsstaetten im Ostraum im Jahre 1940."

28. NI-11324, Oswald Pohl to Heinrich Himmler, RFSS, 4 Sept. 1941, "Abgabe von treuhänderisch verwalteten Ziegeleien."

29. From December 1940, cited after Georg, *Die wirtschaftlichen Unternehmungen der SS*, 79. Von Neurath's first name is alternatively spelled with a "C" and a "K," even in contemporary sources such as various *Who's Who* volumes.

30. Heineman, *Hitler's First Foreign Minister*, 191, 186–212 in general. See also Duff, *A German Protectorate*, 98–102.

31. Alfons Leitl, "Professor Hermann Gretsch/Eine Ausstellung seiner Arbeiten," *Die Bauwelt* 32 (1941): 4.

32. I thank Jan-Eirk Schulte for this biographical information which he received from the Kulturamt of the Landeshauptstadt Stuttgart and the *Westdeutsche Wirtschafts-chronik* (Stuttgart, 1954), 1:843.

33. Affidavit of Kurt Brune, Defense Document Books of Hans Hohberg; Karl Bestle and Karl Niemann, 30 June 1943, "Protokol der Aufsichtsratssitzung"; Kurt May, 23 June 1941, "Jahresrechnung des Vorstandes der Drucker AG, Dampfsägewerke und Holz-warenfabriken in Brünn zur Jahresrechnung 1940"; undated, unsigned, "Technische Grundlagen" and "Protokoll der Aufsichtsratssitzung der Deutschen Edelmöbel Aktiengesellschaft am 29 Jun. 42," T-976/22. Compare Hounshell, *From the American System to Mass Production*, 145–46, 310–14.

34. May, "Jahresrechnung"; undated, unsigned, "Technische Grundlagen" and "Protokoll"; Affidavit of Kurt Brune.

35. Ibid.

36. Georg, *Die wirtschaftlichen Unternehmungen der SS*, 81, and Dr. Höring, Report of 16 Nov. 1941, T-976/22: 556–756. Later Pohl acquired up to 94 percent of Drucker's stock.

37. Georg, *Die wirtschaftlichen Unternehmungen der SS*, 80–81. The DWB recapitalizing the German Masterpiece with over 800,000 Reichsmarks: Volk to Pohl, 1 Sept. 1941, "Umorganization der Ämter," T-976/35; Affidavit of Kurt Brune, Defense Document Books of Hans Hohberg; "Technische Grundlagen" and "Protokoll der Aufsichtsrats-sitzung der Deutschen Edelmöbel Aktiengesellschaft am 29 Jun. 42," T-976/22. The chief of the SS Reich Security Main Office, Reinhard Heydrich, replaced Neurath as *Protektor* on 27 Sept. 1941.

38. See Hohberg's correspondence in BAK NS3/1471, esp. 17 June 1940, "Entwurf eines Vertrages zur Herstellung eines Organverhältnisses," and Anmerkung Salpeter, 18 June 1940. The imposition of standard, individual contracts started in March 1941.

39. Dr. Leo Volk, signed off by Pohl, 4 Sept. 1941, "Stichworte zu den Ausführungen des Gruppenführers für die Ansprache bei der Amtschefsitzung," T-976/35; Volk to Pohl, 1 Sept. 1941, "Umorganization der Ämter," T-976/35. Noteworthy is the absence of any armaments factories or mention of self-sufficiency for the W-SS, which the SS was neither striving for nor interested in.

40. See note 39.

41. Unsigned [probably Salpeter], preparations from 14 June 1940, "Aktenvermerk über die Amtsbesprechung," T-976/35. See HAHB unrecognized to Bauleitung KL

Flossenbürg, 6 July 1939, BAK NS4/34 Fl; likewise HAHB to Bauleitung Flossenbürg, 20 Jan. 1942, BAK NS4/61 Fl.

42. Dipl. Ing. Hermann Gretsch, "Zeitgemäßes Wohnen," *Die Bauzeitung* 38 (1941): 425.

43. Trommler, "Von Bauhausstuhl zur Kulturpolitik," 86–110.

44. Gretsch, "Zeitgemäßes Wohnen," 426. See also the photos in this article, pp. 427–30.

45. Fromm, *Arbeiter und Angestellte*, 142–50. Fromm's statistical samples of NSDAP members were so small as to render his data unrepresentative, but he also noted, curiously, an overrepresentation of Nazis among admirers of *neue Sachlichkeit*.

46. Rössler and Schleiermacher, "Der Generalplan Ost," 10; Kaplan, "Traditions Transformed."

47. Leitl, "Professor Hermann Gretsch," 4. On Gretsch as designer-engineer, compare Heskett, "Design in Interwar Germany," 269–71. Note that what counted as National Socialist design was never as homogeneous as many postwar historians would make it seem. Heskett shows that the Zeppelin "LZ 129 Hindenburg," for example, had tube-steel chairs in its lounge. Nevertheless, designers argued that National Socialism *should* enforce homogeneous design.

48. Hermann Gretsch, "Die Stellung des Entwerfers in der Wirtschaft," *Bauen, Siedeln, Wohnen* 19 (1939): 751.

49. Trommler, "The Avant-Garde and Technology," 397–416, and "The Creation of a Culture of Sachlichkeit," 465–85; Jenkins, "The Kitsch Collections and *The Spirit in the Furniture*," 123–41.

50. Gretsch is quoted citing Gottfried Semper (from the 1850s), in Leitl, "Professor Hermann Gretsch," 4.

51. Pohl to Hans Baier, 18 Mar. 1937; Pohl to Baier, 6 Sept. 1939; and Pohl's report to Himmler, prepared with Baier, 10 June 1937, "Nachwuchs für SS-Verwaltungsführer," BDC SS Personal-Akte Hans Baier. See also Wegner, *Hitlers politische Soldaten*, 150–61; Baier to Pohl, 6 Nov. 1936, BDC SS Personal-Akte Hans Baier; Staatsministerium für Wirtschaft to Baier, 6 Feb. 1940, BDC Personal-Akte Hans Baier. Baier inexplicably seems to have used his personnel file as a repository of all his private correspondence.

52. Baier's letter to a friend, 26 May 1937, BDC Personal-Akte.

53. Quotation from Baier to the Wirtschaftsamt der Stadt Dachau, 15 Apr. 1942. Pohl was not above rebuking his friend, albeit gently. See Pohl to Baier, 22 Oct. 1938, BDC Personal-Akte Hans Baier.

54. Quotation from Baier to Pohl, 30 Mar. 1937; see also Baier to Pohl, 6 Nov. 1936, and Lebenslauf, all from BDC Personal-Akte Hans Baier. Baier's father was a jeweler and had sent his son to one of Germany's modern secondary schools (the *Realgymnasium*).

55. Baier to Pohl, 30 Mar. 1937, BDC Personal-Akte Hans Baier.

56. Baier to Pohl, 6 Nov. 1936, BDC SS Personal-Akte Hans Baier.

57. Ibid. and NO-1577, Affidavit of Hans Baier.

58. Pohl to Baier, 6 Sept. 1939, BDC Personal-Akte Hans Baier.

59. Wolf Lubbe to Hans Baier, 12 Mar. 1942, BDC Personal-Akte Hans Baier.

60. Baier to HSSPF Rössener, 12 Feb. 1942, BDC Personal-Akte Hans Baier.

61. Pohl's report to Himmler, prepared with Baier, 10 June 1937, "Nachwuchs für SS-Verwaltungsführer," BDC SS Personal-Akte Hans Baier.

62. Ibid.

63. Baier to Pohl, 4 Sept. 1938, BDC Personal-Akte Hans Baier. Perhaps Pohl urged Baier to learn something about industry. In 1942 Baier did look into courses in production engineering at the Technical University of Munich, but this never found its way into SS administrative training.

64. NO-542, Salpeter, "Tasks, Organization and Finance Plan of Office III (W)." For similar problems of coordination: Werkleiter Walther to Geschäftsleitung DESt, 1 Sept. 1941, "Abstellung von Bewachungsmannschaften," T-976/25; Mummenthey, 27 Nov. and 20 Dec. 1940, "Monatsbericht," T-976/26.

65. NO-3698, Georg Loerner to RFSS, 14 Sept. 1940, "Branch offices of the Main Office for Budgets and Building for prisoner allocation"; Leckebusch to Chef of Amt I, VuWHA, 6 Jan. 1940, BDC Wilhelm Burböck. These had most likely been initiated at the end of 1939. NO-2315, Pohl, 5 Sept. 1941, "Häftlingseinsatz"; NO-718, Burböck to all *Schutzhaftlager* "E," 28 Nov. 1941, "Assignment of internees' detachments; Regulation IKL"; PS-3677, Burböck, 7 Nov. 1941, "Allgemeine Dienstanweisung für die Schutzhaftlagerführer 'E.' " Jan-Erik Schulte chides me for attributing this innovation to Burböck instead of Glücks ("Verwaltung des Terrors," 396), and his sedulous research has uncovered an order by Glücks on 7 Oct. 1941 establishing the *Schutzhaftlagerführer* "E." Given Glücks's own general lack of initiative or administrative talent I find it extraordinarily unlikely that this initiative came from anyone other than Burböck. Clearly Burböck could not order the Kommandanten directly to do anything, and the authority had to come from Glücks as Inspector of Concentration Camps (just as Gerhard Maurer relied on Glücks to sign orders after Burböck).

66. Koehl, *The Black Corps*, 143; Kimpel, "Agrarreform und Bevölkerungspolitik," 124–45; Dienstleistungszeugnis by Herbert Backe, 11 May 1938, BDC Wilhelm Burböck. On Backe, see Dallin, *German Rule in Russia, 1941–1945*, 35–40; Gerlach, *Krieg, Ernährung, Völkermord*, 10–84.

67. "Personalien," signed Leckebusch, undated [from ca. Mar. 1940], BDC Wilhelm Burböck; Leckebusch to Chef of Amt I, VuWHA, 6 Jan. 1940, BDC Wilhelm Burböck. For analysis of the status of low-level civil servants like Burböck, see Speier, *German White-Collar Workers*, 48–54.

68. NO-1712, Wilhelm Burböck to Dienststellen des HAHB, 27 Nov. 1940, "Neuer Organizationsplan für die Hauptabteilung I/5"; PS-3677, Burböck, 7 Nov. 1941, "Allgemeine Dienstanweisung für die Schutzhaftlagerführer 'E.' "

69. PS-3677, Burböck, 7 Nov. 1941, "Allgemeine Dienstanweisung für die Schutzhaftlagerführer 'E' "; Glücks to all Kommandanten, 14 Oct. 1941, "Abstellung von Häftlingskommandos," BAP, PL5: 42055.

70. Herbert, *Fremdarbeiter*, 270–71. There is no evidence that Burböck sought to use Hollerith machines at this time. Compare Black, *IBM and the Holocaust*, 351–74.

71. NO-3651, Burböck to CCs and Chef HAHB, 17 February 1941, "Taking of photographs by the commanders of working parties." See Anson Rabinbach, *The Human Motor*, 100–119.

72. NO-3668, Wilhelm Burböck to Dienststelle des HAHB, 6 Nov. 1940, "14-tägiger Bericht der Aussenstellen." Mummenthey, 27 Nov. and 20 Dec. 1940, "Monatsbericht," T-976/26.

73. NO-2126, Affidavit of Phillipp Grimm.

74. Orth, "Die Kommandanten der nationalsozialistischen Konzentrationslager," 761–63, 774–75; Segev, *Die Soldaten des Bösen*, 175–85; Kogon, *Der SS-Staat*.

75. Quotation from NO-3657, Grimm to administration of Buchenwald, 24 Oct. 1940. See also NO-2126, Affidavit of Phillipp Grimm; NO-2120 and NO-2105, Phillipp Grimm to Staf. Koch, Kommandant Buchenwald, 28 Oct. and 13 Dec. 1940, "Transfer of Prisoners."

76. NO-2120.

77. Orth, *Die Konzentrationslager-SS*, 63–64.

78. Ibid. and NO-2105.

79. Wilhelm Burböck to all KLs, 11 Dec. 1941, "Monatsberichte," BDC SS Personal-Akte Burböck. See also NO-3696, NO-3661, Burböck to all KLs, 14 Nov. 1940 and 11 Apr. 1941, "Einsatzführerbesprechung"; NO-3668, Wilhelm Burböck to Dienststelle des HAHB, 6 Nov. 1940, "14-tägiger Bericht der Aussenstellen"; Glücks to all KLs, 12 Feb. 1942, "Herabminderung der Häftlingszahl für Lagerbetriebe," BAP, PL5: 42056; Glücks to Lagerkommandanten, 20 Feb. 1942, "Arbeitseinsatz," BDC SS Hängeordner: 1825.

80. NO-3715, Affidavit of Franz Auer. His estimate is almost certainly too high, as the largest employer of prisoners at this time was the Deutsche Ausrüstungswerke, at a little over 7,000. Georg, *Die wirtschaftlichen Unternehmungen der SS*, 61; NO-3665, Burböck to KLs, 23 July 1941, "Strength Report"; NO-1712, Wilhelm Burböck to Dienststellen des HAHB, 27 Nov. 1940, "Neuer Organizationsplan für die Hauptabteilung I/5." Compare with the complaints of Mummenthey, 27 Nov. and 20 Dec. 1940, "Monatsbericht," T-976/26.

81. NO-2636, Testimony of Ferdinand Roemhild. See also NO-2637, Postwar testimony of Dr. med. Fritz Mennecke.

82. PS-1151, Liebehenschel to KL Kommandanten, 10 Dec. 1941, "Former correspondence of 12 Nov. 41"; PS-1234, Kommandant Groß-Rosen to Liebehenschel, 16 Dec. 1941, "Selection of Inmates" and accompanying list of prisoners; Czech, "KL Auschwitz as an Extermination Camp," 29–50; Pressac, *Die Krematorien von Auschwitz*, 38–50.

83. PS-1151. This document is also reproduced in Kogon et al., *Nationalsozialistische Massentötung durch Giftgas*, 76; NO-907, Menneke to his wife, 25 Nov. 1941.

84. PS-1151.

85. Ibid.

86. Ibid.

87. Orth, *Das System der nationalsozialistischen Konzentrationslager*, 114–21, 131–37.

88. Cited in ibid., 117.

89. Hacking, *The Social Construction of What?*, 145.

90. Burleigh, *Death and Deliverance*, 226, 221–30 generally.

91. NO-907, Menneke to his wife, 25 Nov. 1941. Selected letters and other information about Mennecke are also published in translation in Aly, Chroust, and Pross, *Cleansing the Fatherland*, 238–93.

92. Quotations from Wise, Introduction, 5–6, and Alder, "A Revolution to Measure," 41. See also Wise, "Precision," 92; Rusnock, "Quantification, Precision, and Accuracy," 18; Porter, "Precision and Trust," 180–81; Heilbron, "Introductory Essay" and "The Measure of Enlightenment." On the Enlightenment in Germany, see Koselleck, *Preußen*

zwischen Reform und Revolution, 163–217, 663–71. On technological control during this period, see Brose, *The Politics of Technological Change in Prussia*.

93. Bauman, *Modernity and the Holocaust*, 98. See also Aly, *Endlösung*, 15–17, 100, 133, 236–50.

CHAPTER FOUR

1. Halder's notes on Adolf Hitler's "Kommissar Befehl," early Mar. 1941, cited in Shirer, *The Rise and Fall of the Third Reich*, 830. See also Bartov, *Hitler's Army*, and Mayer, *Why Did the Heavens Not Darken?*

2. Aly and Heim, *Vordenker der Vernichtung*, 365. See also Gerlach, "Wirtschaftsinteressen, Besatzungspolitik und der Mord an den Juden."

3. Shirer, *The Rise and Fall of the Third Reich*, 833.

4. Mayer, *Why Did the Heavens Not Darken?*; Haffner, *The Meaning of Hitler*.

5. Heinrich Himmler, 30 Jan. 1942, "Richtlinien für die Planung und Gestaltung der Städte in den eingegliederten deutschen Ostgebieten," BDC Sammelliste 54.

6. For documentation within the SS, see Greifelt? (signature unrecognized), 4 Jan. 1939, "Vermerk," BAK NS19/1084; Seidler, *Fritz Todt*, 214–16, 239–72; Eichholtz and Lehmann, *Geschichte der deutschen Kriegswirtschaft*, 41–118.

7. Reichsminister im Auftrag, Dr. Timm to Pohl (as Ministerialdirektor), 2 Jan. 1940; Schnellbrief from Schulze-Fielitz (OT), 5 Dec. 1939; and Chef HAHB to Neubauleitung Flossenbürg, 27 Nov. 1939, BAK NS4/34 Fl. I would like to thank Paul Jaskot for pointing me to the BAK NS4 Flossenbürg collection and sharing his own research notes with me. Wegner, *Hitlers politische Soldaten*, 100–103; Koehl, *The Black Corps*, 166–68; Broszat, "Nationalsozialistische Konzentrationslager," 82–86.

8. Greifelt?, 4 Jan. 1939, "Vermerk," BAK NS19/1084. As armaments minister, Todt supported the SS in running conflicts with the Wehrmacht; for example, see Seidler, *Fritz Todt*, 252; Ulrich Greifelt, i. A. Himmler, to Wolff, Chef des Pers. Stab, 5 May 1939, "Vierjahresplan," BAK NS19/1084; Amtskommissar Stadt Auschwitz to ZBL Auschwitz, 7 July 1942, USHMM RG-11.001M.03: 23 (502-1-72); Derpa, Regierungsbaurat, Kattowitz to Bürgermeister of Auschwitz, 3 Jan. 1941, and Besprechung am 3.12.42 in Kattowitz, USHMM RG-11.001M.03: 23 (502-1-78).

9. Pohl to Wolff, 11 Sept. 1939, BAK NS19/1084. Meeting announced by Pohl to Wolff, 9 June 1939, "Einreihung der Bauten," BAK NS19/1084.

10. Lörner to Neubauleitung Auschwitz, 18 Mar. 1941, "Prüfungsbericht," USHMM RG-11.001M.03: 18 (502-1-1); List to all Neubauleitungen, 20 Nov. 1940, USHMM RG-11.001M.03: 18 (502-1-1); Bischoff to Bauinspektion Ost, 16 Apr. 1942, "Baupolizeiliche Behandlung," USHMM RG-11.001M.03: 23 (502-1-82). See also Seidler, *Fritz Todt*, 214–15.

11. Chef Amt II, Stbaf. Heidelberg to Neubauleitung Flossenbürg, 4 July 1940, BAK NS4/31. I first found the term "Neubauleitung" in Flossenbürg documents from December 1938. It does not appear in the Auschwitz collection until 1940. Karl Bischoff to Bauinspektion Ost, 16 Apr. 1942, "Baupolizeiliche Behandlung," USHMM RG-11.001M.03: 23 (502-1-82). The Reich Finance Ministry may have also made this shambles an object of its criticism; see NO-4007, Affidavit of Hubert Karl.

12. This has been pointed out to me by Michael Neufeld, for which I thank him.

13. Breitman, *The Architect of Genocide*, 171.

14. Van Pelt and Dwork, *Auschwitz, 1270 to the Present*, 140. The book was Walter Geisler's *Der deutsche Osten als Lebensraum für alle Berufsstände* (Berlin: Volk und Reich Verlag, 1942), 12.

15. Müller, *Hitlers Ostkrieg*, 86–89. The General Plan–Ost was written by Konrad Meyer but was predated by the RSHA's own version of the New Order in January 1940. Roth, " 'Generalplan Ost'—'Gesamtplan Ost,' " 25–82. Lublin and Globocnik are dealt with throughout Pohl, *Von der "Judenpolitik" zum Judenmord*. Van Pelt and Dwork, *Auschwitz, 1270 to the Present*, 236–55, 308–15, 327.

16. On the network of Superior SS and Police Führer (HSSPF), see Birn, *Die Höheren SS- und Polizeiführer*, 157–65; Müller, *Hitlers Ostkrieg*, 87, 220.

17. Pohl, *Von der "Judenpolitik" zum Judenmord*, 33–40, 99; Breitman, *The Architect of Genocide*, 94–104. I thank Joseph Poprzeczny for alerting me to Globocnik's pet name with Himmler.

18. Pohl, *Von der "Judenpolitik" zum Judenmord*, 99.

19. Marszalek, *Majdanek Konzentrationslager Lublin*, 17–18, quotation from 18.

20. Witte, "Slowakische Juden im Distrikt Lublin"; Pohl, *Von der "Judenpolitik" zum Judenmord*, 158–63.

21. Evans, *In Hitler's Shadow*, 88. Compare Maier, "A Holocaust Like All the Others?," 82.

22. NI-1553, CIOS Consolidated Advance Field Team VII, interview with Diplom-Ingenieur Kurt Gerstein.

23. Pohl, *Von der "Judenpolitik" zum Judenmord*, 48, 53, 63, in general 47–87; Aly, *Endlösung*, 207–21; Safrian, *Die Eichmann Männer*, 23–67.

24. Van Pelt and Dwork, *Auschwitz, 1270 to the Present*, 140, quoting Konrad Meyer.

25. Pohl, *Von der "Judenpolitik" zum Judenmord*, 40.

26. Breitman, *The Architect of Genocide*, 129–30, 136–37; Pohl, *Von der "Judenpolitik" zum Judenmord*, 57, 79–90; Black, "Odilo Globocnik—Himmlers Vorposten im Osten," 103–15. See also Odilo Globocnik, "SS- und Polizeistützpunkte," BDC Sammelliste 54. The report is not dated but is initialed by Himmler, 18 July 1941. Wasser, *Himmlers Raumplanung im Osten*, 47–71.

27. Allgemeine Anordnung des RKF, 26 Nov. 1940, cited after Aly, *Endlösung*, 208. Van Pelt and Dwork, *Auschwitz, 1270 to the Present*, notes throughout that the Nazis conceived themselves as the *culmination* of an imagined past, not as reactionaries who wished to return to it. In general, see Pohl, *Von der "Judenpolitik" zum Judenmord*.

28. Heinrich Himmler, 30 Jan. 1942, "Richtlinien für die Planung und Gestaltung der Städte in den eingegliederten deutschen Ostgebieten," BDC Sammelliste 54; Wasser, *Himmlers Raumplanung im Osten*, 83–91. On modern design, planning, and traditionalism, see Stein and Franzke, "German Design and National Identity, 1890–1914," 67–69; Kaplan, "Traditions Transformed."

29. The first quotation is from Odilo Globocnik, "SS- und Polizeistützpunkte," BDC Sammelliste 54; the second is from Wasser, *Himmlers Raumplanung im Osten*, 45 (see also 119–21).

30. Cited after van Pelt and Dwork, *Auschwitz, 1270 to the Present*, 140.

31. The first quotation is from Hanelt, 9 Aug. 1941, "Notiz für Brigadeführer [Globocnik], Stabsbesprechung am 6 Aug. 41"; the second is from an unsigned source, most

likely Hanelt, "Besprechung zwischen Brigadeführer Globocnik, Herrn Stadthauptmann Sauermann, Herrn Seebert, SS Hauptsturmführer Lerch, and Untersturmführer Hanelt am 3 Feb. 41," in USHMM RG-15.027.M:1:6 (Records of SS and Polizeiführer im Distrikt Lublin). See also Affidavit of Willy Mutert, 26 Oct. 1961, ZSL IV 4B AR-Z 82/1968, Band III; Hellmut Müller to RuSHA, 15 Oct. 1941. I thank Joseph Poprzeczny for supplying Hanelt's first name, Gustav, which does not appear in the sources cited here. Compare Herbert, "Rassismus und rationales Kalkül," 26; Browning, "Vernichtung und Arbeit," 37–52; and Frei, "Wie Modern war der Nationalsozialismus?," 368–74. See also Gerlach, "Wirtschaftsinteressen, Besatzungspolitik und der Mord an den Juden"; Marszalek, *Majdanek Konzentrationslager Lublin*, 15–34.

32. Aly and Heim, *Vordenker der Vernichtung*, 337–43.

33. Aly, "Theodor Schieder, Werner Conze oder die Vorstufen der physischen Vernichtung," 173.

34. Odilo Globocnik, "SS- und Polizeistützpunkte," BDC Sammelliste 54; Wasser, *Himmlers Raumplanung im Osten*, 83–91; Heinrich Himmler, 15 May 1942, "Auszug aus den Wirtschafts- und Verwaltungsanordnungen vom 15. Mai 1942 Nr. 3; Befehl über die Einrichtung von SS- und Polizeistützpunkten," BDC SS Personal-Akte Oswald Pohl, Band 1.

35. Himmler's Vermerk, 20–21 July 1941, BDC Sammelliste 54; NI-11314, Oswald Pohl, 4 Nov. 1941, "Aktenvermerk"; Himmler, "Auszug"; Testimony of August Frank, Protocol: 2364.

36. Himmler's Vermerk, 20–21 July 1941. See also Kammler's plans for *Bauinspektion* discussed later.

37. Himmler's Vermerk, 21 July 1941, and Eggeling, record of telephone conversation with Kammler and Karl Bischoff on 26 Sept. 1941, Vermerk, 1 Oct. 1941, USHMM RG-11.001M.03: 24 (502-1-85).

38. Marszalek, *Majdanek Konzentrationslager Lublin*, 19; van Pelt and Dwork, *Auschwitz, 1270 to the Present*, 300–315.

39. Himmler to Pohl, 19 Dec. 1941, BAK NS3/52, cited in Orth, *Das System der nationalsozialistischen Konzentrationslager*, 154–55. Documents regarding specific KLs: NO-2147, Maurer and Vogel, Reports of 11 and 17 Dec. 1941 and 9 Jan. 1942, esp. "Bericht über die am 10 Dec. 41 erfolgte Besichtigung des Zivil-Gefangenenlagers Stutthof und der sich dort befindlichen Werkstätten"; I. A. Maurer to Chef Amt II, W1, W4, W5, IKL Glücks, 24 Jan. 1942, "KL Stutthof," BAP, PL5: 42055; Pohl and Kammler to RFSS, 4 Dec. 1941, "Vorläufiges Friedensbauprogramm des HAHB, Amt II–Bauten. 2 Zusammenstellungen des Friedensbauprogrammes mit Übersichtsplan" and map, BAK NS19/2065; Wasser, *Himmlers Raumplanung im Osten*, 86, 129–30, 318–19; BDC Personal-Akte Joachim Caesar. Dr. Joachim Caesar transferred from the Race and Settlement Main Office for this purpose in early 1942. His wife was also a Ph.D. chemist, and both worked at the concentration camp.

40. Höss testimony, cited in Orth, *Das nationalsozialistische Konzentrationslager*, 78; Rudolf Höss, "Meine Psyche, Werden, Leben, u. Erleben," IfZ 13/1: 25; van Pelt and Dwork, *Auschwitz, 1270 to the Present*, 134; Wolschke-Bulmahn, "Biodynamischer Gartenbau, Landschaftsarchitektur und Nationalsozialismus," 593–94; Kranz, "Das KL-Lublin," 364–74; van Pelt and Dwork, *Auschwitz, 1270 to the Present*, 188–211.

41. Breitman, *The Architect of Genocide*, 136. Breitman renders this quotation in slightly different translation. Of note, he rendered "Das Wohn-Bau Programm," not inaccurately, as "housing construction." I have altered the second sentence slightly. "Wohnen" is not translatable directly into English. The original German reads "Das Wohn-Bau Programm, das die Voraussetzung fuer eine gesunde und soziale Grundlage der Gesamt-SS wie des gesamten Fuehrercorps ist, ist nicht denkbar." My sincere thanks to Richard Breitman for providing this information. See also Wasser, *Himmlers Raumplanung im Osten*, 86 and 318, reproduced from the city archive of Lublin, "Bauanzeige," [early 1942]; "Stadthauptmann Lublin to Gouverneur des Distrikts Lublin," 25 Nov. 1942; and "Be- und Entwässerung des Kriegsgefangenenlagers Lublin sowie Errichtung einer Faulgasgewinnungsanlage."

42. Wolff to Gen. Kastner-Kirdorf, Chef d. Luftwaffenpersonalamtes, 6 Mar. 1941, BDC SS Personal-Akte Hans Kammler; "Gutachten" from 31 May 1941 signed by the Reichsminister der Luftfahrt und Oberbefehlshaber der Luftwaffe, Luftwaffenverwaltungsamt, BDC SS Personal-Akte Hans Kammler. Kammler must have agreed to work for the HAHB immediately (for his official personnel card lists him as an officer of the Amt II already in August 1940).

43. Testimony of Oswald Pohl, Protocol: 1281–82.

44. Speer, *Slave State*, 12.

45. Affidavit of Heinrich Werner Courte, Defense Document Book of Max Kiefer.

46. BDC SS Personal-Akte Hans Kammler, esp. Lebenslauf. Kammler must have been aware of the SS's Berlin, Krumme Lanke settlement in this capacity, for it lay in his region of Zehlendorf.

47. Hans Kammler, "Zur Bewertung von Geländeerschliessungen für die grossstädtische Siedlung" (Engineering diss., Hanover, 1931), "Bibliographischer Anhang."

48. Lebenslauf, BDC Personal-Akte Hans Kammler.

49. See the discussion of Speer's and Joachim Fest's construction of the Nazi technocrat as a literary trope in Van der Vat, *The Good Nazi*, 330–32; Agoston, *Teufel oder Technokrat?*

50. Edgar Hotz and Hans Kammler, *Grundlagen der Kostenrechnung und Organization eines Baubetriebs für den Wohnungs- und Siedlungsbau in Stadt und Land* (Berlin: Verlagsgesellschaft R. Müller, 1934), 1; Kammler, "Zur Bewertung von Geländeerschliessungen," 37. See the anonymous report, 22 May 1943, "Besprechung mit dem Amtsgruppenchef C. Besuch am 21.5.43," USHMM RG-11.001M.03: 20 (502-1-25).

51. Amt II to Neubauleitung Flossenbürg, 10 Aug. 1940, BAK NS4/59. This slur was apparently common in the building trades. Wolschke-Bulmann, "Biodynamischer Gartenbau," 640.

52. Hotz and Kammler, *Grundlagen*, 1.

53. Ibid., vii.

54. Neumann, *Behemoth*, 472.

55. Durth, *Deutsche Architekten*, 23–40, 65–86, 102–3; Ludwig, *Technik und Ingenieure*, 20–35, 105–9; Wengenroth, "Zwischen Aufruhr und Diktatur," 215–23; Gispen, *New Profession, Old Order*, 48–51; Bolenz, *Vom Baubeamten*, 33–39, and regarding housing, 85–86.

56. Durth, *Deutsche Architekten*, 23–40, 65–86; Ludwig and König, *Technik, Inge-*

nieure und Gesellschaft; Ludwig, *Technik und Ingenieure*, 20–35, 105–9; Hortleder, *Das Gesellschaftsbild des Ingenieurs*, 111. James, *The German Slump*, 148, points out that the lack of any vision of the future was also a general sentiment among businessmen and financiers. Prinz, *Vom Mittelstand zum Volksgenossen*, 134–36.

57. James, *The German Slump*, 370–72, and in general 343–419; Overy, " 'Blitzkriegs-wirtschaft'?," 370–435. For an excellent example of one consumer-durable industry, see Sudrow, "Das 'deutsche Rohstoffwunder' und die Schuhindustrie," 63–92. Prinz, *Vom Mittelstand zum Volksgenossen*, 175–81.

58. Hotz and Kammler, *Grundlagen*, 41–47.

59. Quotations here and earlier from ibid., 69. For elucidation of traditional administration, see Koselleck, *Preußen zwischen Reform und Revolution*, 663–71.

60. Testimony of Karl Fanslau, Protocol: 2562; Testimony of Oswald Pohl, Protocol: 1303–5, 1743; NO-1922, Affidavit of Max Kiefer; Affidavit of Heinz Schürmann, Defense Document Books of Franz Eirenschmalz; NO-2613, Affidavit of Franz Eirenschmalz; Kammler, end of Dec. 1941, "Bericht des Amtes II–Bauten über die Arbeiten im Jahre 1941," USHMM RG-11.001M.03: 19 (502-1-13); Kammler, "Hauptamt Haushalt und Bauten Amt II–Bauten. Geschäftsverteilungsplan vom 20.6.41," USHMM RG-11.001M:19 (502-1-9).

61. Elektro- und Radiogroßhandlung Gebrüder Schwiger GmbH to Mayrl, 27 Jan. 1939, BAK NS4/48 Fl.

62. Kammler to Hauptabteilungen, Zentralbauleitungen, 22 Aug. 1941, "Mittelbe-willigungen," USHMM RG-11.001M.03: 18 (502-1-1).

63. For examples of Kammler's supervision in action, see NO-1922, Affidavit of Max Kiefer. See also Kammler's first reorganization of the New Construction Directorates: Kammler, 20 June 1941, "Organization. Amtsbefehl 3.," USHMM RG-11.001M.03: 19 (502-1-11); Kammler to sämtliche SS Baudienststellen, 5 Nov. 1942, "Bauunterhalts- und Baubedarfsanträge," BAK NS4/59, BAK NS4/29 Fl; Kammler to Zentralbauleit-ungen, Zentrale Bauinspektion Lublin, 21 Aug. 1941, "Richtlinien über Vergabewesen und Rechnungslegung für das Bauwesen im Dienstbereich," USHMM RG-11.001M.03: 18 (502-1-1); Kammler to Zentralbauleitungen, Bauleitungen und Hauptabteilungen HAHB, 22 July 1941, "Formulare im Bauwesen der SS," USHMM RG-11.001M.03: 18 (502-1-1); Obstbaf Kögel, undated, "Telefonverzeichnis," BAK NS4/5 Fl.

64. Kammler to Höss, 18 June 1941, "KL Auschwitz Baumaßnahmen 2. und 3. Kriegs-wirtschaftsjahr," USHMM RG-11.001M.03: 19 (502-1-11).

65. NO-1292, Kammler to Glücks, 10 Mar. 1942, "Einsatz von Häftlingen, Kriegs-gefangenen, Juden usw. für die Durchführung des Bauprogrammes des SS-WVHA, Amtsgruppe C 1942."

66. See Kammler to Kommandanten des KL Auschwitz, 9 June 1942, "Bauvorhaben im 3. Kriegsjahr," USHMM RG-11.001M.03: 19 (502-1-9); Kammler, 25 June 1943, "Auf-stellung einer Baubetriebsdienststelle bei der Verwaltung des KL Auschwitz," and Kammler, Chef Amtsgruppe C6, 30 Mar. 1943, "Vorläufige Dienstanweisung für Bau-Betriebsdienststellen der W-SS und Polizei," USHMM RG-11.001M.03: 18 (502-1-1).

67. Himmler to Richard Glücks, 26 Jan. 1942, BDC Hängeordner 643; see also Arthur Liebehenschel to all KLs, 19 Jan. 1942, "Überstellung von Juden," printed in Harry Stein, *Juden in Buchenwald* (Weimar: Gedenkstätte Buchenwald, 1992), 119. This decree orders

the immediate transfer of the "number of Jews able to work to the POW Camp Lublin as reported by teletype." It mentions a teletype from 8 December. Note this predated serious talks of KL armaments works with Speer or Walther Schieber. I thank Peter Witte for providing this document. Testimony of Oswald Pohl, Protocol: 1303–5. See also Wasser, *Himmlers Raumplanung im Osten*, 86; Freund, Perz, and Stuhlpfarrer, "Der Bau des Vernichtungslagers Auschwitz-Birkenau," 187–214; Gerlach, "Wirtschaftsinteressen, Besatzungspolitik und der Mord an den Juden," 276; Kranz, "Das KL-Lublin," 367, 372.

68. Karny, " 'Vernichtung durch Arbeit.' Sterblichkeit in den NS-Konzentrationslagern," 133–58; Rückerl, *NS-Vernichtungslager*, 126–29; Browning, "The Semlin Gas Van," 57–67, 68–87; Gerlach, "Die Wannsee-Konferenz," 7–44.

69. Compare Pohl and Kammler to RFSS, 4 Dec. 1941, "Vorläufiges Friedensbauprogramm des HAHB," BAK NS19/2065; Kammler, 14 Aug. 1941, "Polizeistützpunkt Typ A" and "Typ B," countersigned by Pohl the next day and also, on 18 August, by Kurt Daluege, WAP Lublin Inventarz Zentralbauleitung Lublin: folders 237 and 238.

70. Kammler to Himmler, 10 Feb. 1942, "Aufstellung von SS-Baubrigaden für die Durchführung von Bauaufgaben des RFSS im Kriege und Frieden," BAK NS19/2065; Heinrich Himmler, 30 Jan. 1942, "Richtlinien für die Planung und Gestaltung der Städte in den eingegliederten deutschen Ostgebieten," BDC Sammelliste 54.

71. Himmler's notes from 12 Mar. 1942 on Kammler to Himmler, "Aufstellung."

72. Regierung des Generalgouvernements, Hauptabteilung Innere Verwaltung, Dr. S/H, Vermerk, 27 Mar. 1942, and Untersturmführer Dr. W. Gradmann, Vermerk, 19 Mar. 1942, printed in Madajczyka, *Samojszczyzna—Sonderlaboratorium SS*, 66, 53–60.

73. Kammler to Himmler, 10 Feb. 1942, "Aufstellung von SS-Baubrigaden für die Durchführung von Bauaufgaben des RFSS im Kriege und Frieden," BAK NS19/2065.

74. Quotations from Himmler reply to Pohl and Kammler, 31 Jan. 1942, BAK NS19/2065. Note the total absence of preoccupation with SS armaments factories. Pohl's and Kammler's disputes with Himmler over minor details are dealt with at greater length in Allen, "Engineers and Modern Managers in the SS," 324–29.

75. Oswald Pohl, and signed by August Frank, 27 Jan. 1942, "Befehl des Hauptamtschefs: Neuorganisation," T-976/36; NO-495, Pohl, 19 Jan. 1942, "Organisation of the Administration"; organizational chart dated 15 Jan. 1942. See also NO-555, Karl Niemann, Mar. 1944, "Business Report of the Deutsche Ausrüstungswerke GmbH for the year 1943."

76. BDC SS Personal-Akte August Frank; BDC SS Personal-Akte Dipl. Kauf. Georg Lörner; Testimony of August Frank, Protocol: 2326–28, 2356–60. The Office Group A allowed other branches of SS service to raid its staff. Even its chief, Frank, left to work at the behest of the Wehrmacht, where he revamped the army's entire payroll system in 1944 (another indication that the relationship between mid- and even top-level officials of Reich authorities were less split by "polycratic" conflict than held together by mutual exchange). The Wehrmacht also thanked Frank profusely. NO-498, Pohl, 19 Nov. 1942, "Organization order for external branches of the WVHA, which are *not* under the jurisdiction of an SS-Wirtschafter."

77. Note the absence of discussion of armaments factories or even the upcoming Office Group D–Labor Action in Pohl, "Befehl des Hauptamtschefs" of 19 Jan. 1942 (T-976/36), and in his similar directive to the SS companies of 27 Jan. 1942 (NO-495). To

believe that the WVHA was reorganizing to pursue armaments factories at this time one would have to believe that SS bureaucrats were "thoughtless" indeed, so "thoughtless" as never to mention this supposedly essential goal in their most important internal correspondence. Breitman, *The Architect of Genocide.*

78. Quotation from Kammler, 17 Nov. 1941, "Organization der SS Baudienststellen" (prepared for Pohl), USHMM RG-11.001M.03: 19 (502-1-12), which contained "Dienstanweisungen für den Leiter einer Bauinspektion," "Dienstanweisung für den Leiter einer ZBL," and "Dienstanweisung für den Leiter der Bauleitung."

79. Werner Jothan to all Bauleitungen, 2 Aug. 1944, "Geschäftsbetrieb bei den ZBL mit den angegliederten Bauleitungen und Abteilungen," USHMM RG-11.001M.03: 24 (502-1-84).

80. Kammler, evaluation of Franz Eirenschmalz, 17 Nov. 1944, BDC SS Personal-Akte Franz Eirenschmalz.

81. Kammler evaluation, 26 Apr. 1944, BDC SS Personal-Akte Robert Riedl.

82. Quotation from BDC SS Personal-Akte Karl Hoffmann. Walter Dejaco, who oversaw the construction of crematoria in Auschwitz, sat for five months in an Austrian prison. Like Gerhard Weigel, he engaged in street fights for the movement. The architect Rudlieb Görcke helped found the Schlageter Bund, a Nazi club, in 1928. Leo Schlageter, the club's icon, had been shot by the French for sabotage in the Ruhr; he was a Nazi martyr figure like Horst Wessel. Also, Wilhelm Lenzer, leader of a Building Inspection, led a Hitler Youth troop. Others merely drew attention for less overt acts of sympathy, like Professor Hans Schleif, who conspicuously insisted on appearing in SS uniform before his colleagues. See BDC SS Personal-Akte Walter Dejaco, Rudlieb Görcke, Wilhelm Lenzer. See also Führer of the Nationalsozialistischer Dozentenbund und der Dozentenschaft der Friedrich Wilhelms Universität to the Rektor, 21 June 1937, BDC Reichs- und Preußisches Ministerium für Wissenschaft, Erziehung und Volksbildung-Akte Hans Schleif.

83. BDC Personal-Akte Heinrich Courte.

84. BDC RuSHA Akte Heinrich Courte.

85. Hortleder, *Das Gesellschaftsbild des Ingenieur,* 111; Hartung, "Bauästhetik im Nationalsozialismus," 71–84; Zeller, "Landschaften des Verkehrs," 323–40; Seidler, *Fritz Todt.*

86. Regarding Bischoff's altercation with his immediate superior Gustav Rall, see the correspondence Bischoff to Kammler, Rall to Bischoff, and Kammler to Rall, July–Sept. 1942, USHMM RG-11.001M.03: 23 (502-1-82).

87. See Allen, "The Puzzle of Nazi Modernism."

88. On the ongoing feud between the municipality of Auschwitz and the SS building authority at the concentration camp, see Polenz, 3 Dec. 1942, "Besprechung am 3.12.42 in Kattowitz," USHMM RG-11.001M.03: 23 (502-1-78); Amtskommissar Butz der Stadt Auschwitz to ZBL Auschwitz, 7 July 1942, USHMM RG-11.001M.03: 23 (502-1-72); and same to ZBL Auschwitz, 13 July 1942, with Bischoff's reply, 13 July 1942, USHMM RG-11.001M.03: 23 (502-1-76).

89. BDC Personal-Akte Karl Bischoff. He was *Zellenleiter* from 1936 to 1938 in Munich. Smelser, *Robert Ley,* 155, and on DAF ideology of organization, 98–179 in general.

90. Ortsgruppenleiter der NSDAP bei Kommandantur Auschwitz to Bischoff, 24 Apr. 1942, and reply of 29 Apr. 1942, USHMM RG-11.001M.03: 25 (502-1-99).

91. The first quotation is from Karl Bischoff to HUTA AG, 18 Mar. 1943, "Verhaengung von Strafen ueber Zivilarbeiter," USHMM RG-11.001M.03: 22 (502-1-58); the second is from Werner Jothann to Gestapo Auschwitz, 2 Aug. 1944, "Wiederbeibringung von 2 auslaendischen Arbeitern," USHMM RG-11.001M.03: 22 (502-1-70).

92. Seibel, "Staatsstruktur und Massenmord," 539–69.

CHAPTER FIVE

1. Herbert, *Fremdarbeiter*, 138–51; Eichholtz and Lehmann, *Geschichte der deutschen Kriegswirtschaft*, 1–47, 179–226; Janssen, *Das Ministerium Speer*, 89–90; Ludwig, *Technik und Ingenieure*, 383.

2. Herbert, *Fremdarbeiter*, 135–49; Eichholtz and Lehmann, *Geschichte der deutschen Kriegswirtschaft*, 187–88; Streit, *Keine Kameraden*, 207–16; Gerlach, *Krieg, Ernährung, Völkermord*, 33–35, 10–84 generally.

3. Seidler, *Fritz Todt*, 239–72; Ludwig, *Technik und Ingenieure*, 189–98; Eichholtz and Lehmann, *Geschichte der deutschen Kriegswirtschaft*, 41–118; Milward, *The German Economy at War*, 54–71; Overy, *Why the Allies Won*, 180–244.

4. Fröbe, "KZ-Häftlinge als Reserve qualifizierte Arbeitskraft," 657–58.

5. Hayes, *Industry and Ideology*; Orth, *Das System der nationalsozialistischen Konzentrationslager*, 175–79.

6. Hayes, *Industry and Ideology*, 350–54; NI-1240, Göring to Himmler, 18 Feb. 1941. These prisoners were intended only for construction at the time. Setkiewicz, "Häftlingsarbeit im KZ Auschwitz III–Monowitz," 586–87. See also Pingel, *Häftlinge unter SS-Herrschaft*, 145–48.

7. Mommsen and Grieger, *Das Volkswagenwerk*, 499–507, 543; Grieger, " 'Vernichtung durch Arbeit' in der deutschen Rüstungsindustrie," 48–50; Budraß and Grieger, "Die Moral der Effizienz," 99–101. Compare Adolf Hitler to Himmler, Porsche, Jakob Werlin, 11 Jan. 1942, "Führer Anordnung," T-175/70. The text reads, "I approve the suggestion of party member Professor Dr. Porsche and my deputy, party member Werlin, to transfer the construction, finishing work, and operation of this foundry to the Reichsführer SS and Chief of German Police, who will make available workers from the concentration camps for the project." NO-421, Walther Schieber, 17 Mar. 1942, "Vermerk: Ausnutzung des KZ-Lagers Neuengamme," states that the dealings over joint armaments ventures with the SS at that time had already begun "a few months ago," but mentions no exact date or meeting; these could, however, be only the Porsche or Steyr-Daimler-Puch negotiations. Compare Chef des Waffen- und Geräteamtes der Waffen-SS to RFSS, 9 May 1940, "Beschaffung von Waffen und Gerät für die Waffen-SS" and "Denkschrift," T-175/104, which discuss the congeniality of Todt toward the W-SS.

8. Budraß and Grieger, "Die Moral der Effizienz," 97–99; Mommsen and Grieger, *Das Volkswagenwerk*, 500–501, 496–515 in general; NO-1287, Hohberg, Memorandum of 31 Jan. 1942, "Foundry Volkswagenwerk Fallersleben"; Dr. Hohberg to Grf. Pohl, Professor Porsche, Obf. Kammler, 29 Jan. 1942, "Erlass des Führers vom 11 Jan. 42 re: Einrichtung einer Giesserei in Fallersleben (VW)," T-976/36; NO-595, Kammler, Memorandum of 3 Feb. 1942. Mommsen's analysis here bears some scrutiny. The original agreement between Himmler and Porsche codified in Hitler's decree of 11 January did call for the SS to run the Fallersleben foundry. However, the SS's representatives Pohl,

Kammler, and Hans Hohberg quickly abandoned this when Porsche insisted on undisputed control of factory operations. The SS would supply only labor, construction supervision, and security for the prisoners. Hans Mommsen maintains that Pohl "followed the point of view that the SS should take over not only the finishing construction but also the operation of the light-metals foundry. Porsche disputed this with great emphasis" (Mommsen and Grieger, *Das Volkswagenwerk*, 499–500). Both Grieger and Mommsen cite Hohberg's *Vermerk* of 29 January 1942 to support this. The exact language is therefore instructive. "SS Gruppenführer Pohl presented [*erläutert*] the instructions of the Führer, in whose authority the Schutzstaffel is supposed to undertake the finishing construction and the operation of a light-metals foundry at the Volkswagenwerke in Fallersleben." This was hardly an aggressive move. Pohl merely presented or explained (*erläutern*) the Führer's decree, as one would expect anyone to do who opened such business discussions. The memo, taken by the SS's own personnel, does not mention that Pohl thereafter continued to insist that the SS own the means of production and operate the foundry independently. If this had truly been the SS's goal, would SS internal memoranda keep it secret or fail to express disappointment when it had to be abandoned? In other memos, about settlements, for example, the SS was quite candid about its goals. Porsche, as both Mommsen and Grieger acknowledge, offered the 4,000 amphibious vehicles at this meeting. They insist that this was done as a political maneuver, a *Gegenleistung*, in order to stave off Pohl's demands. If this were so, Pohl did not need much convincing. He knew full well that the SS did not have the personnel, the skill, or the interest to run a light-metals foundry. Again Hohberg's record: "It came to our attention that there is an unclarity in the position of the tasks before us. Professor Porsche expressed that the Schutzstaffel should make available workers and quantities of materials, but the operation of the light-metals foundry is to be the work of the Volkswagen. Regarding the text of the [Führer's] order there has already been a conflict with Dr. Ley [head of the DAF]." The subjunctive in the original German indicates that Porsche, not the SS, had run into conflict with Ley. Pohl sought immediately to assuage matters: "SS-Gruppenführer Pohl made known that first of all clarity shall be reached about this, about who is to operate the works. Gruppenführer Pohl will make it his concern to secure an eventual new draft of the Führer's order." These were hardly the words of someone insisting that the SS expand its industrial capacity into light-metals foundries against the wishes of German armaments moguls like Porsche. The tone was one of solicitous cooperation. Perhaps as yet unknown evidence exists in the VW archives, the archives of the DAF, or other sources which prove that the SS wanted and insisted on running the foundry. Unless the authors produce such evidence, their interpretation on this point cannot be considered convincing.

9. Perz, *Projekt Quarz*, 81–84, and "Der Arbeitseinsatz im KZ Mauthausen," 535–36. Orth, *Das System der nationalsozialistischen Konzentrationslager*, 147–48, speculates that the first impulse to use labor at Steyr-Daimler-Puch came from Ziereis, but evidence cited by Perz from early 1942 documents only Meindl's initiative in the matter.

10. Orth, *Das System der nationalsozialistischen Konzentrationslager*, 175–79; Brenner, "Der 'Arbeitseinsatz' der KZ-Häftlinge in den Außenlagern des Konzentrationslagers Flossenbürg," 682–706.

11. Spoerer, "Profitierten Unternehmen von KZ-Arbeit?," 84. Compare Schröder,

"Das Erste Konzentrationslager in Hannover," 52–54, as well as Dokumentenanhang, 3-5, pp. 590–93, from the same volume, which mentions above all the initiative of Walther Schieber.

12. In terms of the criminal actions of Nazi industry, the SS has been used as exactly the kind of alibi that Gerald Reitlinger first put his finger on in *The SS, Alibi of a Nation*. See also Janssen, *Das Ministerium Speer*, 99. One of the few to recognize the ideological motivations of the SS in their full breadth has been Roth, " 'Generalplan-Ost'—'Gesamt-plan Ost,' " 80 and n.

13. Seidler, *Fritz Todt*, 365–91.

14. Speer, Chronik der Dienststellen, Feb.–Mar. 1942, BAK R3/1736, hereafter cited as Chronik; Janssen, *Ministerium Speer*, 60–96; Eichholtz and Lehmann, *Geschichte der deutschen Kriegswirtschaft*, 57–58, 200–204; Ludwig, *Technik und Ingenieure*, 420–62.

15. Janssen, *Ministerium Speer*, 63.

16. NO-2448, Gottlob Berger to Himmler, 22 Apr. 1942; NO-421, Walther Schieber, 17 Mar. 1942, "Vermerk: Ausnutzung des KZ-Lagers Neuengamme"; Janssen, *Das Minis-terium Speer*, 97–98; NO-2468, Schieber to Saur, 20 Mar. 1942. Later, the Armaments Ministry added a flack gun at Auschwitz, a signaling device at Ravensbrück, and pistol assembly at Neuengamme. Some stress the legitimate grounds of German belief in final victory in 1944. See Salewski, "Die Abwehr der Invasion als Schlüssel zum 'Endsieg,' " and, most brilliantly, Weinberg, "German Plans for Victory, 1944–1945." But compare Wegner, "Defensive ohne Strategie," and Schabel, "Wenn Wunder den Sieg bringen sollen."

17. Speer, *Slave State*, 174; see also 3, 28. On the publication of this book, see Van der Vat, *The Good Nazi*, 364–65.

18. Speer, *Slave State*, 4.

19. Ibid., 8.

20. Speer, *Der Sklavenstaat*, 102–21; Ludwig, *Technik und Ingenieure*, 205; Eichholtz and Lehmann, *Geschichte der deutschen Kriegswirtschaft*, 85, 95–97, 147–48; Janssen, *Ministerium Speer*, 41; BDC SS Personal and RuSHA-Akten Walther Schieber; NI-568 and NI-9292, Interrogation of Walther Schieber. See the warm praise for Schieber's efforts in the Chronik, 4 Sept. 1942. For Speer's own contemporary Nazi punditry, Speech of 10 Jan. 1944, Tagung der Gau- und Kreispropagandaleiter in Berlin, "Wieder-aufbauplanung und Rüstung" in the Chronik. Speer also lied about (among other things) his close cooperation with the *Sicherheitsdienst* of the SS, which he engaged to enforce policy and conduct espionage for his ministry: Chronik, Sept.–Oct. 1943. Later Schieber was investigated by the SS for industrial graft. Some historians, notably Gregor Janssen, have used this as evidence that the SS tried to "infiltrate" Speer's offices by torpedoing Schieber. However, the incident was precipitated by the actual graft of Schieber's brother in Thüringen. The investigation of Schieber was likely promoted by Fritz Sauckel and was used against Speer not by Himmler but by Bormann. Schieber actually requested an SS investigation *by the SS* to clear his own name, and Speer cooperated with Himmler in the matter. Schieber wished to prove that he had acted "honorably" as "an SS man" (his words). See the correspondence in 1944 in BAK R3/1631, especially Schieber's resignation letter of 31 Oct. 1944 and Speer's acceptance 10 Nov. 1944. Schieber stayed in Speer's service as a consultant.

21. Schieber's resignation letter of 31 Oct. 1944, BAK R3/1631. See also Walther Schieber, "Kartoffelkraut-Solanum," *Der Vierjahresplan* 4 (1940): 695.

22. Schieber to Gruppenführer Staatsrat Pg. Hennicke, 14 Feb. 1938, T-175/30.

23. Himmler to Pohl, 23 Mar. 1942, BAK NS19/2065; NO-2167, Glücks to all KL Schutzhaftlagerführer "E" (Einsatz), 20 Feb. 1942, BDC SS Hängeordner: 1825; Pohl to Himmler, 14 Feb. 1942, T-175/129.

24. Compare Herbert, "Arbeit und Vernichtung," 218–19. Similarly, Koehl, *The Black Corps*, 171. Compare Testimony of Pohl, Protocol: 1279–80. The order of 3 March is referred to by Pohl, in 13 Mar. 1942, "Order Nu. 10," Defense Document Books of Oswald Pohl, and confirmed by the order of the Chef of the FHA, Jüttner, NO-3169, 16 Mar. 1942, "Transfer of authority concerning concentration camps." The first organizational chart of Office D is dated from 3 March 1942, with Himmler's signature, showing that Pohl took immediate action, at least on paper. Speer's Chronik records that Speer first met with Fritz Sauckel on 12–13 March 1942, that is, after Himmler had already long been in contact with the Armaments Ministry regarding concentration camp labor. Herbert actually cites documents that mention polycratic *cooperation* at this time (in n. 90, NO-3169, Chef d. FHA, 16 Mar. 1942). See also NO-569, Saur, conference memo, 16 Mar. 1942, "Transfer of armament production to concentration camps." The list of participants is given in NO-2468, Niederschrift of 17 Mar. 1942, "Verlegung von Rüstungsfertigung in Konzentrationslager": Glücks, Schieber, Oberstlt. v. Nicolai, Major Schaede, Direktor Dr. Stallwaag, and Saur. Pohl was not present. The SS was being solicited as a partner here and reacted hastily to oblige.

25. Pohl to Himmler, 30 Apr. 1942, "Eingliederung der IKL," R-129; Oswald Pohl, Protocol: 1798–1800.

26. Pohl to IKL, all Lagerkommandanten, 11 Mar. 1942, and Glücks, "Stabsbefehl Nr. 1" as Amtsgruppenchef D in WVHA, came on 16 Mar. 1942, but was merely a copy of Pohl's organizational plan with Glücks's signature on it, BAP, PL5: 42055, a typical sign of the latter's laziness. Pohl, "Order Nu. 10," Defense Document Books of Oswald Pohl; Pohl to Amt D, Lagerkommandanten, Werkleiter, Amt W, 30 Apr. 1942, T-967/36; PS-1234, Arthur Liebehenschel to Kommandant Groß-Rosen, 26 Mar. 1942; Affidavit of Anton Kaindl, Defense Document Books of Rudolf Scheide; BDC SS Personal-Akten Gerhard Maurer and Enno Lolling and Maurer affidavit in Defense Document Books of Oswald Pohl; NO-719, Himmler to Pohl, 29 May 1942, "Incorporation of the Inspectorate of Concentration Camps into the SS WVHA," *Trials of War Criminals*, 5:301–2.

27. Quotation from NO-2549, Scheiber, 20 Mar. 1942, "Verlegung von Rüstungsfertigung in Konzentrationslager." See also Affidavit of Anton Kaindl, Defense Document Books of Rudolf Scheide.

28. NO-2549, Scheiber, 20 Mar. 1942, "Verlegung von Rüstungsfertigung in Konzentrationslager."

29. Speer, *Slave State*, 5. See also NI-9292, Interrogation of Walther Schieber.

30. R-129, Pohl's report to Himmler, 30 Apr. 1942. The meeting was held 23–24 April 1942.

31. NO-1994, Pohl to Himmler, 28 July 1942, and Brandt to Pohl, 23 Aug. 1942, "Reassignment and detachment of commanders of the concentration camps," IMT: 306;

Segev, *Die Soldaten des Bösen*, Künstler: 88–89, Roedl: 164–69; Orth, *Die Konzentrationslager-SS*, 205–13.

32. Orth, *Die Konzentrationslager-SS*, quotation from 167; on Suhren generally, 165–70; on Hoppe, 115–24, 217–20; Hassebroek, 118–24; Kramer, 103–4; Gideon, 211–13. See also Segev, *Soldaten des Bösen*, 205–15, 233; Richardi, *Schule der Gewalt*, 136–38. On Kaindl, see Tuchel, *Konzentrationslager*, 258–59, 263–64, and BDC Personal-Akte Anton Kaindl and Affidavit, Defense Document Books of Rudolf Scheide.

33. "Dienstvorschrift für die Gewährung von vergünstigungen an Häftlinge gültig ab 15 May 43," BAP, PL5: 42056.

34. Maurer's memorandum to Kommandanten, 27 July 1943, "Bewachung der Häftlinge," BAP PL5: 42053, under the header of Office D1, thus probably in cooperation with Arthur Liebehenschel. This included new "Lagerordnung für Häftlinge," NO-517.

35. Birn, *Die Höheren SS- und Polizeiführer*, 89–124, 316–22; Browning, "Vernichtung und Arbeit," 37–52; Pohl, *Von der "Judenpolitik" zum Judenmord*, 157–65; NO-2128 a and b, Oswald Pohl to Heinrich Himmler, 27 and 23 July 1942, "Enforcement Regulations for the RFSS-Decree of 18 Jun. 42"; "Durchführungsbestimmungen zum RFSS-Befehl vom 18.6.42," BAK 19/14.

36. The average age of the SS economic Führer was forty-six, and only one was born after the turn of the century. Three were over fifty. NO-2128 a and b, Oswald Pohl to Heinrich Himmler, 27 and 23 July 1942, "Enforcement Regulations for the RFSS-Decree of 18 Jun. 42." The six: Eduard Bachl, Otto Bonnes, Konrad Breuer, Rudolf Klotz, Erich Schellin, Josef Spacil. Later in the war, Pohl would replace some of them with a new set of handpicked men chosen from his business enterprises. See BDC SS Personal-Akte Max Horn and Hanns Bobermin. This was far too late. Birn, *Die Höheren SS- und Polizeiführer*, 124, claims that the position of the SS Wirtschafter within the HSSPF gave Pohl leverage, but her more general claim that the Main Offices of the SS could exercise only limited authority within the domain of the HSSPF proves more correct in this case than she thinks. In practice, the SS Wirtschafter did not give Pohl a strong position, although they were intended to.

37. Erich Schellin, Lebenslauf, BDC SS Personal-Akte. He accused "Social Democr. Potato Buyers" of ruining a former business.

38. Report of Gerichtsoffizier of the HAHB, 8 July 1941; Pohl to Schellin, 30 July 1941; but see Pohl's glowing accounts of Schellin's work from 3 Sept. 1937, all in BDC SS Personal-Akte Schellin.

39. Strafverfahren report of Dr. Herbert Seifert, 8 June 1943, BDC SS Personal-Akte Schellin.

40. I translate based on my estimation of how such insults might be phrased in American vulgarity. The first insults were "Wasserköpfe" and "Dummköpfe." The more salacious insult is one that does not make sense even in the original German: "Sie [haben] jeden Abend solch junges Mädchen bei sich . . . sagen Sie mal, Sie haben wohl einen Schwanz wie eine Kuh, dass Ihnen alle Frauen nachlaufen, oder?" Because "cows" don't have penises, it is hard to know whether this is a misquote or whether Schellin was as incompetent at anatomy and deficient in wit as he was in other matters. Einstellungsverfügung from 16 Sept. 1943, BDC SS Personal-Akte Schellin.

41. Schellin to Gruf. von Herff, 28 Sept. 1943, BDC SS Personal-Akte Schellin.

42. SS-Richter's letter to SS and Polizeigericht Krakau, 9 Sept. 1943, BDC SS Personal-Akte Schellin.

43. NO-557, Gerhard Maurer to Pohl, 15 Apr. 1941, "Monatsbericht für Monat März 1941." See also Georg, *Die wirtschaftlichen Unternehmungen der SS*, 61; BDC Personal-Akte Maurer and undated, postwar "Lebenslauf," BAP PL5: 42063; NO-495, Pohl's Organizational Chart, 3 Mar. 1942; Pohl to Salpeter as Chef Amt III A, 31 Jan. 1940, and Hohberg to Pohl, 18 Sept. and 31 Jan. 1940, "Übernahme der wirtschaftlichen Betriebe der SS in Dachau in die Deutschen Ausrüstungswerke," all in BAP PL5: 42055; NI-15149, 1 Apr. 1941, "Bericht über eine Besprechung im KZ Auschwitz 28 Mar. 41"; Affidavits of Gerhard Maurer and Karl Sommer, BAP PL5: 42063; BDC Personal-Akte Gerhard Maurer and undated, "Lebenslauf," BAP PL5: 42063; Dr. Wilhelm Schneider, 6 Sept. 1939, "Urkundenrolle," NS3/435; NO-1202, Affidavit of Hans Moser; NI-1065, Affidavits of Karl Sommer and Gerhard Maurer in Defense Document Books of Karl Sommer; Paul Zapke, 1 Aug. 1952, "In der Reichsschutzsache Gerhard Maurer," BAP PL5: 42064. On Pohl's rubber-stamping of Maurer's initiatives: NO-1290, Oswald Pohl, 22 Jan. 1943, "Working Time of the Prisoners." See BDC Personal-Akte Karl Sommer and NO-1578, NO-2739, Affidavits of Karl Sommer.

44. Affidavits of Gerhard Maurer, BAP PL5: 42063; Maurer to all KL Kommandantur, 24 June 1942, "Häftlingseinsatztagung 1 Jun. 42," BAP PL5: 42055. For Burböck's organization, see NO-718, Burböck to all Schutzhaftlagerführer "E," 28 Nov. 1941, "Assignment of internees' detachments; Regulation IKL dated 14 Oct. 41 Par. E," and compare chapter 3. Although Glücks ordered the dissolution of the regional *Schutzhaftlagerführer "Einsatz"* at the end of 1941, these were now reinstated as Labor Action Führer. Maurer to all Kommandanten and Arbeitseinsatzführer, 21 Nov. 1942, "Übersicht über Anzahl und Einsatz der Häftlinge im KL," BAP PL5: 42055; Testimony of Oswald Pohl, Protocol: 1315; NO-1566, Affidavit of Hanns Bobermin; Interrogation of Gerhard Maurer, BAP PL5: 42063. Compare Allen, "The Banality of Evil Reconsidered," 253–94.

45. Maurer to all KL Kommandantur, 24 June 1942, "Häftlingseinsatztagung 1 Jun. 42," BAP PL5: 42055.

46. Tuchel, *Konzentrationslager*, 28; see also 286–89. See also Pingel, *Häftlinge unter SS-Herrschaft*, 12–33, 150. Compare Alfred Chandler's hallmark example of modern management, the innovation of William P. Shinn for Andrew Carnegie, in *The Visible Hand*, 266. Similarly, Beniger, *The Control Revolution*, 224–25.

47. Peukert, "The Genesis of the 'Final Solution' from the Spirit of Science," 241; Biagioli, "Science, Modernity, and the Final Solution," 185–205. Also Burleigh, *Death and Deliverance*, 238–66.

48. Hans Mirbeth to Schutzhaftlager Au III, 10 Jan. 1944, "Bericht über das Kommando Golleschau," PMO, D-AuIII/Golleschau, Band 3.

49. Karl Mummenthey, 27 Nov. 1940, "Monatsbericht," BAK NS3/1346.

50. NO-1285, Oswald Pohl to RFSS, 16 Mar. 1943, "SV.-Häftlinge."

51. Maurer to all KL Kommandantur, 24 June 1942, "Häftlingseinsatztagung 1 Jun. 42," BAP PL5: 42055. Maurer actually used the phrase "Nicht arbeits- und nicht einsatzfähige Häftlinge," not "arbeitsunfähig." See also Lifton, *The Nazi Doctors*, 172; Kaul, *Ärzte in Auschwitz*, 6–83, 168; Paczula, "Organisation und Verwaltung des ersten Häftlingskrankenbaus in Auschwitz," 159–71.

52. Béon, *Planet Dora*, 73. See also Paczula, "Organisation und Verwaltung des ersten Häftlingskrankenbaus in Auschwitz."

53. NI-10815, Glücks to all KL Lagerärzte, 26 Dec. 1942, "Ärztliche Tätigkeit in den KL." See also Karny, "'Vernichtung durch Arbeit.' Sterblichkeit in den NS-Konzentrationslagern," 133–58.

54. Kommandoführer to Schutzhaftlagerführer KL Au. III, 20 Apr. 1944, "Meldung wegen Wachvergehen," PMO, D-AuIII/Golleschau, Band 3.

55. NO-1935, Maurer to Buchenwald, 7 Dec. 1942, "Häftlingsüberstellung (Bauhilfsarbeiter) zum KL Auschwitz." Maurer badgered commanders and Labor Action officers throughout the war: Maurer to Höss, Kommandant Auschwitz, 4 Sept. 1943, "Abgabe von Juden-Häftlingen," BAP PL5: 42056; Maurer to all Kommandanten and Arbeitseinsatzführer, 21 Nov. 1942, "Übersicht über Anzahl und Einsatz der Häftlinge im KL," BAP PL5: 42055; NO-1977, Maurer to Kommandanturen-Arbeitseinsatz, 29 Jan. 1945, "Forderungsnachweise, Zusammenstellung und Übersicht"; NO-1547, Gerhard Maurer to all KL Kommandanten, 3 Jan. 1944, "Neueinlieferung von Häftlingen; Schutzhäftlinge aus den besetzten Ostgebieten"; Setkiewicz, "Häftlingsarbeit im KZ Auschwitz III–Monowitz," 594–96.

56. Pingel, *Häftlinge unter SS-Herrschaft*, 182–83; Karny, "'Vernichtung durch Arbeit.' Sterblichkeit in den Konzentrationslager," 140–45, 150–52; Orth, *Das System der nationalsozialistischen Konzentrationslager*, 201, 218; PS-1469, Oswald Pohl to Heinrich Himmler, 30 Sept. 1943, "Todesfälle in den KL"; NO-020, Oswald Pohl to Heinrich Himmler, 5 Apr. 1944, "Konzentrations- und Arbeitslager," and reply, 22 Apr. 1944. The Office Group D installed a chief doctor to coordinate all camp medical duties, Enno Lolling, a mediocre man known to have a history of morphine addiction. Here, as with all operations in the Office Group D, Maurer directed matters, not old-hand IKL personnel. NO-065 and NO-407, Affidavits of Oswald Pohl, and NO-444, Affidavit of Dr. Rudolf Brandt; BDC SS Personal-Akte Enno Lolling. See Kaul, *Ärzte in Auschwitz*, 76–83, 168, 170; Lifton, *Nazi Doctors*, 384–416. Also Kremer, "Diary of Johann Paul Kremer."

57. Affidavits of Gerhard Maurer, BAP PL5: 42063; Affidavit of Friedrich Köberlein, Defense Document Books of Georg Lörner. These affidavits were confirmed by Maurer, Pohl, and Glücks in the newly promulgated "Dienstvorschrift für die Gewährung von vergünstigungen an Häftlinge gültig ab 15 May 43," BAP, PL5: 42056.

58. Sofsky, *Ordnung des Terrors*, 33.

59. *Zeitschrift des Vierjahresplan* 3 (1939): 225; *Das Schwarze Korps* (4 May 1939).

60. Erich Buchmann and Fritz Sauckel, *Die Wilhelm-Gustloff-Stiftung. Ein Tatsachen- und Rechenschaftsbericht ueber Sozialismus der Gesinnung und der Tat in einem Nationalsozialistischen Musterbetrieb des Gaues Thueringen der NSDAP* (Weimar: Gauleiter und Reichsstatthalter in Thueringen, 1938), quotation from 10; in general, 6–14.

61. Erich Buchmann, *Von der juedischen Firma Simson zur Nationalsozialistischen Industriestiftung Gustloff-Werke* (Erfurt: U. Bodung-Verlag, 1944), 19.

62. NO-505, Glücks to RSHA, 6 Apr. 1942, "Rüstungsfabrikation in Buchenwald und Neuengamme"; NO-2468, Schieber to Saur, 20 Mar. 1942.

63. RFSS to Pohl, 7 June 1942, reprinted in Bartel et al., *Buchenwald*, 236–37. See also NO-2549, Scheiber, 20 Mar. 1942, "Verlegung von Rüstungsfertigung in Konzentrationslager"; NO-505; NO-2448, Gottlob Berger to Himmler, 22 Apr. 1942.

64. Quotation from Kommandant Pister to D2, 14 June 1942, "Arbeitszeiten der Häftlinge," in Bartel et al., *Buchenwald*, 236–37. NO-598, Heinrich Himmler to Pohl, 7 July 1942, and reply, 8 Sept. 1942; Speer, 28 June 1942, Chronik.

65. Himmler to Speer, 5 Mar. 1943, T-175/73.

66. Ibid. Note the total absence of any plot to "take over" the armaments works in an internal letter the same day to Oswald Pohl, Himmler to Pohl, 5 Mar. 1943, T-175/73.

67. Himmler to Pohl, 5 Mar. 1943, T-175/73.

68. Compare Janssen, *Das Ministerium Speer*, 97; see also 98, 164. Himmler's formulation included "self-responsibility" but not "independence" (*selbstverantwortlich* but not *selbstständig*). He did not, for instance, couple his complaints with additional demands for weapons for the W-SS.

69. Himmler to Speer, 5 Mar. 1943, T-175/73.

70. Albert Speer to Himmler, 25 Mar. 1943, and Armaments Ministry report, 17 Mar. 1943, "Gewehr-Fertigung im KL Buchenwald," T-175/73.

71. NO-2468, Schieber to Saur, 20 Mar. 1942. See Orth, *Das System der national-sozialistischen Konzentrationslager*, 186; Allen, "Engineers and Modern Managers in the SS," 392–97.

72. Telegram, Pister to Persönlicher Stab RFSS, 16 Aug. 1943, T-175/73.

73. Rudolf Brandt to Pohl, 17 Aug. 1943, and draft letter from Himmler to Pohl, 17 Aug. 1943, T-175/73.

74. Telegram, Pister to Brandt, RFSS, 17 July 1943, T-175/19.

75. Ibid.

76. Himmler to Pohl, 5 Dec. 1941, cited in Karny, "SS-Wirtschafts-Verwaltungs Hauptamt," 154.

77. Himmler to Reichsverkehrsministerium, 14 Apr. 1942, "Ihr Schrieben von 7 Mar. 42," in APMM Fot. Nr. 177. I thank Tomasz Kranz for this reference. See also the ministry's letter, 7 Mar. 1942, in same collection, and Kranz, "Das KL-Lublin," 368-69.

78. Regierung des Generalgouvernements, Hauptabteilung Innere Verwaltung, Dr. S/H, Vermerk, 27 Mar. 1942, in Madajczyka, *Samojszczyzna—Sonderlaboratorium SS*, 66. See also Untersturmführer Dr. W. Gradmann, Vermerk, 19 Mar. 1942, in ibid., 53–60.

79. Boelcke, *Deutschlands Rüstung im Zweiten Weltkrieg*, 187–88. Compare Janssen, *Das Ministerium Speer*, 100–101.

80. Chronik, 15 Sept. 1942. No mention of any 3–5 percent cut of production appears in either the Armaments Ministry records or those of the SS.

81. Hans Frank, Hauptabteilungsleitersitzung, 22 June 1942, in Präg and Jacobmeyer, *Das Diensttagebuch des deutschen Generalgouverneurs in Polen*, 516–17. See also Chronik, 19 and 25 Mar. 1944.

82. HAHB Goetze, Aktenvermerk, 11 Nov. 1941, USHMM RG-11.001M.03 (502-1-148). See also Ludwig, *Technik und Ingenieure*, 473–80; Wegner, *Hitlers politische Soldaten*, 112–28, 263–81.

83. Despite Schulte's well-researched emphasis of the eastern settlements elsewhere in his work, he relegates his one reference to them during the negotiations of 1942 to a footnote in "Verwaltung des Terrors," 207 n. 123. Schulte omits completely Pohl's discussion of "peace building" in September with Speer (227–28).

84. NI-15392, Pohl to Himmler, Memorandum of meeting on 15 Sept. 1942 with Speer,

Pohl, Schieber, Saur, Ministerialrat Steffen, Ministerialrat Briese, Kammler, 16 Sept. 1942, "Rüstungsarbeiten. Bombenschäden." Compare Schulte, "Rüstungsunternehmen oder Handwerksbetrieb?," 567.

85. NI-15392, Pohl to Himmler, Memorandum of meeting on 15 Sept. 1942.

86. Freund, Perz, and Stuhlpfarrer, "Der Bau des Vernichtungslagers Auschwitz-Birkenau," 194.

87. Saur to Pohl, 19 Sept. 1942, BDC SS Hängeordner: 4048; Fröbe, "KZ-Häftlinge als Reserve qualifizierte Arbeitskraft," 641.

88. NI-15392, Pohl to Himmler, Memorandum of meeting on 15 Sept. 1942.

89. Ibid. A report one year later listing all of the SS's *independent* industrial endeavors includes not a single armaments firm: NO-551, unsigned, 30 Sept. 1943, "Die Wirtschafts-unternehmungen der Schutzstaffel."

90. Naasner, *Neue Machtzentren*, 305.

91. Janssen, *Das Ministerium Speer*, 164.

92. Chronik, 28 Sept., 5 Oct., and 8 Dec. 1943. Of note, my research has yielded only two exceptions to the SS's otherwise consistent pattern of avoiding SS-owned arma-ments factories; each stems from a single document, both of which occur *after* Speer's offer of September 1942. Neither is representative of general SS policy. The first is Pohl to Chef der Ordnungspolizei Kurt Daluege, 17 Feb. 1943, T-175/67, instructing him *to stop plans* to erect an independent mine manufacturing plant within a concentration camp. The second is an order from Himmler to Jüttner, 8 Feb. 1944, T-175/80: 600549, concern-ing a new grenade launcher captured from Soviet troops, which appealed to Himmler's techno-blinkered temperament: "It is to be striven for that we should take over such production in a KL factory in as great a quantity as possible. Our goal should be to get as great a number of grenade launchers for our troops from the amount to be delivered." I have found no record of any implementation of this order on the part of SS industries. Neither contradicts the fact that actual SS industries in no way ran as self-contained, vertically integrated, independent armaments factories. There was undoubtedly enthu-siasm within the SS to participate in armaments projects during the war, to make a contribution, and the like, but one can believe that the SS tried to force its way into armaments production only if we consider the SS's industries as *mere* objects of political history, and thus ignore the history of industry, technology, and genuine SS intentions. Cooperation was almost always more prevalent than conflict. Consider that courtesy visits and gift exchanges were typical of Speer, and are a constant preoccupation of the Chronik (e.g., 19 and 25 Mar. 1944). Biographers have repeatedly sought to put Speer in a more realistic light: Schmidt, *Albert Speer*; Van der Vat, *The Good Nazi*. Speer's self-aggrandizement and vanity were also disclosed as early as 1965 by Milward, *The German Economy at War*, 54–56. Compare Speer, *Slave State*, 102–21.

CHAPTER SIX

1. Prinz, *Vom Mittelstand zum Volksgenossen*, 133–36; Walker and Renneberg, "Natur-wissenschaftler, Techniker und der Nationalsozialismus," 21–22.

2. Mommsen and Grieger, *Das Volkswagenwerk*, 500–505.

3. Reported by Pohl to Himmler, 28 Apr. 1942, "Fertigstellung und Einrichtung der Leichtmetallgiesserei Volkswagenwerk," T-175/139. Grieger, "'Vernichtung durch Arbeit'

in der deutschen Rüstungsindustrie," 50, and Mommsen and Grieger, *Das Volkswagen-werk*, 506, both claim that the resulting factory halls were "armaments ruins." They neglect to cite evidence that claims that, in fact, the SS satisfied its part of the bargain. Both Mommsen and Grieger make much of Speer's decision to abandon the Light Metals Foundry project to stave off the SS, an unlikely strategy because his offices were seeking Himmler's cooperation at the very same time. See Dr. Hans Hohberg to Oswald Pohl, Professor Porsche, and Obf. Kammler, 29 Jan. 1942, "Erlass des Führers vom 11 Jan. 42 re: Einrichtung einer Giesserei in Fallersleben (VW)," T-976/36, which clearly states: "Professor Porsche expressed that the Schutzstaffel should place workers and raw mate-rials at the disposal of the project, but that the operation of the light-metals foundry is the task of the Volkswagen works." See also Kammler's memorandum, NO-595, 3 Feb. 1942, "Giesserei Volkswagenwerke." From the outset, no takeover was ever intended, although a mutually beneficial partnership was.

4. Mommsen and Grieger, *Das Volkswagenwerke*, 515.

5. Speer to Volkswagenwerk, copy to Himmler, 23 Mar. 1942, "Bauvorhaben Leicht-metallgiesserei," T-175/139; Pohl to Himmler, 28 Apr. 1942, "Fertigstellung und Ein-richtung der Leichtmetallgiesserei Volkswagenwerk," T-175/139; Budraß and Grieger, "Die Moral der Effizienz," 97–101. Compare Mommsen and Grieger, *Das Volkswagen-werke*, 497–515; Orth, *Das System der nationalsozialistischen Konzentrationslager*, 170, 207–8.

6. NO-3032, Pohl (apparently drafted by Kammler if one judges from the letterhead) to RFSS, 11 July 1942, "Building of a temporary [*befehlsmässig*] gun factory in Weimar-Buchenwald in conjunction with the Gustloff-Werke"; Oswald Pohl to Himmler, 20 June 1943, and Fernschreiben, Brandt to Kammler, 16 June 1943, T-175/73.

7. Kammler to RFSS Persönlicher Stab, 12 June 1943; Kammler to Persönlicher Stab, 16 June 1943, "Reichsbahnbau Weimar-Buchenwald"; Kammler to Persönlicher Stab, 24 May 1943, "Eisenbahnbau Weimar-Buchenwald" with enclosed draft to Referent of Gauleiter Fritz Sauckel; Sauckel to Himmler, 24 May 1943; and Speer to Himmler, 30 May 1943, T-175/73: 0787–0868.

8. Kammler to RFSS, 12 June 1943.

9. Kammler to RFSS, 24 May 1943. The finishing date was 16 June 1943.

10. Only the Organization Todt matched the SS's proficiency in the management of forced labor on large-scale construction projects. See Raim, *Die Dachauer KZ-Außen-kommandos*.

11. NO-2128 a, Pohl to Himmler, 27 July 1942, "Enforcement Regulations for the RFSS Decree of 18 Jun. 42." Compare Himmler, 18 June 1942, "Neugliederung der Wirtschafts-und Verwaltungsdienststellen bei den Höheren SS- und Polizeiführern in den besetzten Gebieten einschließlich Generalgouvernement," T-976/36.

12. Herr Lothert, 31 Jan. 1943, "Besprechung am 30 Jan. 43 mit dem SS-Wirtschafter, Gruppe C–Bauwesen," and Busch, 12 Apr. 1942, "Besprechung beim SS-Wirtschafter, Gruppe C5–Bauwesen am 3 Dec. 42," formerly SF-10/16334 Militärgeschichtliches Ar-chiv Potsdam; Hans Schleif, 11 Dec. 1944, "Begründung," BDC Personal-Akte Fritz Blaschek; Kammler to all Amtsgruppen WVHA and the Höheren SS und Polizeiführer in Reichsgebiet and in the General Government, 16 Oct. 1942, "Sparsamster Einsatz der Häftlinge bei den Bauvorhaben," T-976/36.

13. NO-498, Pohl, 19 Nov. 1942, "Organization order for external branches of the WVHA, which are 'not' under the jurisdiction of a SS-Wirtschafter"; Kammler, end of Dec. 1941, "Bericht des Amtes II–Bauten über die Arbeiten im Jahre 1941," USHMM RG-11.001M.03: 19 (502-1-13); Kammler, "Hauptamt Haushalt und Bauten Amt II–Bauten. Geschäftsverteilungsplan vom 20 Jun. 41," USHMM RG-11.001M:19 (502-1-9); Heinrich Himmler, 15 May 1942, "Auszug aus den Wirtschafts- und Verwaltungsanordnungen vom 15. Mai 1942 Nr. 3; Befehl über die Einrichtung von SS- und Polizeistützpunkten," BDC Wirtschaftsverwaltungsanordnungen, Mitteilungen für die W-SS; Wasser, *Himmlers Raumplanung im Osten*, 84, 318 Dokumentenanhang.

14. "Mindener Bericht" (written by WVHA officers in Allied captivity in Minden, probably Mummenthey, Volk, and Hoffmann), reprinted in Naasner, *SS-Wirtschaft und SS-Verwaltung*, 159; hereafter cited as "Mindener Bericht." Regarding Loibl, see chapter 1. See also Nelson, *Small Wonder*.

15. Hölsken, *Die V-Waffen*, esp. 10–33. Speer, *Inside the Third Reich*, 366: "Our most expensive project was also our most foolish one. Those rockets, which were our pride and for a time my favorite armaments projects, proved to be nothing but a mistaken investment."

16. For Allied strategy in industry and battlefield tactics, see Overy, *Why the Allies Won*. Compare Schabel, "Wenn Wunder den Sieg bringen sollen"; Neufeld, *The Rocket*, 170–72; Salewski, "Die Abwehr der Invasion als Schlüssel zum 'Endsieg,'" 221–23.

17. Freund, *Arbeitslager Zement*, 25; Bornemann, *Geheimprojekt Mittelbau*, 29. The island had been a favorite duck-hunting spot in his family. Hölsken, *Die V-Waffen*, 18–19; Neufeld, *The Rocket*, 41–72, 167; Wagner, "Verlagerungswahn und Tod," 32–101.

18. Neufeld, *The Rocket*, 112.

19. Letter to OKW, 3 Feb. 1942, "Vorbereitung der Massenherstellung der Peenemünder Geräte," BA MA RH8/1211.

20. Neufeld, *The Rocket*, 171–75; "Ein Mann schafft Lokomotiven. Die Kriegslok meistert ein Problem-Erfolg straffer Gemeinschaftsarbeit," *Völkischer Beobachter*, 16 Aug. 1943; "Ritterkreuz des KVK Direktor G. Degenkolb," *Die Lokomotive* 40 (1943): 192–93. I thank Alfred Gottwaldt for copies of this material.

21. Ordway and Sharpe, *The Rocket Team*, 61.

22. Signature unrecognized, 6 Aug. 1943, "Niederschrift über die Besprechung am 4 Aug. 43 beim Heimat-Artillerie-Park 11," BA MA RH 8/v 1254; "Entstehungsgeschichte des Versuchsserienwerkes Peenemünde, 1943," 23–26 Feb. 1943, BA MA RH/1210, hereafter cited as "Entstehungsgeschichte"; Bornemann, *Geheimprojekt Mittelbau*, 32–33; Freund, *Arbeitslager Zement*, 37; Neufeld, *The Rocket*, 143–47, 173–76. Degenkolb also wanted to expand production in the Demag Lokomotive Works in Falkensee in the summer of 1943. Wagner, "Verlagerungswahn und Tod," 68.

23. Neufeld, *The Rocket*, 189; Bornemann, *Geheimprojekt Mittelbau*, 35.

24. Cited after Hölsken, *Die V-Waffen*, 41 (who quotes Dornberger, *V-2—Der Schuß ins Weltall*, 196); see 37–41 in general. See also Neufeld, *The Rocket*, 176–78; Freund, *Arbeitslager Zement*, 42.

25. Quotation from Arthur Rudolph, 16 Apr. 1943, "Besichtigung des Häftlingseinsatzes bei den Heinkel-Werken, Oranienburg, am 12 Apr. 43," BA MA RH8/1210. See also Freund, *Arbeitslager Zement*, 43–45; Neufeld, *The Rocket*, 186–89. The prisoners are

listed as "1000 Russen, 800 Franzosen, 800 Polen." Signature unrecognized, 6 Aug. 1943, "Niederschrift über die Besprechung am 4 Aug. 43 beim Heimat-Artillerie-Park 11," BA MA RH 8/v 1254; "Entstehungsgeschichte," Apr. 1943.

26. Neufeld, *The Rocket*, 197–201; "Entstehungsgeschichte," 18 Aug. 1943; Wagner, "Verlagerungswahn und Tod," 69–71; Freund, *Arbeitslager Zement*, 52–53; Bornemann, *Geheimprojekt Mittelbau*, 40–43.

27. Quotation from Eichholtz and Lehmann, *Geschichte der deutschen Kriegswirtschaft*, 156.

28. Perz, *Projekt Quarz*, 138; Fröbe, "KZ-Häftlinge als Reserve qualifizierte Arbeitskraft," 636–81.

29. Neufeld, *The Rocket*, 197–201; Freund, *Arbeitslager Zement*, 58–62; Bornemann, *Geheimprojekt Mittelbau*, 38–43; Neufeld, *The Rocket*, 199; Reissinger (first name unknown), 23 Aug. 1943, "Kurzübersicht über die Weiterführung des A4-Programms," BA MA RH8/1211.

30. Freund, *Arbeitslager Zement*, 58–60; Bornemann, *Geheimprojekt Mittelbau*, 40.

31. Raim, *Die Dachauer KZ-Außenkommandos*, 20–60. The SS was still holding out for that future day when the peace building program could finally be mobilized. See With to RFSS, 22 May 1943, "Vereinheitlichung des Bauwesens bei allen Wehrmachtsteilen einschliesslich Waffen-SS," and Kammler's reply, 1 July 1943, BAK NS19/2065. Kammler's reasoning was, "The Reichsführer SS finds it necessary to carry out the giant program after the end of the war as a higher building authority [*Bauhoheit*] . . . the preconditions for a quick victory will be accomplished at the same time that the infrastructure of the W-SS is therewith strengthened for the peace program." See also Perz, *Projekt Quarz*, 140–41.

32. Himmler to Speer, 21 Aug. 1943, BAK NS19/2055.

33. See Himmler's urgent interest in fuel injection for another example. Himmler to Milch, 3 Feb. 1943, and Milch to Himmler, 18 Feb. 1943, BAK NS19/1197.

34. Speer to Himmler, 22 Dec. 1943, BAK R3/1583.

35. Neufeld, *The Rocket*, 202; Freund, *Arbeitslager Zement*, 61–62; Perz, *Projekt Quarz*, 144; Bornemann, *Geheimprojekt Mittelbau*, 43; Orth, *Das System der nationalsozialistischen Konzentrationslager*, 244–46; Wagner, "Das Außenlagersystem des KL Mittelbau-Dora," 711–16. The most thorough treatment of the Mittelwerk-Mittelbau complex is Wagner, "Verlagerung und Tod," which also stresses the dense networks of industrial cooperation and interdependence surrounding Nordhausen.

36. Freund, *Arbeitslager Zement*, 38–39; Bornemann, *Geheimprojekt Mittelbau*, 51; "Niederschrift über die Gesellschafter-Versammlung," 24 Sept. 1943; Handelsregister Notiz; and Vermerk of Adolf Schnurre, Anwalt (prepared registration), 5 Oct. 1943, BAK R121/544; Professor Dr. Hettlage, 23 Sept. 1943, "Vermerk über Besprechung, betr. Mittelwerk GmbH, am 21 Sept. 43," BAK 121/405; Kunze, 14 Dec. 1943, "Niederschrift über die 1. Sitzung des Beirates der Mittelwerk GmbH, 10 Dec. 43," BAK R121/405. Degenkolb not only chose Hans Kammler to join him on Mittelwerk's board of directors, but he also appointed the concentration camp Kommandant of Dora, Otto Förschner, to be one of the four factory directors (*Betriebsführer*) at Mittelwerk. No evidence suggests that the SS sought to "infiltrate" Mittelwerk through Kammler and Förschner. Quite the con-

trary, Förschner, submissive to the wishes of Kammler, distinguished himself among KL Kommandant by obeying the wishes of production engineers and was accordingly despised as a weakling among other IKL old hands. See Segev, *Die Soldaten des Bösen*, 38. See another document with signature unrecognized, 6 Aug. 1943, "Niederschrift über die Besprechung am 4 Aug. 43 beim Heimat-Artillerie-Park 11," BA MA RH 8/v 1254. Mittel- werk's Vorstand eventually offered Förschner a bonus of 10,000 Reichsmarks, which raised the ire of Himmler, who resented seeing an SS Kommandant receive perquisites of corporate life. Himmler removed Förschner shortly thereafter because he feared the "infiltration" of private business into the SS. See Kammler to Rüstungskontor GmbH, 2 Oct. 1943, "Vorstandsmitglied SS," BAK R121/405; Speer to Himmler, 29 Jan. 1945, BAK R3/15583; BDC SS Personal-Akte Otto Förschner.

37. Betriebsordnung 1, 13 Dec. 1943, and Geschäftsordnung, also prepared on 13 Dec. 1943, R121/405.

38. Schmid-Loßberg to Degenkolb, 5 Jan. 1944, BAK R121/544. See also Neufeld, *The Rocket*, 179–80.

39. Kunze, 14 Dec. 1943, "Niederschrift über die 1. Sitzung des Beirates der Mittelwerk GmbH, 10 Dec. 43," BAK R121/405.

40. Neumann, *Behemoth*.

41. Ferencz, *Less than Slaves*.

42. Miller, *The Way of Death*, esp. "The Economics of Mortality," 657–92; quotation from 681.

43. Ordway and Sharpe, *The Rocket Team*, 71, similarly 62.

44. Speer to Kammler, 17 Dec. 1943, BDC SS Personal-Akte Hans Kammler. See also Chef des Amtes Bau, Stobe-Dethleffsen to Kammler, 21 Dec. 1943, "Bombensichere Ausweichbauten," BAP 46.03/68; Speer's Chronik der Dienststellen, Nov.–Dec. 1943, BAK R3/1738, hereafter cited as Chronik.

45. Neufeld, *The Rocket*, 260–65; Béon, *Planet Dora*, 11–12 (among other references); Orth, *Das System der nationalsozialistischen Konzentrationslager*, 309.

46. Jägerstabbesprechung, 26 May 1944, BA MA RL3/7.

47. Remdt and Wermusch, *Rätsel Jonastal*, 47. The survivor was Fred Wander.

48. Jägerstabbesprechung, 2 May 1944, BA MA RL3/6.

49. Béon, *Planet Dora*, 30.

50. Testimony of Josef Ackermann, prisoner–medical assistant at Dora-Buchenwald, Protocol: 952–56; Raim, *Die Dachauer KZ-Außenkommandos*, 220–39.

51. Fauser, "Zur Geschichte des Außenlagers Langenstein-Zwieberge," 39.

52. Lenzer to BI, ZBL, 20 May 1942, and Chef Amt II, Stbaf. Heidelberg to Neu- bauleitung Flossenbürg, 4 July 1940, BAK NS4/31; Kammler to Kommandanten des KL Auschwitz, 9 June 1942, "Bauvorhaben im 3. Kriegsjahr," USHMM RG-11.001M.03: 19 (502-1-9); Lenzer to BI, ZBL, Amtsgruppe C, Amtsgruppe D, 17 June 1942, "Anforderung und Berechnung von Häftlingen für die Bauleitungen," USHMM RG-11.001M.03: 18 (502-1-3); Kammler to Rüstungskontor GmbH, 2 Oct. 1943, "Vorstandsmitglied SS," BAK R121/405; Kammler, 25 June 1943, "Aufstellung einer Baubetriebsdienststelle bei der Verwaltung des KL Auschwitz," and Kammler and Chef Amtsgruppe C6, 30 Mar. 1943, "Vorläufige Dienstanweisung für Bau-Betriebsdienststellen der W-SS und Polizei," USHMM RG-11.001M.03: 18 (502-1-1).

53. Bornemann, *Aktiver und Passiver Widerstand im KZ Dora*, 24; Fröbe, "KZ-Häftlinge als Reserve qualifizierte Arbeitskraft."

54. The SS engineering corps also trained concrete workers among prisoners: NO-1386, Schmauser (??) to Himmler, 20 Apr. 1942, "Zurückziehung der bei den Dienststellen des RKF, Kattowitz und beim Sonderbeauftragten des RFSS für fremdvölkischen Arbeitseinsatz, Sosnowitz," BAP, PL5: 42054, and NI-11235, Affidavit of Heinrich L. Koradhan. Colonel W. R. J. Cook et al., Underground Factories in Central Germany, CIOS File No. XXXII-17 (London: HM Stationery Office, Deutsches Museum): 12–13, 34–40, 144–50; Jan Jecha, survivor of Terezín and Auschwitz, interview with the author, 16 Nov. 1994; Svetozar Gucek, survivor of Leitmeritz, interview with the author, 16 Nov. 1994; Svetozar Gucek, "Bau des Konzentrationslagers Leitmeritz und der Fabrik Richard I," and Bohumil Kos, survivor of Leitmeritz, "Anfänge des Kommandos Richard im Theresienstädter Gefängnis," both papers delivered, 16 Nov. 1994, at the Internationalen Konferenz zum 50. Jahrestag der Errichtung des KZ in Litomerice (Leitmeritz), copies in Terezin Ghetto Museum. See also Béon, *Planet Dora*, 84.

55. Dieckmann, "Existenzbedingungen und Widerstand im KL Dora-Mittelbau," 201–8; Wagner, "Verlagerungswahn und Tod," 298–325.

56. Cited after Ordway and Sharpe, *The Rocket Team*, 70. Compare Van der Vat, *The Good Nazi*.

57. Bornemann, *Aktiver und Passiver Widerstand im KZ Dora*, 70–71; NO-1564, Affidavit of Erwin Tschentscher; Testimony of Alberto Bertin in Fauser, "Zur Geschichte des Außenlagers Langenstein-Zwieberge," 41, 45–46.

58. Wagner, "Das Außenlagersystem des KL Mittelbau-Dora," 716–17.

59. Wagner, "Verlagerungswahn und Tod," 77–78; Karny, "'Vernichtung durch Arbeit' in Leitmeritz," 52–53.

60. NO-405, Vogel to Hans Baier, 7 Apr. 1944.

61. Colonel W. R. J. Cook et al., Underground Factories in Central Germany, CIOS File No. XXXII-17 (Deutsches Museum): 12–13, 34–40, 84, 144–50; interviews with Jan Jecha and Svetozar Gucek, and contributions of Svetozar Gucek, "Bau des Konzentrationslagers Leitmeritz und der Fabrik Richard I," and Bohumil Kos, survivor of Leitmeritz, "Anfänge des Kommandos Richard im Theresienstädter Gefängnis"; Speer, "Besprechung beim Führer vom 21–23 Sep. 44," 24 Sept. 1944, BAK R3/1510; Karl Bischoff, 17 Feb. 1944, "Aktenvermerk," USHMM RG-11.001M.03: 23 (502-1-81). Evidence that the techniques of Kahla were used by the SS, prisoner testimony quoted by Perz, *Projekt Quarz*, 195–96. The interior of Mittelwerk was planned as a grid with connecting tunnels shooting between parallel main galleries. Excavation began at twenty separate pitheads for cross galleries and four separate pitheads for service tunnels. Thus the layout of Mittelwerk's grid effectively maximized the tunnel face that SS Construction Directorates could attack simultaneously.

62. Raim, *Die Dachauer KZ-Außenkommandos*, 28–32; Janssen, *Das Ministerium Speer*, 185–86; Perz, *Projekt Quarz*, 142–43; Overy, *Why the Allies Won*, 133.

63. Hughes, *Rescuing Prometheus*, 3–14.

64. "Besprechung im Sonderzug, Messerschmitt-Regensburg," 10 Mar. 1944, BA MA RL3/1.

65. Jägerstabbesprechung, 31 Mar. 1944, BA MA RL3/4, and "Schnellbericht," 6 Mar.

1944, BA MA RL3/10. This was in spite of the objections of the RLM. Chronik, 5 and 10 Jan. 1944. Mark Spoerer estimates the GDP of Germany at about 150 billion Reichsmarks in 1942. I thank him for this information.

66. PS-1584(I), Hermann Göring to Heinrich Himmler, 15 Feb. 1944, "Aufstellung der 7. Staffel/Fliegergruppe zB V. 7"; Perz, *Projekt Quarz*, 145–48; Freund, *Arbeitslager Zement*, 84; Bornemann, *Geheimprojekt Mittelbau*, 77–78; Chronik, May 1944; "Vermerk über die Besprechung am 16.3.44, Verlagerung von Fertigung untertage," and MR Richter, 11 May 1944, "Unterirdische Verlagerungen," BAK R7/1173; Jägerstabbesprechung, 15 Apr. 1944, BA MA RL3/5; Jägerstabbesprechung, 31 Mar. 1944, BA MA RL3/4; "Anordnung über die Errichtung des 'Jägerstabes,'" 13 Mar. 1944, BDC Ordner 274; Anonymous, 1 May 1944, "Durchführung der baulichen Arbeiten für die unterirdische Verlagerung"; Meffert, 1 Feb. 1944, "Kommission für unterirdische Verlagerungen"; and Stobe-Dethleffsen to Kammler, 21 Dec. 1943, "Bombensichere Ausweichbauten," BAP 46.03/68.

67. First quotation is from Pingel, *Häftlinge unter SS-Herrschaft*, 62; second, Himmler to Pohl, 23 Mar. 1942, BAK NS19/2065.

68. PS-1584 (III), Himmler to Göring, 9 Mar. 1944, "Einsatz von Häftlingen in der Luftfahrtindustrie." Spoerer, "Profitierten Unternehmen von KZ-Arbeit?," 61–95, found only one firm that was "forced" to take prisoners out of thirty-three case studies. Examples of typical enthusiastic cooperation: Karl Hermann Frank to Speer, 28 Apr. 1944, BAK R3/1578; Generaldirektor Steyr-Daimler-Puch AG to RFSS, 14 July 1943, "Errichtung eines KL in Wr. Neudorf," T-175/19; Generalbevollmächtigter für Sonderfragen der chemischen Erzeugung to RFSS, 27 July 1943, BDC SS Hängeordner: 7139; Von Kruedener (Hauptmann under Milch) to Pohl, 6 Oct. 1944, BDC SS Hängeordner: 4234; Speer to Himmler, 27 Nov. 1944, BAK R3/1583; Speer to Himmler, 27 May 1944, T-175/146. Ferdinand Porsche also returned to ask for prisoners again: NO-067, Himmler to Pohl, 4 Mar. 1944. Paul Budin, Generaldirektor of Hasag and Sondervollmacht über Hochlauf Panzerfaust in RMRK to Himmler, 17 Oct. 1944, PS-4021.

69. Examples of Himmler's importunity: Pohl to Himmler, 24 Jan. 1944, Höhlenbau in den SS-eigenen Steinbrüchen," T-175/21, and Brandt to Pohl, 19 Mar. 1944, BAK NS19/2602. Also Himmler to Milch, 3 Feb. 1943, BAK NS19/1197.

70. "Besprechung bei Allach," 10 Mar. 1944, BA MA RL3/1. See also Perz, *Projekt Quarz*, 149–50, and Bornemann, *Geheimprojekt Mittelbau*, 82–83.

71. "Niederschrift der Besprechungen während des 'Unternehmens Rubertus,'" 8 Mar. 1944, BA MA RL3/1.

72. Ibid.

73. Jägerstabbesprechung, 12 Mar. 1944, BA MA RL3/2.

74. See Kater, *The Nazi Party*, 169–233.

75. "Schlußbesprechung in der Reichsbahn-Direktion Stuttgart," 10 Mar. 1944, BA MA RL3/1.

76. Jägerstabbesprechung, 5 July 1944, BA MA RL3/9.

77. Chronik, 26 May 1944.

78. Jägerstabbesprechung, 1 May 1944, BA MA RL3/6.

79. Jägerstabbesprechung, 31 Mar. 1944, BA MA RL3/4.

80. Braun, "Fertigungsprozesse im deutschen Flugzeugbau 1926–45."

CHAPTER SEVEN

1. Himmler to Pohl, 9 Sept. 1942, BAK NS19/14.

2. See chapter 5.

3. NI-15392, Pohl to Himmler, memorandum of meeting on 15 Sept. 1942 with Speer, Pohl, Schieber, Saur, Ministerialrat Steffen, Ministerialrat Briese, Kammler, 16 Sept. 1942, "Rüstungsarbeiten. Bombenschäden."

4. NO-555, Karl Niemann, Mar. 1944, "Business Report of the Deutsche Ausrüstungs-werke GmbH for the year 1943." On IG Farben at Auschwitz, see also NO-1216, travel report of Dr. May (with Hohberg) to Pohl, 1–8 June 1942, T-976/32; Affidavit of Dipl. Kaufmann Heinz Savelsberg, Defense Document Books of Hans Hohberg; NO-553, Karl Niemann, Mar. 1943, "Geschäftsbericht der DAW, 1942." At this time the DAW was still making "Siedlermöbel."

5. Knoll, "Die Porzellanmanufaktur München-Allach," 124; Hohberg to Frank, 3 Oct. 1942, "Finanzierung DESt," T-976/25; NO-549, monthly reports of DAW from 16 Oct.–17 Dec. 1942. Regarding the ODBS, see Allen, "The Puzzle of Nazi Modernism," 538–58. Regarding textiles, see Lechler to Pohl, 6 Nov. 1942, T-976/15. See also the Manifesto for SS Wirtschaft by Leo Volk: NO-1016, 11 July 1944, "Organization und Aufgaben der Amtsgruppe W." Compare Schulte, "Rüstungsunternehmen oder Handwerksbetrieb?," 558–83.

6. Orth, *Das System der nationalsozialistischen Konzentrationslager*, 53, 165, 192.

7. For statistics on the SS work force at the DAW, see NO-553, Karl Niemann, Mar. 1943, "Geschäftsbericht der DAW, 1942." Turnover went from 5,366,547 Reichsmarks in 1941 to over 23,000,000 Reichsmarks in 1942. For statistics on Buchenwald, see Karny, "SS-Wirtschafts-Verwaltungs Hauptamt," 161.

8. "Protokoll Aufsichtsratssitzung der D. Drucker AG," 10 Mar. 1942, T-976/22.

9. Bericht, 31 Mar. 1942, "Wiederaufbau der Möbelfabrik Drucker AG im Auftrag durch WVHA Stab Hauptamtschef," T-976/22. See also Protokoll der Aufsichtsrats-sitzung der Deutschen Edelmöbel Aktiengesellschaft am 29 June 1942, T-976/22, and "Protokoll Aufsichtsratssitzung der D. Drucker AG," 10 Mar. 1942, T-976/22. In keeping with the SS's fascination with technical wonders, the factory had introduced a newly patented synthetic glue for Messerschmitt tailpieces. The synthetic glue had impressed the Luftwaffe from the beginning: Kurt May, 23 June 1941, "Jahresrechnung des Vor-standes der Drucker AG, Dampfsägewerke und Holzwarenfabriken in Brünn zur Jahr-esrechnung 1940, T-976/22; Hohberg to Fachamt des Deutschen Handwerks, 17 July 1942, "Interesse für verschiedene Verfahren," BAK NS3/4.

10. Affidavit of Werner Kahn, Defense Document Books of Karl Mummenthey; Affidavit of Dr. Martin Löffler, Defense Document Books of Hans Hohberg; Dr. Höring, Report of 16 Nov. 1941, T-976/22. On "Aryanization" in general, see Hayes, "Big Business and Aryanization," 254–81.

11. NO-553, Karl Niemann, Mar. 1943, "Geschäftsbericht der DAW, 1942"; NO-3742, Karl Bestle and Karl Niemann to Magistrat der Hauptstadt Prag, 20 Oct. 1942, "Gewer-beanmeldung"; and BDC Personal-Akte Joseph Opperbeck.

12. An example, by no means exceptional, is Paul Wartenburg, a young SS manager. He headed a division within the ODBS and was caught embezzling funds in Warsaw. He was, like May, investigated on criminal charges. Although the investigation did uncover

serious discrepancies, it found no burning sins against SS ideology. The result was that nothing happened to Wartenburg. SS Richter Weber to PHA from SS und Polizeigericht Breslau, 21 Jan. 1944, BDC SS Personal-Akte Paul Wartenberg.

13. "Mindener Bericht," 120–21, 155–56, 167, 172–73, 179–87.

14. Pohl, "Die großen Zwangsarbeitslager der SS- und Polizeiführer," 415–20.

15. PS-1553, Affidavit of Kurt Gerstein; PS-3868, Affidavit of Rudolf Höß. See also Browning, "The Development and Production of the Nazi Gas Van," 57–67; Adalbert Rückerl, *NS-Vernichtungslager*, 61, 72–74; Friedlander, *The Origins of Nazi Genocide*, 296–99; Burleigh, *Death and Deliverance*, 232–34; Breitman, *The Architect of Genocide*, 190, 200–201, 235. See Pingel's treatment of the contradiction between "Ideologie und Ökonomie," *Häftlinge unter SS-Herrschaft*, 139–44; Pohl, *Von der "Judenpolitik" zum Judenmord*, 33–87.

16. NO-1906, Memorandum signed by Oswald Pohl and Georg Lörner, 14 July 1943. Notably absent from this list are any independent SS armaments industries. For evidence that OSTI was connected closely with settlements, see NO-1267, Hans Hohberg to Horn, 1 May 1943, which mentions the Siedler-Wirtschaftsgemeinschaft Lublin.

17. Pohl, "Die großen Zwangsarbeitslager der SS- und Polizeiführer," 429.

18. NO-519, Horn to Pohl, 24 Jan. 1944.

19. Pohl, *Von der "Judenpolitik" zum Judenmord*, 158–62. Horn's first reports are dated 13 and 26 Feb. 1943: Horn, 13 Feb. 1943, "Queries regarding the Osti Ltd," Prosecution Document Book 3; NO-1265, Horn to Hohberg, 26 Feb. 1943. See also Max Horn, *Festschrift zum 70. Geburtstag. Der Wirtschaftsprüfer als Unternehmer* (Berlin: Technische Universität-Berlin, 1974). Horn had worked in the *gemeinnützige* business sector, as many *Wirtschaftsprüfer* did.

20. NO-1271, Memorandum of Albert Kruse (prisoner in the Schreibstube of OSTI), 21 June 1944, and NO-1906, Memorandum signed by Oswald Pohl and Georg Lörner, 14 July 1943; NO-599, Pohl (drafted by Maurer), 7 Sept. 1943, "Taking over of Jewish Labor Camps from SS and Police Chiefs of the GG"; Dr. Horn, 13 Mar. 1944, Bericht, BDC SS Hängeordner: 1748. Pohl instructions recorded in NO-1270, Memorandum of Dr. Max Horn, 13 Feb. 1943, "Queries regarding Osti Ltd."

21. NO-1270.

22. NO-519, Horn to Pohl, 24 Jan. 1944.

23. These marginalia are found in the correspondence cited in the quotations of this paragraph.

24. NO-519, Horn to Pohl, 24 Jan. 1944. The action was code-named after Reinhard Heydrich, the chief of the Reich Security Main Office who had been assassinated by Czech partisans. Rückerl, *NS-Vernichtungslager*, 37; Pohl, *Von der "Judenpolitik" zum Judenmord*, 152–61; Witte, "Slowakische Juden im Distrikt Lublin"; NO-1040, Hohberg to Chef Amt A2 and Hauptsturmführer Millmer, 3 June 1943; NO-1266, Hohberg, 26 June 1943, "Darlehen aus dem Reinhardfonds"; and NO-1269, Horn to Bankkonto Emissionsbank, 17 Aug. 1943. The funds were handled by the Dresdner Bank. NO-056, Globocnik to Himmler, 4 Nov. 1943; NO-064, Globocnik to Himmler, 5 Jan. 1944; NO-057, 18 Jan. 1944, "Wirtschaftlicher Teil der Aktion Reinhardt." Regarding the amount of material which changed hands: NO-060, Globocnik to Himmler, 4 Nov. 1943,

gives detailed list of items. On fenced dental gold, see Peter Hayes, "Degussa and the Persecution of the Jews," paper delivered at the Lessons and Legacies Conference, Boca Raton, Florida, 9 Nov. 1998. See also Pohl, *Von der "Judenpolitik" zum Judenmord*, 172; Georg Wippern, Chef der Verwaltung "Operation Reinhard," 27 Oct. 1962, ZSL IV 4b AR-Z 82/1968, Band IV; Eizenstat et al., *U.S. and Allied Efforts to Recover and Restore Gold and Other Assets Stolen*, 160–69; Perz and Sandkühler, "Auschwitz und die 'Aktion Reinhard,'" 289–301.

25. Paxton and Marrus, *Vichy France*, 75–115, 152–61.

26. Eizenstat et al., *U.S. and Allied Efforts to Recover and Restore Gold and Other Assets*, 160–69.

27. Horn's report, 14 Feb. 1944, BDC Hängeordner: 1750; NO-2187, Dr. Horn to Pohl, 13 Mar. 1944; Pohl, *Von der "Judenpolitik" zum Judenmord*, 157, 170–71.

28. Cited after Pohl, *Von der "Judenpolitik" zum Judenmord*, first quotation from 171; second, 166. Regarding military actions at this time, Overy, *Why the Allies Won*, 85–100.

29. Pohl, "Die großen Zwangsarbeitslager der SS- und Polizeiführer," 428, and *Von der "Judenpolitik" zum Judenmord*, 170–74, note that some few work Jews escaped, seemingly at random. On the other hand, others working for private armaments firms (thus directly for the war effort) under Speer's ministry were killed. NO-057, 18 Jan. 1944, "Wirtschaftlicher Teil der Aktion Reinhardt," mentions the requests of Speer and Speer's Rüstungsinspektionen for labor, "The demand for these places was very great." But in regard to promises to provide General Schindler, Rüstungsinspektion Krakau, with laborers, the report states laconically, "The agreement could no longer be upheld."

30. NO-2187, Dr. Horn to Pohl, 13 Mar. 1944. Liquidations at DWB factories in Lemberg were almost parallel. Sandkühler, "Das Zwangsarbeitslager Lemberg-Janowska 1941–1944," 606–35.

31. Even the Wannsee Conference protocol had discussed the preservation of some few Jews for slave labor. See BDC SS Personal-Akte Schellin; NO-057, 18 Jan. 1944, "Wirtschaftlicher Teil der Aktion Reinhardt"; Orth, *Das System der nationalsozialistischen Konzentrationslager*, 213–14.

32. NO-2187, Dr. Horn to Pohl, 13 Mar. 1944.

33. Ibid.

34. Browning, "Bureaucracy and Mass Murder," 142. The pattern at Lublin also repeated itself at Litzmannstadt. See the correspondence between Pohl and Himmler, 5, 15, and 24 July 1943, BDC SS Hängeordner: 1328, and NO-482, Heinrich Himmler to WVHA, HSSPF, RSHA, 5 July 1943. Horn to Pohl, 14 Feb. 1944, BDC Hängeordner: 1750, shows that whatever was left of the original OSTI industries was returned to the SSPF. See also Browning, "Beyond 'Intentionalism' and 'Functionalism,'" 211–33, on Globocnik's role, 220; Pohl, *Von der "Judenpolitik" zum Judenmord*, 165–70.

35. Lechler to Pohl, 6 Nov. 1942, T-976/15, and NO-510, undated memorandum of Weber, early 1944; Drobisch and Wieland, *System der NS-Konzentrationslager*, 82–87.

36. This report can be found unsigned and undated in BAK NS19/2302. It is undeniably by Schondorff, who was the only brick engineer of sufficient rank to be in correspondence with the Personal Staff of the RFSS. A note in BAK NS19/46 refers to Schon-

dorff's "Bericht über Metallwarenfabrik- Ziegel- und Betonsteinfabrikarbeiter," dated 16 Dec. 1943, and states that it had been lent out. The missing report is almost certainly the one located in BAK NS19/2302 and cited here.

37. Unsigned, undated report, BAK NS19/2302.

38. Ibid. See also Karny, "'Vernichtung durch Arbeit.' Sterblichkeit in den NS-Konzentrationslagern," 141. As Karny notes, the absolute number of deaths in the KZs never sank over the years. Only the percentage of deaths among camp populations sank under the control of the WVHA. The sinking death rate reflected the massive expansion of the KZ population, not the decreasing presence of death. Compare Pingel, *Häftlinge unter SS-Herrschaft*, 181–87.

39. In reality productivity in SS companies varied wildly, from as low as 17 percent for civilian workers at an SS cement factory near Auschwitz, to 20 percent at a cinder-block factory in Linz, to 70 percent and above for some DAW factories. See Dr. Hohberg to Pohl, Kammler, Mummenthey, and Maurer, 6 Aug. 1942, "Schlackenwerk Linz," T-976/24. On the cement factory, see Allen, "The Puzzle of Nazi Modernism," 562–63. On the DAW, see Opperbeck to Hauptamtschef WVHA, 18 Oct. 1943, T-976/32; Naasner, *Neue Machtzentren*, 346.

40. See Pohl to RFSS, 18 Apr. 1944, "Lehmbauweise Posen und Mauern mit heisser Kelle"; Pohl to RFSS, 25 Jan. 1944, "Lehmbauweise Posen"; Schondorff to Pohl, 22 Jan. 1944, "Verfestigung von Lehm zu wasserunlöslichen Massen"; Himmler to Pohl, 3 Mar. 1944, "Lehmbauweise Posen"; Himmler to Pohl, 17 Dec. 1943; Pohl and Schondorff, Report of 18 Apr. 1944, BAK NS19/3115; Budraß and Grieger, "Die Moral der Effizienz," 126. For further SS efforts toward rationalization of slave labor, see Allen, "Engineers and Modern Managers in the SS," 397–416.

41. Quotation from Wirtschaftshändlerjahrbuch 1941, submitted by Hohberg in his defense, Defense Document Books of same.

42. "Mindener Bericht," 78. The authors of the "Mindener Bericht" also insisted that "those with insight into the DWB exerted themselves to develop the concern ever more in purely economic directions" (76). They were already playing to the well-established impression in the Anglo-American world that a clear line divided "pure" business rationality from ideological considerations, as if the "business" itself had not involved manufacturing in a racial-supremacist utopia, slave labor, and other crimes. On Mummenthey's biography, see Allen, "Engineers and Modern Managers in the SS," 507–16.

43. Mummenthey to Pohl and Hohberg, 17 Sept. 1941, BAK NS3/1344.

44. Pohl's handwritten marginalia and Hohberg to Mummenthey, 25 Sept. 1941, BAK NS3/1344. See also Mummenthey to Pohl, 18 July 1940, "Organizationsplan und Geschäftsordnung der Deutschen Erd- und Steinwerke," T-976/25.

45. "Gutachten über die Golleschauer Portland-Zementfabrik," BAK NS3/164. See also Hans Baier, 22 Apr. 1944, "Prüfung der wirtschaftlichen Betriebe der Schutzstaffel (historische Entwicklung)," T-976/1, which also discusses the early history of Hohberg, Wenner, and Salpeter's early reorganization of the DESt and DWB.

46. Hohberg to Pohl, 27 Feb. 1941, "Zementfabrik Golleschau und Zementfabrik Saturn," BAK NS3/164. See also Hohberg letter to ODBS, 3 May 1943, and Bobermin to DWB, 27 Apr. 1943, BAK NS3/1369.

47. NO-1954, Oswald Pohl to Hohberg, 21 Sept. 1942, "Neuregelung der Dienststelle 'Wirtschaftsprüfer' in meinem Stabe."

48. NO-1035, Hohberg to Pohl, 21 July 1942.

49. Hohberg to Finanzamt Börse, 5 May 1943, "Gesellschaftssteuerbescheid vom 9.9.42," BDC SS Hängeordner: 3463. See also Hans Baier, 22 Apr. 1944, "Prüfung der wirtschaftlichen Betriebe der Schutzstaffel (historische Entwicklung)," T-976/1.

50. Testimony of Pohl, Protocol: 1644–45; Affidavit of Richard Ansorge, Defense Document Books of Hohberg. That Hohberg sometimes signed as "Chef," NO-1040, Hohberg to Chef Amt A2 and Hauptsturmführer Millmer, 3 June 1943; Hohberg to Finanzamt; RWM Staatssekretär Dr. Landfried, 3 June 1943, and Pohl to Deutsche Revisions- und Treuhand AG, 14 Sept. 1943, "Vermerk," T-976/1.

51. Leo Volk, 22 Apr. 1944, "Vermerk für Oberführer Baier," T-976/1. See also the ridicule expressed about Hohberg and accusations of betrayal made by Hans Baier, 22 Apr. 1944, "Prüfung der wirtschaftlichen Betriebe der Schutzstaffel (historische Entwicklung)," T-976/1.

52. NO-1016, 11 July 1944, "Organization und Aufgaben der Amtsgruppe W."

53. Leo Volk, 22 Apr. 1944, "Vermerk für Oberführer Baier," T-976/1.

54. See note 51 and Stab W to Dr. Karoli, undated draft of letter from late 1943; Pohl to Dr. Richard Karoli, 31 Oct. 1944, "DWB-Konzern-Prüfung"; and Baier, Vertraulicher Aktenvermerk, 12 Jan. 1945, T-976/1.

55. Mommsen, "Die Realisierung des Utopischen," 213–14. Compare Zuboff, *In the Age of the Smart Machine*, 97–223, 224–44; Jaeger, "Unternehmensgeschichte in Deutschland seit 1945," 126; Mai, "Die Ökonomie der Zeit," 319–20.

56. Goldhagen, *Hitler's Willing Executioners*.

57. Brustein, *The Logic of Evil*, and Sofsky, *Die Ordnung des Terrors*, esp. 32–33. See Allen, "Technocrats of Extermination."

58. If self-help books for managers take (and took) issues such as corporate culture seriously, historians of modern organization during the Nazi period should do so as well. For an example of a contemporary modern corporation that stumbled because of a lack of consensus, see the observations of Hirsh, *Technology and Transformation*, esp. his well-formulated thesis on p. 3. Berghoff, "Unternehmenskultur und Herrschaftstechnik," 172–80. On midlevel managers in general, see Beniger, *The Control Revolution*; Zunz, *Making America Corporate*; Chandler, *Scale and Scope*, which takes a comparative perspective. Regarding the German case, see Kocka, *Die Angestellten*. Regarding the Weimar and Nazi period, see Prinz, *Vom Mittelstand zum Volksgenossen*. Compare Mommsen, "Konnten Unternehmer im Nationalsozialismus apolitisch bleiben?," 69–72, 130.

59. Cited after Naasner, *Neue Machtzentren*, 394.

60. Dr. Volk to Technisches Amt Viii FHA, 31 May 1943, "Erfinderschutz bei SS und Polizei," T-976/14. See also Aktenvermerk in Stab W, 20 Dec. 1943; Chef des Technischen Amtes to Pohl, 17 Apr. 1944, "Erfinderschutz und Erfinderbetreuung," T-976/14; Walker, *German National Socialism and the Quest for Nuclear Power*, 136–37.

61. Reported in HSSPF Südwest to Reichsstatthalter Stuttgart und Gauleiter Baden, 24 Jan. 1944, "Unterbringung empfindlicher Rüstungsbetriebe," T-175/21. See also

Dr. Fitzner to Unstf. Daniels, personal staff RFSS, 10 Jan. 1944, T-976/14; "Mindener Bericht," 187–89.

62. NO-3793, Jacobi to Stab W, 4 Dec. 1944, "Lagerbericht"; Pohl to Grf. Grimm, 26 May 1944, BDC Hängeordner: 4244; NO-4071, 2 May 1944, Dr. Hoffmann to Mummenthey. Schieferöl GmbH included the foundation of a new satellite camp of Natzweiler through Maurer. The camp was supposed to hold 1,000 prisoners but was overfilled with 1,500, mostly Polish, prisoners. Conditions were atrocious and upward of 15 prisoners died per day in December 1944: Bauleitung der W-SS Erzingen Kreis Balingen Württemburg to Amt D2, 31 May 1944, BDC Hängeordner: 4242; "Mindener Bericht," 185–86; Pohl to Geilenberg, 6 Sept. 1944, BDC SS Hängeordner: 4240; Dr. Hoffmann; Jacobi; Volk, 10 Oct. 1944, "Aktenvermerk über die Entwicklung und den Stand der Schieferölgewinnung," BDC SS Hängeordner: 3453; NO-3893, Report of a meeting on 14 Oct. 1944 between Pohl, Krauch, and Volk. For further details of the Schieferöl GmbH, see Allen, "Engineers and Modern Managers in the SS," 537–42.

63. NO-3901, Bericht, 29 Nov. 1944, meeting of 21 Nov. 1944. See also NO-4077, Jacobi to Stab W, 6 Dec. 1944; Baier to Mummenthey, 22 Aug. 1944, BDC SS Hängeordner: 4241.

64. H. Jacobi to Hoffmann, 28 Mar. 1945 and also 29 Mar. 1945, BDC SS Hängeordner: 4228.

65. Ibid.

66. Ibid.

67. Orth, *Das System der nationalsozialistischen Konzentrationslager*, 192; Ludwig, *Technik und Ingenieure*, 487–502; Fröbe, "Der Arbeitseinsatz von KZ-Häftlingen," 358–63; Herbert, *Fremdarbeiter*, 251–63; Speer, Chronik, May–June 1944, BAK R3/1739.

68. Erhard Milch to Maurer, 13 Apr. 1943, T-175/80. In particular cases, Speer also sought the direct intervention of Maurer. See Speer, 8 June 1944, "Generalkommissar für die Sofortmassnahmen," T-175/70: 87633–36. Industries also sought his cooperation: Maurer to Hohberg, 5 Apr. 1943, "18-Fertigung bei den Reichswerken Hermann Goering in Druette bei Braunschweig," and Pohl to Paul Pleiger, "Vertrag mit den HGW in Druette b/Braunschweig" 5 June 1943, BAP, PL5: 42064.

69. NI-315, Budin's letter included in Pohl to Himmler, 17 Oct. 1944. See other examples: Mummenthey to Pohl, 1 Sept. 1944, "Giesserei und Temperei," and Mummenthey to Pohl, 26 Aug. 1944, "Angelegenheit Tweer," BAK NS3/1344. This was the typical, not the exceptional, relationship of private industry to slave-labor brokers. Spoerer, "Profitierten Unternehmen von KZ-Arbeit?," 61–95.

70. Budraß and Grieger, "Die Moral der Effizienz," 126. Actual productivity rates hovered around 90 percent that of civilian workers, still an incredibly high figure.

71. NO-1202, Affidavit of Hans Moser; Fröbe, "Der Arbeitseinsatz von KZ-Häftlingen," 361–65; Budraß and Grieger, "Die Moral der Effizienz," 110–32; Dieckmann, "Existenzbedingungen und Widerstand in KL Dora-Mittelbau," 69–71; Janssen, *Das Ministerium Speer*, 102; NI-638, Reichsminister fuer Ruestungs- und Kriegsproduktion (unsigned) to all Ruestungskommissionen, Inspecteure, Kommandeure, Hauptausschuesse, Ringe und Produktionsausschuesse, Reichsvereinigung Eisen, GBA, WVHA, 9 Oct. 1944, "Anforderung und Einsatz von KZ-Haeftlingen"; NO-1977, Maurer to Kommandanturen-Arbeitseinsatz, 29 Jan. 1945, "Forderungsnachweise, Zusammenstellung und Übersicht."

72. Neufeld, *The Rocket*, 216. For examples of praise for Kammler, see Göring to Himmler, 27 June 1944, and Dorsch to Kammler, 16 June 1944, NS19/2065.

73. Neufeld, *The Rocket*, 180–82.

74. Baudrexl, "Als Techniker in der deutschen Rüstung," 18. Compare Neufeld, *The Rocket*, 213–20.

75. Ordway and Sharpe, *The Rocket Team*, 193; see also 62.

76. Mommsen, "Erfahrungen mit der Geschichte der Volkswagenwerk GmbH," 47; similarly, *Der Mythos von der Modernität*, 25.

77. Overy, *Why the Allies Won*, 239–40. Compare Neufeld, *The Rocket*, 176–82, 239; Bornemann, *Geheimprojekt Mittelbau*, 124. Rocket team member Gerhard Stegmaier informed on his colleagues. Arrests among members of the rocket team had occurred before on the same grounds—those of the primacy of security. No less, in these earlier cases voluntary informants within the high-tech project itself had triggered the suspicions of Himmler's secret police, which Himmler then sought to use to his advantage. Typically the SS operated by acting on the information, dubious or not, of willing informants. There is no reason to suspect that the SS acted toward the rocket team any differently than it acted against the German population in general. See Gellately, *The Gestapo and German Society*, 13. The research and development "scientists" were becoming increasingly marginalized within Mittelwerk's own civilian management before the SS's "takeover." The end of "Entstehungsgeschichte des Versuchsserienwerkes Peenemünde," 1944, BA MA RH8/1210, relates that von Braun's experiment stand was packed up and assigned to one of the factory centers (Zement at the RAX Works). Jens-Christian Wagner has pointed out to me that von Braun remained in control of final testing, however; notably, this meant final judgment over whether "sabotage" had occurred. Both Speer and Degenkolb supported the effort to curb research and development in favor of production.

78. Fritz Kranfuß to RFSS, 3 Nov. 1944, BAK NS19/1677.

79. Brandt to Kranfuß, 7 Nov. 1944, BAK NS19/1677.

80. NO-2144, Heinrich Himmler to Kammler, 5 Aug. 1944; Anonymous, undated [from July or Aug. 1944], "Aktennotiz 260," NS4 Anhang/31; Himmler to Kammler, Jüttner, Grothmann, 6 Aug. 1944, NS19/2055. See the similar case of the Nazi powder rocket: Hübner, BzbV Heer to Oberstlt. Borttscheller, 9 Jan. 1945; and Kammler to Beauftragte zur besonderen Verfügung des Heeres (powder rocket), 9 Jan. 1945, "Rheinbote," BA MA RH8/1306; Neufeld, *The Rocket*, 230, 240–48; Bornemann, *Geheimprojekt Mittelbau*, 103–8; NO-2615 and NO-2611, Weigel and Kammler to Pohl/Himmler, 29 July 1944, "Bericht über den Einsatz der 5. und 1. SS-Baubrigade beim Bau von Stellungen und Nachschubanlagen für V1 u. V2 im Westen." Jens-Christian Wagner puts the ratio at two prisoners to every one civilian worker in the spring of 1944 (and higher than this before). Albin Sawatzki had planned on a ratio of 8:1. As Wagner notes, many of the civilians were not "workers" in the conventional sense but office personnel, technicians, low-level managers, foremen, and the like. In May 1944, for example, almost 5,000 prisoners worked on rocket assembly; 2,500 civilians supervised them or worked by their side. Meanwhile the Special Staff–Kammler was driving an additional 10,000 prisoners in tunneling and construction. See "Verlagerungswahn und Tod," 179–84, esp. table 5. I thank Wagner for providing me this table.

81. From 31 Mar. 1945, *Die Tagebücher von Joseph Goebbels*, IfZ.

82. Neufeld, *The Rocket*, 257 and 239–65 in general; NO-2615 and NO-2611; Duwe to all Einsatzgruppen der OT, Amtsgruppe C5, WVHA, Noell, 9 Sept. 1944, "Baumassnahmen der W-SS, 'Unternehmen X,'" Vha, Fond "OT," 31/5; unrecognizable signature to Professor Osenberg, 13 Feb. 1945, "Besprechung mit General Dornberger am 10 Feb. 45," BAK R26 III/52; Bornemann, *Geheimprojekt Mittelbau*, 103–25; Speer, "Besprechung beim Führer am 12 Oct. 44," and "Besprechung beim Führer vom 21–23 Sep. 44," BAK R3/1510.

83. Speer to Himmler, 27 Nov. 1944, BAK R3/1583.

84. Jägerstabsitzung, 26 May 1944, BA MA RL3/7. See also Karny, "'Vernichtung durch Arbeit' in Leitmeritz," 37–61; Ludwig, *Technik und Ingenieure*, 463–73.

85. Jägerstabbesprechung, 1 July 1944, BA MA RL3/9.

86. James, "Die Rolle der Banken im Nationalsozialismus," 33–34; Ludwig, *Technik und Ingenieure*, 495–99; Jägerstabbesprechung, 31 Mar. 1944, BA MA RL3/4, and 5 July 1944, BA MA RL3/9. Compare Orth, *Das System der nationalsozialistischen Konzentrationslager*, 248–54.

87. "Niederschrift über die Jägerstabbesprechung," 18 Mar. 1944, BA MA RL3/2.

88. Piper, *Die Zahl der Opfer von Auschwitz*, table D; Hilberg, *The Destruction of the European Jews*, 805–16. On labor shortages, see the Jägerstabbesprechung, 14 and 15 Apr. 1944, BA MA RL3/5, and Jägerstabbesprechung, 26 May and 9 June 1944, BA MA RL3/7.

89. Jägerstabbesprechung, 1 July 1944, BA MA RL3/9; Fröbe, "Hans Kammler," 316; Raim, *Die Dachauer KZ-Außenkommandos*, 44–56; Perz, *Projekt Quarz*, 208–9, 245–50. On labor shortfalls, see Jothan to BI Schlesien, 25 Mar. 1944, "Bericht über den Stand der Bauarbeiten im KL Auschwitz einschl. Häftlingseinsatz," USHMM RG-11.001M.03: 24 (502-1-83); Bischoff, 7 May 1943, "Bericht über die Schwierigkeiten, welche beim Ausbau des KGL auftreten," USHMM RG-11.001M.03: 24 (502-1-83); Albrecht, "Military Technology and National Socialist Ideology," 88–125; Himmler to PHA, 3 Feb. 1945, BDC SS Personal-Akte Hans Kammler; unrecognizable signature to Professor Osenberg, 13 Feb. 1945, "Besprechung mit General Dornberger am 10 Feb. 45," BAK R26 III/52; Speer, "Besprechung beim Führer am 28 Nov. 44," BAK R3/1510; Speer to Himmler, 27 Nov. 1944, BAK R3/1583; Berger to Himmler, 9 Nov. 1944, "Flugzeugbauprogramm 162," T-175/122; Verteiler, Oberkommando des Heeres Leeb, 27 Feb. 1945, BA MA RH8/1941. On utopian building, see Kammler to Brandt, 12 Dec. 1944, BAK NS19/3346; Remdt and Wermusch, *Rätsel Jonastal*, 76–77; Orth, *Das System der nationalsozialistischen Konzentrationslager*, 305–8.

90. On the Auschwitz engineers, see Pressac, *Die Krematorien von Auschwitz*, 173–80. On Glücks, see NO-1202, Affidavit of Hans Moser. On the bizarre festival at Dachau, see NO-2331, Affidavit of Gerhard Wiebeck.

EPILOGUE

1. Protocol: 1543; Affidavit of Adolf Reiner and Otto Georges, Defense Document Books of Karl Mummenthey. See also Defense Document Books of August Frank and Hans Baier. For a similar case to Baier, see those of Franz Eirenschmalz.

2. Affidavit of Otto Georges.

3. Testimony of Eugen Kogon, Protocol: 893–95.

4. Biagioli, "Science, Modernity, and the Final Solution," 185–205; Hecht and Allen, "Authority, Political Machines, and Technology's History."

5. Allen, "Modernity, the Holocaust, and Machines without History."

6. Scott, *Seeing Like a State*.

7. Schabel, "Wenn Wunder den Sieg bringen sollen," 397.

8. Albrecht, "Military Technology and National Socialist Ideology."

9. Compare Schulte, "Rüstungsunternehmen oder Handwerksbetrieb?," 558–83.

10. Kempner, *SS im Kreuzverhör*, 105.

11. Friedrich, *Die kalte Amnestie*.

12. Rostow and Millikan, *A Proposal*, first quotation from 1; second, 39–40. On technical, "scientific" aid, see also 54, 61–63. Rostow had worked within the European Commission for Europe in Geneva and Millikan had worked within various committees of the Marshall Plan. Thus both were involved in far more than intellectual debates about modernization and the reconstruction of Europe; they were active policy makers.

13. Radkau, "Kontinuität über 1945 hinweg," 438; Hutton and Lawrence, *German Engineers*, 115.

14. Gerhard Maurer, undated "Lebenslauf," BAP, PL5: 42063.

15. Orth, "Die Kommandanten der nationalsozialistischen Konzentrationslager," 755.

16. Stern, *Dreams and Delusions*, 39–40.

17. "Urteil des Schwurgerichts Frankfurt," in *Justiz und NS-Verbrechen. Sammlung deutscher Strafurteile wegen nationalsozialistischer Tötungsverbrechen 1945–1966*, vol. 13 (Amsterdam: University Press Amsterdam, 1975), 127.

18. Arendt, *Totalitarianism*, 27.

19. Browning, "The Development and Production of the Nazi Gas Van," 66–67.

20. Bauman, *Modernity and the Holocaust*, 195.

21. Fröbe compares the atom bomb project briefly to similar research and development projects in the Third Reich in "KZ-Häftlinge als Reserve qualifizierte Arbeitskraft," 663 and n.

22. Hüttenberger, "Nationalsozialistische Polykratie," 421–23.

23. Neumann, *Behemoth*, 44.

24. Hilberg, *The Destruction of the European Jews*, 18.

25. Mills, *White Collar*, 234–35.

26. Mommsen, "Konnten Unternehmer im Nationalsozialismus apolitisch bleiben?," 69; Mommsen and Grieger, *Das Volkswagenwerk*, 945.

27. Hilberg, *The Destruction of the European Jews*, 640.

28. Ibid., 4.

29. See the hilarious chapter of Musil, *Man ohne Eigenschaften*, 909–29; Porter, *Trust in Numbers*, 79–84.

30. Parcer and Grotum, *Sterbebücher des KL Auschwitz–Birkenau*, x–xi. Compare Wootton and Wolk, "The Evolution and Acceptance of the Loose-Leaf Accounting System."

31. Compare Zuboff, *In the Age of the Smart Machine*.

32. Harwood, *Styles of Scientific Thought*. See also the lists of professionals in Friedlander, *The Origins of Nazi Genocide*, tables 6.2, 6.4, 6.5.

33. Orth has also treated the continuities of ideology and organization in the IKL throughout *Die Konzentrationslager-SS*.

34. Gellately, " 'A Monstrous Uneasiness,' " 180.

35. Aly, "Theodor Schieder, Werner Conze oder die Vorstufen der physischen Vernichtung," 167. See also Haar, " 'Kämpfende Wissenschaft,' " 215–40.

36. Hans-Ulrich Wehler, discussion at the Ludwig Maximillians Universität München, 14 February 2000.

BIBLIOGRAPHY

Abrahamson, Irving. *Against Silence: The Voice and Vision of Elie Wiesel*. New York: Holocaust Library, 1985.

Ackermann, Josef. *Heinrich Himmler als Ideologe*. Göttingen: Musterschmidt, 1970.

Agoston, Tom. *Teufel oder Technokrat: Hitlers graue Eminenz*. Berlin: Verlag E. S. Mittler & Sohn, 1993.

Albrecht, Ulrich. "Military Technology and National Socialist Ideology." In *Science, Technology and National Socialism*, edited by Mark Walker and Monika Renneberg, 88–125. Cambridge: Cambridge University Press, 1994.

Alder, Ken. *Engineering the Revolution: Arms and Enlightenment in France*. Princeton: Princeton University Press, 1997.

——. "Innovation and Amnesia: Engineering Rationality and the Fate of Interchangeable Parts Manufacturing in France." *Technology and Culture* 38 (1997): 273–311.

——. "A Revolution to Measure: The Political Economy of the Metric System in France." In *The Values of Precision*, edited by M. Norton Wise, 39–71. Princeton: Princeton University Press, 1995.

Allen, Michael Thad. "The Banality of Evil Reconsidered: SS Mid-Level Managers of Extermination through Work." *Central European History* 30 (1997): 253–94.

——. "Engineers and Modern Managers in the SS: The Business Administration Main Office." Ph.D. diss., University of Pennsylvania, 1995.

——. "Flexible Production at Concentration Camp Ravensbrück." *Past and Present* 165 (1999): 182–217.

——. "Ideology Counts: Controlling the Bodies of Concentration Camp Prisoners." In *Cultures of Control*, edited by Miriam Levin, 177–204. Amsterdam: Harwood Academic Publishers, 2000.

——. "Modernity, the Holocaust, and Machines without History." In *Technologies of Power*, edited by Gabrielle Hecht and Michael Thad Allen, 175–214. Cambridge: MIT Press, 2001.

——. "The Puzzle of Nazi Modernism: Modern Technology and Ideological Consensus in an SS Factory at Auschwitz." *Technology and Culture* 37 (1996): 527–71.

——. "Technocrats of Extermination: Engineers, Modern Bureaucracy, and Complicity." In *Lessons and Legacies of the Holocaust*, edited by Ronald Smelser. Evanston: Northwestern University Press, forthcoming.

Aly, Götz. *"Endlösung" Volkerverschiebung und der Mord an den europäischen Juden*. Frankfurt am Main: Fischer, 1995.

——. "Theodor Schieder, Werner Conze oder die Vorstufen der physischen

Vernichtung." In *Deutsche Historiker im Nationalsozialismus*, edited by Winfried
Schulze and Otto Gerhard Oexle, 163–82. Frankfurt am Main: Fischer, 1999.

Aly, Götz, Peter Chroust, and Christian Pross. *Cleansing the Fatherland: Nazi Medicine
and Racial Hygiene*. Baltimore: Johns Hopkins University Press, 1994.

Aly, Götz, and Susanne Heim. *Vordenker der Vernichtung Auschwitz und die deutschen
Pläne für eine neue europäische Ordnung*. Frankfurt am Main: Fischer, 1993.

Appleby, Joyce, Lynn Hunt, and Margaret Jacob. *Telling the Truth about History*. New
York: Norton, 1994.

Arendt, Hannah. *Part Three of Origins of Totalitarianism*. New York: Harcourt Brace
Jovanovich, 1951.

———. *Eichmann in Jerusalem: A Report on the Banality of Evil*. 1963. Reprint, New York:
Penguin Books, 1977.

Armanski, Gerhard. *Maschinen des Terrors: Das Lager (KZ und GULAG) in der
Moderne*. Münster: Verlag Westfälisches Dampfboot, 1993.

Arndt, Ino. "Das Frauenkonzentrationslager Ravensbrück." *Dachauer Hefte* 3 (1987):
125–57.

Bartel, Walter, et al., eds. *Buchenwald Mahnung und Verpflichtung*. Berlin: Kongress
Verlag, 1960.

Bartov, Omer. *Hitler's Army: Soldiers, Nazis, and War in the Third Reich*. Oxford:
Oxford University Press, 1992.

———. *Murder in Our Midst: The Holocaust, Industrial Killing, and Representation*.
Oxford: Oxford University Press, 1996.

———. "Passing into History: Nazism and the Holocaust beyond Memory." *History and
Memory* 9 (1997): 162–88.

Baudrexl, K. Friedrich. "Als Techniker in der deutschen Rüstung." In *Vernichtung durch
Fortschritt am Beispiel der Raketenproduktion im Konzentrationslager Mittelbau*,
edited by Torsten Hess and Thomas Seidel, 16–20. Weimar: Westkreuz Verlag, 1995.

Bauman, Zygmunt. *Modernity and the Holocaust*. Ithaca: Cornell University Press, 1989.

Beniger, James. *The Control Revolution: Technological and Economic Origins of the
Information Society*. Cambridge: Harvard University Press, 1986.

Béon, Yves. *Planet Dora: A Memoir of the Holocaust and the Birth of the Space Age*. New
York: Westview Press, 1997.

Berghoff, Hartmut. "Unternehmenskultur und Herrschaftstechnik Industrieller Pater-
nalismus: Hohner von 1857 bis 1918." *Geschichte und Gesellschaft* 23 (1997): 167–204.

Bernadac, Christian. *Camp for Women: Ravensbrück*. 1968. Reprint, Paris: Ferni
Publishing House, 1978.

Beyerchen, Alan. "Rational Means and Irrational Ends: Thoughts on the Technology of
Racism in the Third Reich." *Central European History* 30 (1997): 386–402.

Biagioli, Mario. "Science, Modernity, and the Final Solution." In *Probing the Limits of
Representation: Nazism and the "Final Solution,"* edited by Saul Friedländer, 185–
205. Cambridge: Harvard University Press, 1992.

Birn, Ruth Bettina. *Die Höheren SS- und Polizeiführer Himmlers Vertreter im Reich und
in den besetzten Gebieten*. Düsseldorf: Droste, 1986.

Black, Edwin. *IBM and the Holocaust: The Strategic Alliance between Nazi Germany and
America's Most Powerful Corporation*. New York: Crown Publishers, 2001.

Black, Peter. "Odilo Globocnik—Himmlers Vorposten im Osten." In *Die Braune Elite II*, edited by Ronald Schmelser and Rainer Zitelmann, 103–15. Darmstadt: Wissenschaftliche Buchgesellschaft, 1993.

Boehnert, Gunnar. "The Third Reich and the Problem of 'Social Revolution': German Officers and the SS." In *Germany in the Age of Total War*, edited by Volker Berghahn and Martin Kitchen, 203–17. London: Barnes & Noble, 1981.

Boelcke, Willi. *Deutschlands Rüstung im Zweiten Weltkrieg: Hitlers Konferenzen mit Albert Speer 1942–1945*. Frankfurt am Main: Akademische Verlagsgesellschaft, 1969.

Bolenz, Eckhard. *Vom Baubeamten zum freiberuflichen Architekten: Technische Berufe im Bauwesen (Preussen/Deutschland, 1799–1931)*. Frankfurt am Main: P. Lang, 1999.

Bornemann, Manfred. *Aktiver und Passiver Widerstand im KZ Dora und im Mittelwerk*. Berlin: Westkreuz Verlag, 1994.

——. *Geheimprojekt Mittelbau: Die Geschichte der deutschen V-Waffen Werke*. Munich: J. F. Lehmanns, 1971.

Bourdieu, Pierre. *Outline of a Theory of Practice*. Cambridge: Cambridge University Press, 1977.

Braun, Hans-Joachim. "Fertigungsprozesse im deutschen Flugzeugbau 1926–45." *Technikgeschichte* 57 (1990): 111–35.

Breitman, Richard. *The Architect of Genocide: Himmler and the Final Solution*. New York: Alfred A. Knopf, 1991.

Brenner, Hans. "Der 'Arbeitseinsatz' der KZ-Häftlinge in den Außenlagern des Konzentrationslagers Flossenbürg—ein Überblick." In *Die Nationalsozialistischen Konzentrationslager—Entwicklung und Struktur*, vol. 2, edited by Ulrich Herbert, Karin Orth, and Christoph Dieckmann, 682–706. Göttingen: Wallstein Verlag, 1998.

Brose, Eric. *The Politics of Technological Change in Prussia: Out of the Shadow of Antiquity, 1809–1848*. Princeton: Princeton University Press, 1993.

Broszat, Martin. "Nationalsozialistische Konzentrationslager." In *Anatomie des SS-Staates*, vol. 2, edited by Helmut Krausnick, Hans Buchheim, Martin Broszat, and Hans-Adolf Jacobsen, 37–73. Munich: Deutscher Taschenbuchverlag, 1967.

——. *Nationalsozialistische Polenpolitik 1939–45*. Stuttgart: Deutsche Verlags-Anstalt, 1961.

Browning, Christopher. "Beyond 'Intentionalism' and 'Functionalism': A Reassessment of Nazi Jewish Policy from 1939 to 1941." In *Reevaluating the Third Reich*, edited by Thomas Childers and Jane Caplan, 211–33. New York: Holmes & Meier, 1993.

——. "Bureaucracy and Mass Murder: The German Administrator's Comprehension of the Final Solution." In *The Path to Genocide: Essays on Launching the Final Solution*, 125–44. Cambridge: Cambridge University Press, 1995.

——. "The Development and Production of the Nazi Gas Van." In *Fateful Months: Essays on the Emergence of the Final Solution*, 57–67. New York: Holmes & Meier, 1985.

——. "The Semlin Gas Van and the Final Solution in Serbia." In *Fateful Months: Essays on the Emergence of the Final Solution*, 68–87. New York: Holmes & Meier, 1985.

——. "Vernichtung und Arbeit: Zur Fraktionierung der planenden deutschen Intelligenz im besetzten Polen." In *"Vernichtungspolitik." Eine Debatte über den Zusammenhang von Sozialpolitik und Genozid im nationalsozialistischen Deutschland*, edited by Wolfgang Schneider, 37–52. Hamburg: Junius Verlag, 1991.

Brustein, William. *The Logic of Evil: The Social Origins of the Nazi Party, 1925–1933*. New Haven: Yale University Press, 1996.

Buchheim, Hans. "Befehl und Gehorsam." In *Anatomie des SS-Staates*, vol. 1, edited by Helmut Krausnick, Hans Buchheim, Martin Broszat, and Hans-Adolf Jacobsen, 215–318. Munich: Deutscher Taschenbuch Verlag, 1979.

———. "Die SS—das Herrschaftsinstrument." In *Anatomie des SS-Staates*, vol. 1, edited by Helmut Krausnick, Hans Buchheim, Martin Broszat, and Hans-Adolf Jacobsen, 1–115. Munich: Deutscher Taschenbuch Verlag, 1979.

———. "The SS—Instrument of Domination." In *Anatomy of the SS State*, vol. 1, edited by Helmut Krausnick, Hans Buchheim, Martin Broszat and Hans-Adolf Jacobsen, 127–302. 1965. Reprint, New York: Walker and Company, 1968.

Budraß, Lutz, and Manfred Grieger. "Die Moral der Effizienz: Die Beschäftigung von KZ-Häftlingen am Beispiel des Volkswagenwerks und der Henschel Flugzeug-Werke." *Jahrbuch für Wirtschaftsgeschichte* 2 (1993): 89–135.

Burleigh, Michael. *Death and Deliverance: "Euthanasia" in Germany, 1900–1945*. Cambridge: Cambridge University Press, 1994.

Burleigh, Michael, and Wolfgang Wippermann. *The Racial State, 1933–1945*. Cambridge: Cambridge University Press, 1991.

Calvert, Monte A. *The Mechanical Engineer in America, 1830–1910: Professional Cultures in Conflict*. Baltimore: Johns Hopkins University Press, 1967.

Caplan, Jane. *Government without Administration: State and Civil Service in Weimar and Nazi Germany*. Oxford: Clarendon Press, 1988.

Chandler, Alfred. *Scale and Scope: The Dynamics of Industrial Capitalism*. Cambridge: Harvard University Press, 1990.

———. *The Visible Hand: The Managerial Revolution in American Business*. Cambridge: Harvard University Press, 1977.

Cohen, Yves. "The Modernization of Production in the French Automobile Industry between the Wars: A Photographic Essay." *Business History Review* 65 (1991): 754–80.

Corni, Gustavo. "Die Agrarpolitik des Faschismus: Ein Vergleich zwischen Deutschland und Italien." *Tel Aviver Jahrbuch für deutsche Geschichte* 17 (1988): 391–423.

———. *Hitler and the Peasants: Agrarian Policy of the Third Reich, 1930–1939*. Munich: Berg, 1990.

Coyner, S. J. "Class Consciousness and Consumption: The New Middle Class during the Weimar Republic." *Journal of Social History* 10 (1977): 310–31.

Czech, Danuta. "KL Auschwitz as an Extermination Camp." In *Selected Problems from the History of KL Auschwitz*, edited by Kazimierz Smolen, 29–50. Oswiecim: Panstwowe Muzeum w Oswiecimiu, 1979.

Dallin, Alexander. *German Rule in Russia, 1941–1945: A Study of Occupation Policies*. London: Macmillan, 1957.

Dawidowicz, Lucy. *The Holocaust and the Historians*. Cambridge: Harvard University Press, 1981.

Dieckmann, Götz. "Existenzbedingungen und Widerstand in KL Dora-Mittelbau unter dem Aspekt der funktionellen Einbeziehung der SS in das System der faschistischen Kriegswirtschaft." D.Phil., Humboldt Universität, 1968.

Dietz, Burkhard, Michael Fessner, and Helmut Maier. " 'Der Kulturwert der Technik'

als Argument der Technischen Intelligenz für sozialen Aufstieg und Anerkennung."
In *Technische Intelligenz und "Kulturfaktor Technik": Kulturvorstellungen von
Technikern und Ingenieuren zwischen Kaiserreich und früher Bundesrepublik
Deutschland*, edited by Dietz, Fessner, and Maier, 1–34. Münster: Waxmann, 1996.

Drobisch, Klaus, and Günther Wieland. *System der NS-Konzentrationslager 1933–39.*
Berlin: Akademie Verlag, 1993.

Duff, Shiela. *A German Protectorate: The Czechs under Nazi Rule.* London: Frank Cass,
1970.

Durth, Werner. *Deutsche Architekten Biographische Verflechtungen 1900–1970.*
Braunschweig: Friedr. Vieweg & Sohn, 1986.

Eichholtz, Dietrich, and Joachim Lehmann. *Geschichte der deutschen Kriegswirtschaft.*
Vol. 2: *1941–43.* Berlin: Akademie-Verlag, 1985.

Eizenstat, Stuart, et al. *U.S. and Allied Efforts to Recover and Restore Gold and Other
Assets Stolen or Hidden by Germany during World War II.* Washington, D.C.: U.S.
Holocaust Memorial Museum, 1997.

Evans, Richard. *In Hitler's Shadow: West German Historians and the Attempt to Escape
from the Nazi Past.* New York: Pantheon, 1989.

Fairbairn, Brett. "History from the Ecological Perspective: Gaia Theory and the
Problem of Cooperatives in Turn-of-the-Century Germany." *American Historical
Review* 99 (1994): 1203–39.

Fauser, Ellen. "Zur Geschichte des Außenlagers Langenstein-Zwieberge." In
Zwangsarbeit und die unterirdische Verlagerung, edited by Torsten Hess, 37–49.
Berlin: Westkreuz Verlag, 1994.

Ferencz, Benjamin. *Less than Slaves: Jewish Forced Labor and the Quest for
Compensation.* Cambridge: Harvard University Press, 1979.

Ford, Henry. *Mein Leben und Werk.* Leipzig: Paul List Verlag, 1923.

Frei, Norbert. "Wie Modern war der Nationalsozialismus?" *Geschichte und Gesellschaft*
19 (1993): 367–87.

Freund, Florian. *"Arbeitslager Zement." Das Konzentrationslager Ebensee und die
Raketenrüstung.* Vienna: Verlag für Gesellschaftskritik, 1989.

Freund, Florian, Bertrand Perz, and Karl Stuhlpfarrer. "Der Bau des Vernichtungslagers
Auschwitz-Birkenau." *Zeitgeschichte* 20 (1993): 187–214.

Friedlander, Henry. "The Nazi Concentration Camps." In *Human Responses to the
Holocaust: Perpetrators and Victims, Bystanders and Resisters*, edited by Michael
Ryan, 33–69. New York: Edwin Mellen Press, 1981.

——. *The Origins of Nazi Genocide: From Euthanasia to the Final Solution.* Chapel Hill:
University of North Carolina Press, 1995.

Friedländer, Saul. Introduction to *Hitler und die Endlösung: "Es is des Führers
Wunsch . . .,"* by Gerald Fleming, xii–xlv. Frankfurt am Main: Ullstein, 1987.

——. "West Germany and the Burden of the Past: The Ongoing Debate." *Jerusalem
Quarterly* 42 (1987): 3–18.

Friedrich, Joerg. *Die kalte Amnestie: NS Täter in der Bundesrepublik.* Frankfurt am
Main: Fischer, 1984.

Fröbe, Rainer. "Der Arbeitseinsatz von KZ-Häftlingen und die Perspektive der
Industrie, 1943–45." In *Europa und der "Reichseinsatz." Ausländische Zivilarbeiter,*

Kriegsgefangene und KZ-Häftlinge in Deutschland 1938–1945, edited by Ulrich Herbert, 351–83. Essen: Klartext, 1991.

———. "Hans Kammler." In *Die SS: Elite unter dem Totenkopf*, edited by Ronald Smelser and Enrico Syring, 298–318. Paderborn: Schöningh, 2000.

———. "KZ-Häftlinge als Reserve qualifizierte Arbeitskraft: Eine späte Entdeckung der deutschen Industrie und ihre Folgen." In *Die Nationalsozialistischen Konzentrationslager—Entwicklung und Struktur*, vol. 2, edited by Ulrich Herbert, Karin Orth, and Christoph Dieckmann, 636–81. Göttingen: Wallstein Verlag, 1998.

Fromm, Erich. *Arbeiter und Angestellte am Vorabend des Dritten Reiches*. Stuttgart: Deutsche Verlags-Anstalt, 1980.

Gellately, Robert. *Backing Hitler: Consent and Coercion in Nazi Germany*. Oxford: Oxford University Press, 2001.

———. *The Gestapo and German Society: Enforcing Racial Policy, 1933–1945*. Oxford: Oxford University Press, 1990.

———. " 'A Monstrous Uneasiness': Citizen Participation and Persecution of the Jews in Nazi Germany." In *Lessons and Legacies: The Meaning of the Holocaust in a Changing World*, edited by Peter Hayes, 178–95. Evanston: Northwestern University Press, 1991.

Georg, Enno. *Die wirtschaftlichen Unternehmungen der SS*. Stuttgart: Deutsche Verlags-Anstalt, 1963.

Gerlach, Christian. *Krieg, Ernährung, Völkermord*. Hamburg: Hamburger Edition HIS Verlag, 1998.

———. "Die Wannsee-Konferenz, das Schicksal der deutschen Juden und Hitlers politische Grundsatzentscheidung, alle Juden Europas zu ermorden." *Werkstattgeschichte* 18 (1997): 7–44.

———. "Wirtschaftsinteressen, Besatzungspolitik und der Mord an den Juden in Weißrußland 1941–1943." In *Nationalsozialistische Vernichtungspolitik 1939–1945: Neue Forschungen und Kontroversen*, edited by Ulrich Herbert, 263–91. Frankfurt am Main: Fischer, 1998.

Giddens, Anthony. *The Consequences of Modernity*. Stanford: Stanford University Press, 1990.

Gillingham, John. *Industry and Politics in the Third Reich: Ruhr Coal, Hitler and Europe*. New York: Columbia University Press, 1985.

Gispen, Kees. "German Engineers and American Social Theory: Historical Perspectives on Professionalization." *Comparative Studies in Society and History* 30 (1988): 550–74.

———. *New Profession, Old Order: Engineers and German Society, 1815–1914*. Cambridge: Cambridge University Press, 1989.

Goffman, Erving. *Asylums: Essays on the Social Situation of Mental Patients and Other Inmates*. New York: Anchor Books, 1961.

Goldhagen, Daniel. *Hitler's Willing Executioners: Ordinary Germans and the Holocaust*. New York: Alfred A. Knopf, 1996.

Green, Nancy. *Ready-to-Wear and Ready-to-Work: A Century of Industry and Immigrants in Paris and New York*. Durham: Duke University Press, 1997.

Grieger, Manfred. " 'Vernichtung durch Arbeit' in der deutschen Rüstungsindustrie." In *Vernichtung durch Fortschritt am Beispiel der Raketenproduktion im Konzentrationslager Mittelbau*, edited by Torsten Hess and Thomas Seidel, 35–62. Bonn: Westkreuz Verlag, 1995.

Grundmann, Friedrich. *Agrarpolitik im Dritten Reich*. Hamburg: Hoffmann und Campe, 1979.

Haar, Ingo. " 'Kämpfende Wissenschaft': Entstehung und Nidergang der völkischen Geschichtswissenschaft im Wechsel der Systeme." In *Deutsche Historiker im Nationalsozialismus*, edited by Winfried Schulze and Otto Gerhard Oexle, 215–40. Frankfurt am Main: Fischer, 1999.

Habermas, Jürgen. *Theorie des kommunikativen Handelns*. Vols. 1, 2. Frankfurt am Main: Suhrkamp, 1981.

Hacking, Ian. *The Social Construction of What?* Cambridge: Harvard University Press, 1999.

Haffner, Sebastian. *The Meaning of Hitler*. Cambridge: Harvard University Press, 1979.

Hartung, Ulrich. "Bauästhetik im Nationalsozialismus und die Frage der Denkmalwürdigkeit." In *Raktionäre Modernität und Völkermord: Probleme des Umgangs mit der NS-Zeit in Austellungen und Gedenkstätten*, edited by Bernd Fankenbach and Franz-Josef Jelich, 71–84. Essen: Klartext-Verlag, 1994.

Harwood, Jonathan. *Styles of Scientific Thought: The German Genetics Community, 1900–1933*. Chicago: University of Chicago Press, 1993.

Hayes, Peter. "Big Business and Aryanization in Germany, 1933–1939." *Jahrbuch für Antisemitismusforschung* 3 (1994): 254–81.

———. *Industry and Ideology: IG Farben in the Nazi Era*. Cambridge: Cambridge University Press, 1987.

———. "Polycracy and Policy in the Third Reich: The Case of the Economy." In *Reevaluating the Third Reich*, edited by Thomas Childers and Jane Caplan, 190–210. New York: Holmes & Meier, 1993.

Hecht, Gabrielle, and Michael Thad Allen. "Authority, Political Machines, and Technology's History." In *Technologies of Power*, edited by Allen and Hecht, 1–28. Cambridge: MIT Press, 2001.

Heilbron, John. "Introductory Essay." In *The Quantifying Spirit in the Eighteenth Century*, edited by Tore Frängsmyr, John Heilbron, and Robin Rider, 1–23. Berkeley: University of California Press, 1990.

———. "The Measure of Enlightenment." In *The Quantifying Spirit in the Eighteenth Century*, edited by Tore Frängsmyr, John Heilbron, and Robin Rider, 207–61. Berkeley: University of California Press, 1990.

Heineman, John. *Hitler's First Foreign Minister: Constantin Freiherr von Neurath, Diplomat and Statesman*. Berkeley: University of California Press, 1979.

Heinemann, Isabel. "Die Kooperation von SS, Polizei und Zivilverwaltung bei der bevölkerungspolitischen Neuordnung." Paper delivered at the special conference Networks of Persecution: The Holocaust as Division-of-Labor-Based Crime, Konstanz, 26 September 2000.

Henke, Josef. "Von den Grenzen der SS-Macht: Eine Fallsstuie zur Tätigkeit des SS-Wirtschaftsverwaltungshauptamt." In *Verwaltung contra Menschenführung im Staat*

Hitlers: Studien zum politisch-administrativen System, edited by Dieter Rebentisch and Karl Teppe, 255–77. Göttingen: Vandenhoeck & Ruprecht, 1986.

Herbert, Ulrich. "Arbeit und Vernichtung: Ökonomisches Interesse und Primat der 'Weltanschauung' im Nationalsozialismus." In *Ist der Nationalsozialismus Geschichte? Zur Historrisierung und Historikerstreit*, edited by Dan Diner, 198–236. Frankfurt am Main: Fischer, 1987.

——. *Best: Biographische Studien über Radikalismus, Weltanschauung und Vernunft 1903–1989*. Bonn: Dietz, 1996.

——. *Fremdarbeiter: Politik und Praxis des "Ausländer-Einsatzes" in der Kriegswirtschaft des Dritten Reiches*. Bonn: Verlag J. H. W. Dietz Nachf., 1986.

——. "Rassismus und rationales Kalkül: Zum Stellenwert utilitaristisch verbrämter Legitimationsstrategien in der nationalsozialistischen 'Weltanschauung.' " In *Vernichtungspolitik: Eine Debatte über den Zusammenhang von Sozialpolitik und Genozid im nationalsozialistischen Deutschland*, edited by Wolfgang Schneider, 25–36. Hamburg: Junius Verlag, 1991.

Herf, Jeffrey. *Reactionary Modernism: Technology, Culture, and Politics in Weimar and the Third Reich*. Cambridge: Cambridge University Press, 1984.

Heskett, John. "Design in Interwar Germany." In *Designing Modernity: The Arts of Reform and Persuasion, 1885–1945*, edited by Wendy Kaplan, 257–85. New York: Thames and Hudson, 1995.

Hilberg, Raul. *The Destruction of the European Jews*. 1961. Reprint, New York: Holmes & Meier, 1985.

——. "Significance of the Holocaust." In *The Holocaust: Ideology, Bureaucracy, and Genocide*, edited by Henry Friedlander and Sybil Milton, 101. Millwood, N.Y.: Kraus International Publications, 1980.

Hirsh, Richard. *Technology and Transformation in the American Electric Utility Industry*. Cambridge: Cambridge University Press, 1989.

Hobsbawm, Eric. "Custom, Wages, and Workload in Nineteenth-Century Industry." In *Workers in the Industrial Revolution: Recent Studies of Labor in the United States and Europe*, edited by Peter Stearns and Daniel Walkowitz, 232–55. New Brunswick, N.J.: Transaction Books, 1974.

Hölsken, Heinz Dieter. *Die V-Waffen, Entstehung-Propaganda-Kriegseinsatz*. Stuttgart: Deutsche Verlagsanstalt, 1984.

Hortleder, Gerd. *Das Gesellschaftsbild des Ingenieurs: Zum politischen Verhalten der Technischen Intelligenz in Deutschland*. Frankfurt am Main: Suhrkamp, 1970.

Höss, Rudolf. "Autobiography." In *KL Auschwitz Seen by the SS: Rudolf Höss, Pery Broad, Johann Paul Kremer*, edited by Jersy Rawicz, 27–102. Oswiecim: Auschwitz-Birkenau State Museum, 1997.

Hounshell, David. *From the American System to Mass Production, 1800–1932: The Development of Manufacturing Technology in the United States*. Baltimore: Johns Hopkins University Press, 1984.

Huber, Gabriele. *Die Porzellan-Manufaktur Allach-München GmbH: Eine "Wirtschaftsunternehmung" der SS zum Schutz der "deutschen Seele."* Marburg: Jonas Verlag, 1992.

Hughes, Thomas Parke. *American Genesis: A Century of Invention and Technology*. New York: Viking, 1989.

——. *Rescuing Prometheus*. New York: Pantheon Books, 1998.

Hundt, Sönke. *Zur Theoriegeschichte der Betriebswirtschaftslehre*. Cologne: Bund-Verlag, 1977.

Hüttenberger, Peter. "Nationalsozialistische Polykratie." *Geschichte und Gesellschaft* 2 (1976): 417–42.

Hutton, Stanley, and Peter Lawrence. *German Engineers: The Anatomy of a Profession*. Oxford: Clarendon, 1981.

Jaeger, Hans. "Unternehmensgeschichte in Deutschland seit 1945: Schwerpunkte—Tendenzen—Ergebnisse." *Geschichte und Gesellschaft* 18 (1992): 107–32.

James, Harold. *The German Slump: Politics and Economics, 1924–1936*. Oxford: Clarendon Press, 1986.

——. "Die Rolle der Banken im Nationalsozialismus." In *Unternehmen im Nationalsozialismus*, edited by Lothar Gall and Manfred Pohl, 25–36. Munich: C. H. Beck, 1998.

Janssen, Gregor. *Das Ministerium Speer: Deutschlands Rüstung im Krieg*. Berlin: Verlag Ullstein, 1968.

Jaskot, Paul. "The Architectural Policy of the SS, 1936–45." Ph.D. diss., Northwestern University, 1993.

——. *The Architecture of Oppression: The SS, Forced Labor and the Nazi Monumental Building Economy*. New York: Routledge, 2000.

Jenkins, Jennifer. "The Kitsch Collections and *The Spirit in the Furniture*: Cultural Reform and National Culture in Germany." *Social History* 21 (1996): 123–41.

Jünger, Ernst. *Der Arbeiter*. In *Werke*. Vol. 6: *Essays II*. 1932. Reprint, Stuttgart: Ernst Klett Verlag, 1964.

Kafka, Franz. *The Trial*. New York: Schocken Classics, 1988.

Kaienburg, Hermann. *"Vernichtung durch Arbeit": Der Fall Neuengamme. Die Wirtschaftsbestrebungen der SS und ihre Auswirkungen auf die Existenzbedingungen der KZ-Gefangenen*. Bonn: Verlag J. H. W. Dietz Nachf., 1990.

Kaplan, Wendy. "Traditions Transformed: Romantic Nationalism in Design, 1890–1920." In *Designing Modernity: The Arts of Reform and Persuasion, 1885–1945*, edited by Kaplan, 19–48. New York: Thames and Hudson, 1995.

Karny, Miroslav. "SS-Wirtschafts-Verwaltungs Hauptamt: Verwalter der KZ-Häftlingsarbeitskräfte und Zentrale des SS-Wirtschaftskonzerns." In *Deutsche Wirtschaft Zwangsarbeit von KZ-Häftlingen für Industrie und Behörden*, edited by Ludwig Eiber and Jan Reemtsma, 153–67. Hamburg: VSA-Verlag, 1991.

——. " 'Vernichtung durch Arbeit' in Leitmeritz: Die SS-Führungsstäbe in der deutschen Kriegswirtschaft." *1999* 4 (1993): 37–61.

——. " 'Vernichtung durch Arbeit': Sterblichkeit in den NS-Konzentrationslagern." In *Sozialpolitik und Judenvernichtung: Gibt es eine Ökonomie der Endlösung?*, edited by Götz Aly and Susanne Heim, 133–58. Berlin: Rotbuch Verlag, 1987.

Kater, Michael. *The Nazi Party: A Social Profile of Members and Leaders, 1919–1945*. Cambridge: Harvard University Press, 1983.

——. "Zum gegenseitigen Verhältnis von SA und SS in der Sozialgeschichte des Nationalsozialismus 1925–39." *Vierteljahresschrift für Sozial- und Wirtschaftsgeschichte* 62 (1975): 339–79.

Kaul, Friedrich. *Ärzte in Auschwitz*. Berlin: VEB Verlag Volk und Gesundheit, 1968.

Kempner, Robert. *SS im Kreuzverhör: Die Elite, die Europa in Schergen Schlug*. Nördlingen: Franz Greno, 1987.

Kershaw, Ian. *Popular Opinion and Political Dissent in the Third Reich: Bavaria, 1933–1945*. Oxford: Oxford University Press, 1983.

Kimpel, Ulrich. "Agrarreform und Bevölkerungspolitik." *Beiträge zur nationalsozialistischen Gesundheits- und Sozialpolitik* 10 (1992): 124–45.

Kissenkoetter, Udo. *Gregor Straßer und die NSDAP*. Stuttgart: Deutsche Verlags-Anstalt, 1978.

Knoll, Albert. "Die Porzellanmanufaktur München-Allach: Das Lieblingskind von Heinrich Himmler." *Dachauer Hefte* 15 (1999): 116–33.

Koch, Peter-Ferdinand. *Himmlers graue Eminenz—Oswald Pohl und das Wirtschaftsverwaltungshauptamt der SS*. Hamburg: Verlag Facta Oblita, 1988.

Kocka, Jürgen. *Die Angestellten in der deutschen Geschichte 1850–1980*. Göttingen: Vandenhoeck & Ruprecht, 1981.

——. "Eisenbahnverwaltung in der industriellen Revolution: Deutsch-amerikanische Vergleiche." In *Historia Socialis et Oeconomica: Festschrift für Wolfgang Zorn zum 65. Geburtstag*, edited by Hermann Kellenbenz and Hans Pohl, 259–77. Stuttgart: Franz Steiner Verlag, 1987.

——. "Scale and Scope, a Review Colloquium." *Business History Review* 64 (1990): 711–16.

——. *Unternehmensverwaltung und Angestelltenschaft am Beispiel Siemens 1847–1914: Zum Verhältnis von Kapitalismus und Bürokratie in der deutschen Industrialisierung*. Stuttgart: Ernst Klett Verlag, 1969.

——. *Unternehmer in der deutschen Industrialisierung*. Göttingen: Vandenhoeck & Ruprecht, 1975.

Koehl, Robert. *The Black Corps: The Structure and Power Struggles of the Nazi SS*. Madison: University of Wisconsin Press, 1983.

——. *RKFDV: German Resettlement and Population Policy, 1939–1945. A History of the Reich Commission for the Strengthening of Germandom*. Cambridge: Harvard University Press, 1957.

Kogon, Eugen. *Der SS-Staat: Das System der deutschen Konzentrationslager*. Munich: Kindler Verlag, 1971.

Kogon, Eugen, Hermann Langbein, Adalbert Rückerl, et al. *Nationalsozialistische Massentötung durch Giftgas*. Frankfurt am Main: Fischer, 1986.

Koselleck, Reinhart. *Preußen zwischen Reform und Revolution: Allgemeines Landrecht, Verwaltung und soziale Bewegung von 1791 bis 1848*. Stuttgart: Ernst Klett Verlag, 1967.

Kracauer, Siegfried. *Die Angestellten*. 1929. Reprint, Frankfurt am Main: Suhrkamp, 1971.

Kranz, Tomasz. "Das KL Lublin—Zwischen Planung und Realisierung." In *Die Nationalsozialistischen Konzentrationslager—Entwicklung und Struktur*, vol. 2, edited by Ulrich Herbert, Karin Orth, and Christoph Dieckmann, 363–89. Göttingen: Wallstein Verlag, 1998.

Kremer, Paul. "Diary of Johann Paul Kremer." In *KL Auschwitz as Seen by the SS*, edited by Jersy Rawicz, 149–215. Warsaw: Interpress Publishers, 1991.

Lane, Barbara Miller. *Architecture and Politics in Germany, 1918–1945*. Cambridge: Harvard University Press, 1968.

Latour, Bruno. *We Have Never Been Modern*. Cambridge: Harvard University Press, 1993.

Lewchuk, Wayne. "Men and Mass Production: The Role of Gender in Managerial Strategies in the British and American Automobile Industry." In *Fordism Transformed: The Development of Production Methods in the Automobile Industry*, edited by Haruhito Shiomi and Kazuo Wada, 219–42. Oxford: Oxford University Press, 1995.

Lifton, Robert Jay. *The Nazi Doctors*. New York: Basic Books, 1986.

Lindenfeld, David. "The Prevalence of Irrational Thinking in the Third Reich: Notes Toward the Reconstruction of Moden Value Rationality." *Central European History* 30 (1997): 365–84.

Ludwig, Karl-Heinz. *Technik und Ingenieure im Dritten Reich*. Düsseldorf: Droste Verlag, 1974.

Ludwig, Karl-Heinz, and Wolfgang König. *Technik, Ingenieure und Gesellschaft: Geschichte des Vereins Deutscher Ingenieure 1856–1981*. Düsseldorf: VDI-Verlag, 1981.

Madajczyka, Czeslawa. *Samojszczyzna—Sonderlaboratorium SS*. Ludowa: Spoldzielnia Wydawnicza, 1977.

Mai, Gunther. "Die Ökonomie der Zeit: Unternehmerische Rationalisierungsstrategien und industrielle Arbeitsbeziehungen." *Geschichte und Gesellschaft* 23 (1997): 311–27.

Maier, Charles. "A Holocaust Like All the Others? Problems of Comparative History." In *The Unmasterable Past: History, Holocaust, and German National Identity*, 66–99. Cambridge: Harvard University Press, 1988.

——. *In Search of Stability: Explorations in Historical Political Economy*. Cambridge: Cambridge University Press, 1987.

——. *The Unmasterable Past: History, Holocaust, and German National Identity*. Cambridge: Harvard University Press, 1988.

Marszalek, Josef. *Majdanek Konzentrationslager Lublin*. Warsaw: Verlag Interpress, 1984.

Mason, Timothy. "Domestic Dynamics of Nazi Conquests: A Response to Critics." In *Reevaluating the Third Reich*, edited by Thomas Childers and Jane Caplan, 161–89. New York: Holmes & Meier, 1993.

——. "Labour in the Third Reich, 1933–39." *Past and Present* 33 (1968): 112–41.

——. "Some Origins of the Second World War." *Past and Present* 29 (1964): 67–87.

Mayer, Arno. *Why Did the Heavens Not Darken?* New York: Pantheon, 1988.

Miller, Joseph. *The Way of Death: Merchant Capitalism and the Angolan Slave Trade, 1730–1830*. Madison: University of Wisconsin Press, 1988.

Mills, C. Wright. *White Collar: The American Middle Classes*. Oxford: Oxford University Press, 1951.

Milward, Alan. *The German Economy at War*. London: Athlone Press, 1965.

Mommsen, Hans. *Beamtentum im Dritten Reich*. Stuttgart: Deutsche Verlags-Anstalt, 1966.

———. "Erfahrungen mit der Geschichte der Volkswagenwerk GmbH im Dritten Reich." In *Unternehmen im Nationalsozialismus*, edited by Lothar Gall and Manfred Pohl, 45–54. Munich: C. H. Beck, 1998.

———. "Konnten Unternehmer im Nationalsozialismus apolitisch bleiben?" In *Unternehmen im Nationalsozialismus*, edited by Lothar Gall and Manfred Pohl, 66–70. Munich: C. H. Beck, 1998.

———. *Mythos von der Modernität: Zur Entwicklung der Rüstungsindustrie im Dritten Reich*. Essen: Klartext, 2000.

———. "Die Realizierung des Utopischen: Die 'Endlösung der Judenfrage' im 'Dritten Reich.' " In *Der Nationalsozialismus und die deutsche Gesellschaft*, 184–232. Reinbek bei Hamburg: Rowohlt, 1991.

Mommsen, Hans, and Manfred Grieger. *Das Volkswagenwerk und seine Arbeiter im Dritten Reich*. Düsseldorf: Econ Verlag, 1996.

Mosse, George. *The Crisis of German Ideology: Intellectual Origins of the Third Reich*. 1964. Reprint, New York: Schocken Books, 1981.

Müller, Rolf-Dieter. *Hitlers Ostkrieg und die deutsche Siedlungspolitik: Die Zusammenarbeit von Wehrmacht, Wirtschaft und SS*. Frankfurt am Main: Fischer, 1991.

Mumford, Lewis. *Technics and Civilization*. 1934. Reprint, New York: Harcourt Brace, 1963.

Musil, Robert. *Mann ohne Eigenschaften*. Frankfurt am Main: Rowohlt, 1987.

Naasner, Walter. *Neue Machtzentren in der deutschen Kriegswirtschaft 1942–45*. Boppard am Rhein: Harald Boldt Verlag, 1994.

———. *SS-Wirtschaft und SS-Verwaltung: "Das SS-Wirtschafts-Verwaltungshauptamt und die unter seiner Dienstaufsicht stehenden wirtschaftlichen Unternehmungen" und weitere Dokumente*. Düsseldorf: Droste, 1998.

Nelson, Walter Henry. *Small Wonder: The Amazing Story of the Volkswagen*. 3d ed. Boston: Little, Brown, 1970.

Neufeld, Michael. *The Rocket and the Reich: Peenemünde and the Coming of the Ballistic Missile Era*. Cambridge: Harvard University Press, 1995.

Neumann, Franz. *Behemoth: The Structure and Politics of National Socialism, 1933–1944*. New York: Harper and Row, 1944.

Nolan, Mary. *Visions of Modernity: American Business and the Modernization of Germany*. Oxford: Oxford University Press, 1994.

Nolte, Ernst. "Vergangenheit die nicht vergehen will." *Frankfurter Allgemeine Zeitung*, 6 June 1986.

Ordway, Frederick, and Mitchell Sharpe. *The Rocket Team*. New York: Thomas Y. Crowell, 1979.

Orth, Karin. "Die Kommandanten der nationalsozialistischen Konzentrationslager." In *Die Nationalsozialistischen Konzentrationslager: Entwicklung und Struktur*, edited by Ulrich Herbert, Karin Orth, and Christoph Dieckmann, 755–86. Göttingen: Wallstein Verlag, 1998.

——. *Die Konzentrationslager-SS: Sozialstrukturelle Analysen und biographische Studien.* Göttingen: Wallstein Verlag, 2000.

——. *Das System der nationalsozialistischen Konzentrationslager—Eine politische Organisationsgeschichte.* Hamburg: Hamburger Edition HIS Verlag, 1999.

Overy, Richard. " 'Blitzkriegswirtschaft'? Finanzpolitik, Lebensstandard und Arbeitseinsatz in Deutschland 1939–42." *Vierteljahreshefte für Zeitgeschichte* 36 (1988): 370–435.

——. "Germany, 'Domestic Crisis' and War in 1939." *Past and Present* 116 (1987): 138–68.

——. *Goering the Iron Man.* London: Routledge & Kegan Paul, 1984.

——. "Mobilization for Total War in Germany, 1939–1941." *English Historical Review* 88 (1988): 613–39.

——. *The Nazi Economic Recovery, 1932–1938.* Cambridge: Cambridge University Press, 1982.

——. *Why the Allies Won.* New York: W. W. Norton, 1995.

Paczula, Tadeusz. "Organisation und Verwaltung des ersten Häftlingskrankenbaus in Auschwitz." In *Die Auschwitz-Hefte*, edited by Jochen August, 159–71. Weinheim: Beltz Verlag, 1987.

Parcer, Jan, and Thomas Grotum. *Sterbebücher des KL Auschwitz-Birkenau.* Oswiecim: Archiwum Panstwowego Muzeum Oswiecim-Brzenzinka, 1996.

Partner, Simon. *Assembled in Japan: Electrical Goods and the Making of the Japanese Consumer.* Berkeley: University of California Press, 1999.

Paxton, Robert, and Michael Marrus. *Vichy France and the Jews.* New York: Basic Books, 1981.

Perz, Bertrand. "Der Arbeitseinsatz im KZ Mauthausen." In *Die Nationalsozialistischen Konzentrationslager—Entwicklung und Struktur*, edited by Ulrich Herbert, Karin Orth, and Christoph Dieckmann, 533–57. Göttingen: Wallstein Verlag, 1998.

——. *Projekt Quarz: Steyr-Daimler-Puch und das Konzentrationslager Melk.* Vienna: Verlag für Gesellschaftskritik, 1991.

Perz, Bertrand, and Thomas Sandkühler. "Auschwitz und die 'Aktion Reinhard' 1942–45: Judenmord und Raubpraxis in neuer Sicht." *Zeitgeschichte* 26 (1999): 283–316.

Petropoulos, Jonathan. *Art as Politics in the Third Reich.* Chapel Hill: University of North Carolina Press, 1996.

Peukert, Detlev. "Alltag und Barbarei: Zur Normalität des Dritten Reiches." In *Ist der Nationalsozialismus Geschichte? Zur Historrisierung und Historikerstreit*, edited by Dan Diner, 51–61. Frankfurt am Main: Fischer, 1987.

——. "The Genesis of the 'Final Solution' from the Spirit of Science." In *Reevaluating the Third Reich*, edited by Thomas Childers and Jane Caplan, 234–52. New York: Holmes & Meier, 1993.

——. *Inside Nazi Germany: Conformity, Opposition, and Racism in Everyday Life.* 1982. Reprint, New Haven: Yale University Press, 1987.

Pingel, Falk. *Häftlinge unter SS-Herrschaft Widerstand, Selbstbehauptung und Vernichtung im Konzentrationslager.* Hamburg: Hoffmann und Campe, 1978.

Piore, Michael, and Charles Sabel. *The Second Industrial Divide: Possibilities for Prosperity.* New York: Basic Books, 1984.

Piper, Franciszek. *Die Zahl der Opfer von Auschwitz*. Oswiecim: Verlag Staatliches Museum, 1993.

Pohl, Dieter. "Die großen Zwangsarbeitslager der SS- und Polizeiführer für Juden im Generalgouvernement 1942–45." In *Die Nationalsozialistischen Konzentrationslager—Entwicklung und Struktur*, edited by Ulrich Herbert, Karin Orth, and Christoph Dieckmann, 415–38. Göttingen: Wallstein Verlag, 1998.

———. *Von der "Judenpolitik" zum Judenmord: Der Distrikt Lublin des Generalgouvernements 1939–1944*. Frankfurt am Main: Peter Lang, 1993.

Porter, Theodore. "Precision and Trust: Early Victorian Insurance and the Politics of Calculation." In *The Values of Precision*, edited by M. Norton Wise, 173–97. Princeton: Princeton University Press, 1995.

———. *Trust in Numbers: The Pursuit of Objectivity in Science and Public Life*. Princeton: Princeton University Press, 1995.

Präg, Werner, and Wolfgang Jacobmeyer, eds. *Das Diensttagebuch des deutschen Generalgouverneurs in Polen 1939–45*. Stuttgart: Deutsche Verlags-Anstalt, 1975.

Pressac, Jean-Claude. *Die Krematorien von Auschwitz: Die Technik des Massenmordes*. Munich: Piper, 1993.

Prinz, Michael. *Vom Mittelstand zum Volksgenossen: Die Entwicklung des sozialen Status der Angestellten von der Weimarer Republik bis zum Ende der NS-Zeit*. Munich: R. Oldenbourg Verlag, 1986.

Rabinbach, Anson. *The Human Motor: Energy, Fatigue, and the Origins of Modernity*. New York: Basic Books, 1990.

Radkau, Joachim. "Kontinuität über 1945 hinweg: Der ökonomische Aspekt als Grundlage von Makrohistorie." In *Deutsche Industrie und Politik von Bismarck bis heute*, edited by Joachim Radkau and G. Hallgarten. Hamburg: Europäische Verlagsanstalt, 1974.

Raim, Edith. *Die Dachauer KZ-Außenkommandos Kaufering und Mühldorf: Rüstungsbauten und Zwangsarbeit im letzten Kriegsjahr 1944/1945*. Landsberg a. Lech: Landsberger Verlagsanstalt Martin Neumeyer, 1992.

Reitlinger, Gerald. *The SS, Alibi of a Nation, 1922–1945*. London: Heinemann, 1956.

Remdt, Gerhardt, and Günter Wermusch. *Rätsel Jonastal: Die Geschichte des letzten Führerhauptquartiers*. Berlin: Christoph Links Verlag, 1992.

Richardi, Hans-Günter. *Schule der Gewalt: Das Konzentrationslager Dachau 1933–34*. Munich: C. H. Beck, 1983.

Ritschl, Albrecht. "Die NS-Wirtschaftsideologie—Modernisierungsprogramm oder reaktionäre Utopie?" In *Nationalsozialismus und Modernisierung*, edited by Rainer Zitelmann and Michael Prinz, 48–70. Darmstadt: Wissenschaftliche Buchgesellschaft, 1991.

Rössler, Mechtild. "Area Research and Spatial Planning from the Weimar Republic to the German Federal Republic." In *Science, Technology, and National Socialism*, edited by Mark Walker and Monika Renneberg, 122–51. Cambridge: Cambridge University Press, 1994.

Rössler, Mechtild, and Sabine Schleiermacher. "Der Generalplan Ost und die Modernität der Großraumordnung: Eine Einführung." In *Der "Generalplan Ost"*:

Hauptlinien der nationalsozialistischen Planungs- und Vernichtungspolitik, edited by Rössler and Schleiermacher, 7–11. Berlin: Akademie Verlag, 1993.

Rostow, Walt Whitman, and Max F. Millikan. *A Proposal: Key to an Effective Foreign Policy*. New York: Harper & Brothers, 1957.

Roth, Karl Heinz. " 'Generalplan Ost'—'Gesamtplan Ost': Forschungsstand, Quellenprobleme, neue Ergebnisse." In *Der "Generalplan Ost": Hauptlinien der nationalsozialistischen Planungs- und Vernichtungspolitik*, edited by Mechtild Rössler and Sabine Schleiermacher, 25–95. Berlin: Akademie Verlag, 1993.

Rückerl, Adalbert. *NS-Vernichtungslager in Spiegel deutscher Strafprozesse: Belzec, Sobibor, Treblinka, Chelmno*. Munich: Deutscher Taschenbuch Verlag, 1977.

Rusnock, Andrea. "Quantification, Precision, and Accuracy: Determinations of Population in the Ancien Régime." In *The Values of Precision*, edited by M. Norton Wise, 17–38. Princeton: Princeton University Press, 1995.

Sachse, Carola. *Siemens, der Nationalsozialismus und die moderne Familie: Eine Untersuchung zur sozialen Rationalisierung in Deutschland im 20. Jahrhundert.* Hamburg: Rasch und Röhring Verlag, 1990.

Safrian, Hans. *Die Eichmann Männer*. Vienna: Europa Verlag, 1993.

Salewski, Michael. "Die Abwehr der Invasion als Schlüssel zum 'Endsieg.' " In *Die Wehrmacht: Mythos und Realität*, edited by Rolf-Dieter Müller and Hans-Erich Volkmann, 210–23. Munich: Oldenbourg, 1999.

Sandkühler, Thomas. "Das Zwangsarbeitslager Lemberg-Janowska 1941–1944." In *Die Nationalsozialistischen Konzentrationslager—Entwicklung und Struktur*, edited by Ulrich Herbert, Karin Orth, and Christoph Dieckmann, 606–35. Göttingen: Wallstein Verlag, 1998.

Saul, John Ralston. *Voltaire's Bastards: The Dictatorship of Reason in the West*. New York: Vintage, 1992.

Schabel, Ralf. "Wenn Wunder den Sieg bringen sollen: Wehrmacht und Waffentechnik im Luftkrieg." In *Die Wehrmacht: Mythos und Realität*, edited by Rolf-Dieter Müller and Hans-Erich Volkmann, 385–404. Munich: Oldenbourg, 1999.

Schmidt, Matthias. *Albert Speer: Das Ende eines Mythos Speers wahre Rolle im Dritten Reich*. Munich: Scherz Verlag, 1982.

Schröder, Hans-Hermann. "Das erste Konzentrationslager in Hannover: Das Lager bei der Akkumulatorenfabrik in Stöcken." In *Konzentrationslager in Hannover: KZ-Arbeit und Rüstungsindustrie in der Spätphase des Zweiten Weltkriegs, Teil I*, edited by Rainer Fröbe et al., 44–130. Hildesheim: August Lax, 1985.

Schulte, Jan-Erik. "Rüstungsunternehmen oder Handwerksbetrieb? Das KZ-Häftlinge ausbeutende SS-Unternehmen 'Deutsche Ausrüstungswerke GmbH.' " In *Die Nationalsozialistischen Konzentrationslager—Entwicklung und Struktur*, edited by Ulrich Herbert, Karin Orth, and Christoph Dieckmann, 558–83. Göttingen: Wallstein Verlag, 1998.

——. "Verwaltung des Terrors: Entwicklung und Tätigkeit der Verwaltungs- und Wirtschaftsämter der SS 1933–1945." D.Phil., Ruhr-Universität Bochum, 1999.

Scott, James. *Seeing Like a State: How Certain Schemes to Improve the Human Condition Have Failed*. New Haven: Yale University Press, 1998.

Segev, Tom. *Die Soldaten des Bösen: Zur Geschichte der KZ-Kommandanten*. Reinbek bei Hamburg: Rowohlt, 1992.

Seibel, Wolfgang. "Staatsstruktur und Massenmord: Was kann eine historisch-vergleichende Institutionenanalyse zur Erforschung des Holocaust beitragen?" *Geschichte und Gesellschaft* 24 (1998): 539–69.

Seidler, Franz. *Fritz Todt: Baumeister des Dritten Reiches*. Frankfurt am Main: Verlag Ullstein, 1988.

Sereny, Gitta. *Albert Speer: His Battle with Truth*. New York: Alfred A. Knopf, 1995.

Setkiewicz, Piotr. "Häftlingsarbeit im KZ Auschwitz III-Monowitz: Die Frage nach der Wirtschaftlichkeit der Arbeit." In *Die Nationalsozialistischen Konzentrationslager— Entwicklung und Struktur*, edited by Ulrich Herbert, Karin Orth, and Christoph Dieckmann, 584–605. Göttingen: Wallstein Verlag, 1998.

Shirer, William. *The Rise and Fall of the Third Reich: A History of Nazi Germany*. New York: Touchstone, 1990.

Siegel, Tilla, and Thomas von Freyberg. *Industrielle Rationalisierung unter dem Nationalsozialismus*. Frankfurt am Main: Campus, 1991.

Sinclair, Bruce. "Inventing a Genteel Tradition: MIT Crosses the River." In *New Perspectives on Technology and American Culture*, edited by Sinclair, 1–18. Philadelphia: American Philosophical Society, 1986.

Slaton, Amy, and Janet Abbate. "The Hidden Lives of Standards." In *Technologies of Power*, edited by Michael Thad Allen and Gabrielle Hecht, 95–143. Cambridge: MIT Press, 2001.

Smelser, Ronald. *Robert Ley, Hitler's Labor Front Leader*. New York: Berg, 1988.

Sofsky, Wolfgang. *Die Ordnung des Terrors: Das Konzentrationslager*. Frankfurt am Main: Fischer, 1993.

Speer, Albert. *Inside the Third Reich: Memoirs*. New York: Macmillan, 1970.

——. *Der Sklavenstaat: Meine Auseinandersetzung mit der SS*. Stuttgart: Deutsche Verlagsanstalt, 1981.

——. *Slave State: Heinrich Himmler's Masterplan for SS Supremacy*. London: Weidenfeld and Nicolson, 1981.

Speier, Hans. *German White-Collar Workers and the Rise of Hitler*. New Haven: Yale University Press, 1986.

Spoerer, Mark. "Profitierten Unternehmen von KZ-Arbeit? Eine kritische Analyse der Literatur." *Historische Zeitschrift* 268 (1999): 61–95.

Stein, Laurie, and Irmela Franzke. "German Design and National Identity, 1890–1914." In *Designing Modernity: The Arts of Reform and Persuasion, 1885–1945*, edited by Wendy Kaplan, 49–77. New York: Thames and Hudson, 1995.

Stern, Fritz. *Dreams and Delusions: National Socialism in the Drama of the German Past*. New York: Vintage Press, 1987.

Streit, Christian. "The German Army and the Policies of Genocide." In *The Policies of Genocide: Jews and Soviet Prisoners of War in Nazi Germany*, edited by Gerhard Hirschfeld, 1–14. London: Allen & Unwin, 1986.

——. *Keine Kameraden: Die Wehrmacht und die sowjetischen Kriegsgefangenen, 1941–45*. Stuttgart: Deutsche Verlags-Anstalt, 1978.

Sudrow, Anne. "Das 'deutsche Rohstoffwunder' und die Schuhindustrie:

Schuhproduktion unter den Bedingungen der nationalsozialistische Autarkiepolitik." *Blätter für Technikgeschichte* 60 (1998): 63–92.

Sydnor, Charles. *Soldiers of Destruction: The SS Death's Head Division, 1933–1945.* 1970. Reprint, Princeton: Princeton University Press, 1990.

Tillion, Germaine. *Ravensbrück.* New York: Anchor Press, 1975.

Trials of War Criminals before the Nuremberg Military Tribunals under Control Council Law No. 10, October 1946–April 1949. Washington, D.C.: U.S. Government Printing Office, 1949–53.

Trischler, Helmut. "Führer-Ideologie im Vergleich zwischen den USA, Groß Brittanien, Deutschland, Österreich, und der Schweiz." *Historische Zeitschrift* 251 (1990): 45–88.

Trommler, Frank. "The Avant-Garde and Technology: Toward Technological Fundamentalism in Turn-of-the-Century Europe." *Science in Context* 8 (1995): 397–416.

———. "Between Normality and Resistance: Catastrophic Gradualism in Nazi Germany." *Journal of Modern History* 64 (1992): 82–101.

———. "The Creation of a Culture of Sachlichkeit." In *Society, Culture, and the State in Germany, 1870–1930,* edited by Geoff Eley, 465–85. Ann Arbor: University of Michigan Press, 1996.

———. "Von Bauhausstuhl zur Kulturpolitik: Die Auseinandersetzung um die moderne Produktkultur." In *Kultur. Bestimmungen in 20. Jahrhundert,* edited by Helmut Brackert and Fritz Wefelmeyer, 86–110. Frankfurt am Main: Suhrkamp, 1990.

Tuchel, Johannes. " 'Arbeit' in den Konzentrationslagern im deutschen Reich 1933–39." In *Arbeiterschaft und Nationalsozialismus in Österreich,* edited by Rudolf Ardelt and Hans Hautmann, 455–69. Vienna: Europaverlag, 1990.

———. *Konzentrationslager: Organisationsgeschichte und Funktion der "Inspektion der Konzentrationslager" 1934–38.* Boppard am Rhein: Harald Boldt Verlag, 1991.

Turner, Henry Ashby. "Big Business and the Rise of Adolf Hitler." In *Nazism and the Third Reich,* edited by Turner, 89–108. New York: Quadrangle Books, 1972.

Van der Vat, Jan. *The Good Nazi: The Life and Lies of Albert Speer.* New York: Houghton Mifflin, 1997.

van Pelt, Robert, and Debórah Dwork. *Auschwitz, 1270 to the Present.* New York: W. W. Norton, 1996.

Wagner, Jens-Christian. "Das Außenlagersystem des KL Mittelbau-Dora." In *Die Nationalsozialistischen Konzentrationslager—Entwicklung und Struktur,* edited by Ulrich Herbert, Karin Orth, and Christoph Dieckmann, 707–29. Göttingen: Wallstein Verlag, 1998.

———. "Verlagerungswahn und Tod: Die Fiktion eines Rüstungszentrums und der KZ-Komplex Mittelbau-Dora 1943–1945." D.Phil., University of Göttingen, 1999.

Walker, Mark. *German National Socialism and the Quest for Nuclear Power, 1939–1949.* Cambridge: Cambridge University Press, 1989.

Walker, Mark, and Monika Renneberg. "Naturwissenschaftler, Techniker und der Nationalsozialismus." In *Ich diente nur der Technik: Sieben Karrieren zwischen 1940 und 1950,* edited by Alfred Gottwaldt, 1–32. Berlin: Nicolaische Verlagsbuchhandlung, 1995.

Wasser, Bruno. *Himmlers Raumplanung im Osten: Der Generalplan Ost in Polen 1940–1944.* Berlin: Birkhäuser Verlag, 1993.

Weber, Max. *Gesammelte Aufsätze zur Religionssoziologie*. Tübingen: Verlag von J. C. B. Mohr, 1922.

Wegner, Bernd. "Defensive ohne Strategie. Die Wehrmacht und das Jahr 1943." In *Die Wehrmacht: Mythos und Realität*, edited by Rolf-Dieter Müller and Hans-Erich Volkmann, 385–404. Munich: Oldenbourg, 1999.

———. *Hitlers politische Soldaten: Die Waffen-SS 1933–1945. Studien zu Leitbild, Struktur und Funktion einer nationalsozialistischen Elite*. Paderborn: Schöningh, 1982.

Weinberg, Gerhard. "German Plans for Victory, 1944–1945." In *Germany, Hitler, and World War II*, 274–86. Cambridge: Cambridge University Press, 1996.

Wengenroth, Ulrich. "Zwischen Aufruhr und Diktatur: Die Technische Hochschule 1918–1945." In *Die Technische Universität München: Annährungen an ihre Geschichte*, edited by Wengenroth, 215–23. Munich: Faktum, 1993.

Wilhelm, Hans-Heinrich. *Die Truppe des Weltanschauungskriegs: Die Einsatzgruppen der Sicherheitspolizei und des SD 1938–1942*. Stuttgart: Deutsche Verlags-Anstalt, 1981.

Winkler, Dörte. *Frauenarbeit im Dritten Reich*. Hamburg: Hoffmann und Campe Verlag, 1977.

Wise, M. Norton. Introduction to *The Values of Precision*, edited by Wise, 3–16. Princeton: Princeton University Press, 1995.

———. "Precision: Agent of Unity and Product of Agreement, Part I—Traveling." In *The Values of Precision*, edited by Wise, 92–102. Princeton: Princeton University Press, 1995.

Witte, Peter. "Slowakische Juden im Distrikt Lublin." Paper presented at the conference Aufbau- und Zerstörungspolitik im deutsch besetzten Osten, Haus der Wannsee-Konferenz, Berlin, 12 June 1995.

Wolschke-Bulmahn, Joachim. "Biodynamischer Gartenbau, Landschaftsarchitektur und Nationalsozialismus." *Das Gartenamt* 9 (1993): 590–95; 10 (1994): 638–42.

Wootton, Charles, and Carel Wolk. "The Evolution and Acceptance of the Loose-Leaf Accounting System, 1885–1935." *Technology and Culture* 41 (2000): 80–98.

Zeitlin, Jonathan, and Steven Tolliday. "Employers and Industrial Relations between Theory and History." In *The Power to Manage? Employers and Industrial Relations in Comparative-Historical Perspective*, edited by Zeitlin and Tolliday, 1–34. New York: Routledge, 1991.

Zeller, Thomas. "Landschaften des Verkehrs Autobahnen im Nationalsozialismus und Hochgeschwindigkeitsstrecken für die Bahn in der Bundesrepublik." *Technikgeschichte* 64 (1997): 323–40.

Ziegler, Dieter. "Die Verdrängung der Juden aus der Dresdner Bank 1933–1938." *Vierteljahrshefte für Zeitgeschichte* 47 (1999): 187–216.

Ziegler, Herbert. *Nazi Germany's New Aristocracy: The SS Leadership, 1925–1939*. Princeton: Princeton University Press, 1989.

Zilt, Andreas. " 'Reactionary Modernism' in der westdeutschen Stahlindustrie? Technik als Kulturfaktor bei Paul Reusch und Hubert Hauttmann." In *Technische Intelligenz und "Kulturfaktor Technik": Kulturvorstellungen von Technikern und Ingenieuren zwischen Kaiserreich und früher Bundesrepublik Deutschland*, edited by Burkhard Dietz, Michael Fessner, and Helmut Maier, 191–202. Münster: Waxmann, 1996.

Zuboff, Shoshana. *In the Age of the Smart Machine: The Future of Work and Power*. New York: Basic Books, 1988.

Zumpe, Lotte. "Arbeitsbedingungen und Arbeitsergebnisse in den Textilbetrieben der SS im Konzentrationslager Ravensbrück." *Jahrbuch für Wirtschaftsgeschichte* 19, no. 2 (1969): 11–51.

——. "Die Textilbetriebe der SS im Konzentrationslager Ravensbrück." *Jahrbuch für Wirtschaftsgeschichte* 19, no. 1 (1969): 11–41.

Zunz, Olivier. *Making America Corporate, 1870–1920*. Chicago: University of Chicago Press, 1990.

INDEX

Birn, Ruth-Betina, 326 (n. 36)

Bischoff, Helmut, 226

Bischoff, Karl, 161–63

Blood and Soil, 10, 62, 98, 142–44; and rational engineering, 54–55. *See also* Ideology

BMW (Bavarian Motor Works), 268

Bobermin, Hanns, 103–4, 257, 282

Boelcke-Kaserne (death camp), 229

Bombing, 1, 183, 188, 197–98, 262, 264; of rocket installations, 214–15; and underground transfer of industry, 214–16, 232–39. *See also* Fighter Staff; Geilenberg, Edmund; Great Building Projects; Special Staff Kammler

Borsig, 212

B Projects, 224, 269

Breitman, Richard, 158

Breuer, Marcel, 109–10

Brick works. *See* German Earth and Stone Works; Schondorff, Erduin

Browning, Christopher, 252, 276

Brustein, William, 79–80

Buchenwald, 44–45, 58, 60, 121–24, 131, 139, 148, 173, 177, 187–88, 190–96, 205–6, 218, 229, 242, 254, 271. *See also* Concentration camps

Budin, Paul, 263

Building Brigades 50, 151–59, 171, 248, 282; and armaments projects, 192, 204–5, 267; Albert Speer's patronage of, 200. *See also* Office Group C

Building Inspections, 146–48, 151–52, 158, 165, 218; and conflict with HSSPFs, 207. *See also* Office Group C

Burböck, Wilhelm, 6, 148–49, 159, 169, 175, 177, 183, 186; and Operation 14 f 13, 118–25. *See also* Office I/5

Bureaucracy, 2–5, 19, 79–80, 113, 252, 275–79; and uniqueness of Holocaust, 5, 272; modern vs. civil service and traditional, 7, 81, 118, 138–39, 145–46, 210–11, 276–81; and power struggle and polycracy, 7–11, 42, 99, 149, 180–82, 190, 217–18, 259, 268; role of ideology and

consensus in, 11–13, 16–17, 30–31, 49–52, 91–96, 136–38, 143–44, 149–50, 158–64, 259–60, 264, 282–84; SS as, 20, 27–29, 47–48, 55–56, 78–82, 107, 147–58; in concentration camps, 38–42, 119–27, 149–50; resistance to, 38–42, 134–35; and slave labor in concentration camps, 46, 72–78, 82, 182–90, 203

Business Administration Main Office (WVHA), 2, 16–18, 165, 175, 252–62, 271–84; and German Workers Front, 9–11; and slave labor, 23–32, 167–79, 182–90; founding of, 154–58; and armaments contracts and Armaments Ministry, 171–206, 216–17, 221, 233, 235, 240–41, 263, 267–68; and General Plenipotentiary for the Reich Labor Action, 191, 215; and rocketry, 208–22, 234–35, 264, 272; and Operation Reinhard, 245–52; and Economics Ministry, 258–59. *See also* names of specific offices

Calvert, Monte, 87

Camp SS, 39, 70, 226. *See also* Death's Head Units; Inspectorate of Concentration Camps; Eicke, Theodor

Canaris, Wilhelm, 24

Cäsar, Joachim, 140

Cell Leaders, 162. *See also* German Workers Front

Central Building Directorates, 147–48, 149–50, 158, 161–63; predecessors, 49–52. *See also* Main Office for Budgets and Buildings; Office II; Office Group C

Chandler, Alfred, 73, 278

Civil engineers, 2, 16, 18, 20, 108, 139, 202–3, 248, 250, 269–70, 274. *See also* Professions—engineers

—in concentration camps, 48–56, 89, 91, 99, 148–50; and slave labor, 203, 205–7, 211–11, 214, 222–32, 234, 239, 282–83

—as modern managers, 49, 108, 145–48; and professional convergence with architects, 48, 142–44; ideology of, 158–64

238–39, 283–84; and technology and organization, 136–38, 143–44, 162–63, 244–52. *See also* Ideology

Rationalization, 45–46, 63–78, 88, 99, 145, 173, 210–2; and slave labor, 222–32, 239, 241, 252–55, 263. *See also* Technocrat; Technology

Ravensbrück, 72–78, 95, 131, 139, 177. *See also* Concentration camps

RAX Works, 211–14

Reich Labor Action, 1, 166, 172, 178, 198–99. *See also* Sauckel, Fritz

Reich Security Main Office (RSHA), 133, 226, 249. *See also* Criminal Police; Gestapo

Reichsführer SS. *See* Himmler, Heinrich

Reichs Kommissar for the Reinforcement of Germandom (RKF), 97–112, 168, 171, 204, 233; and relationship to DWB, 99–100, 112; and HAHB, 129–48, 151–54

Research Institute for Eastern Shelter, 137

Riga, 251. *See also* Concentration camps

Rödl, Arthur, 178

Rössler, Mechtild, 109

Rostow, Walt Whitman, 274

Rudolph, Arthur, 209–10; and slave labor, 213–14, 218, 223, 228–29, 265, 267

SA (Sturmabteilung), 19, 22, 160

Sachse, Carola, 73

Sachsenberg, 37, 59, 139. *See also* Concentration camps

Sachsenhausen, 38, 60, 62, 71, 117, 148, 177, 167, 179, 254. *See also* Concentration camps

Safrian, Hans, 5

Salpeter, Walter, 79–83, 114, 102–3, 255–56, 258, 282, 298 (n. 8); as modern manager, 81–82, 89; and recruitment of SS managers, 85–87, 89; as ideologue, 92–96, 101, 220; and SS settlements, 108

Sauckel, Fritz, 166, 175, 191, 198–99, 205–6, 237

Saul, John Ralston, 86–87

Saur, Karl Otto, 199, 283; as head of Fighter Staff, 233–39, 264, 268

Sawatzki, Albin, 211–12, 218, 221–22, 228–29, 267

Schellin, Erich, 181–82, 207, 246–47

Schieber, Walter, 173–77, 190–94, 196–206, 212, 324 (n. 20)

Schieder, Theodor, 284

Schitli, Wilhelm, 178

Schleiermacher, Sabina, 109

Schondorff, Erduin, 86–90, 92, 96, 112, 114, 131, 143, 145, 180, 182, 206, 225, 253–56, 260, 282–83

Schulte, Jan-Erik, 198

Schwarz, Hans Xaver, 80

Schwarz, Heinz, 85

Science and scientists, 62, 99, 185, 223, 276; and engineering professionalization, 88–89

Security Main Office, 27. *See also* Reich Security Main Office

Seibel, Woflgang, 163

Seidler, Franz, 161

Seifert, Alwin, 62

Settlement, 10, 12, 21, 33, 52–56, 90, 118–19, 204, 241–43, 244; and "New Order," 98–112, 133–40, 150–54, 171, 176, 194–96, 200–201, 270; and racial supremacy, 142–44

Shale Oil GmbH (Schieferöl GmbH), 261–62

Sharpe, Mitchell, 265–67

Shock Troop Adolf Hitler, 21, 38

Siemens and Halske AG, 72

Simson, Arthur, 191

Simson, Julius, 191

Skill, 49–50, 67, 74–78, 89, 117, 226–32. *See also* Labor

Slave labor, 36, 42–48, 66–68, 81–82, 150–55, 253–55; opposition to, 42–43, 57–58; and violence, 43–46, 77–78, 229–39, 263; and SS settlements, 100–102, 222–32, 245–52; and genocide, 117–27; and modernity, 137–40; and armaments,